Agriculture Issues and Policies

nova science publishers

www.novapublishers.com

Agriculture Issues and Policies

Strategies to Achieve Sustainable Development Goals (SDGs): A Road Map for Global Development
Rajani Srivastava, M.Sc., PhD (Editor)
2022. ISBN: 978-1-68507-836-2 (Hardcover)
2022. ISBN: 979-8-88697-027-2 (eBook)

Pistachios: Cultivation, Production and Consumption
Shaziya Haseeb Siddiqui, PhD (Editor)
Shoaib Alam Siddiqui, PhD (Editor)
2022. ISBN: 978-1-68507-949-9 (Hardcover)
2022. ISBN: 979-8-88697-063-0 (eBook)

Ecosystem Services: Types, Management and Benefits
Hanuman Singh Jatav, PhD (Editor)
Vishnu D. Rajput, PhD (Editor)
2022. ISBN: 978-1-68507-614-6 (Hardcover)
2022. ISBN: 978-1-68507-747-1 (eBook)

Underutilized Crops and Their Value Addition
Romesh Kumar Salgotra, PhD (Author)
Monika Sood, PhD (Author)
Surbhi Jasrotia (Author)
2021. ISBN: 978-1-68507-443-2 (Hardcover)
2021. ISBN: 978-1-68507-481-4 (eBook)

Solanum melongena: Production, Cultivation and Nutrition
Abdul Majid Ansari, PhD (Editor)
Wajid Hasan, PhD (Editor)
M. Prakash, PhD (Editor)
2021. ISBN: 978-1-68507-311-4 (Hardcover)
2021. ISBN: 978-1-68507-323-7 (eBook)

More information about this series can be found at
https://novapublishers.com/product-category/series/agriculture-issues-and-policies/

Brojo Kishore Mishra
Editor

Computational Intelligence for Sustainable Development

nova
science publishers
www.novapublishers.com

https://doi.org/10.52305/GYYT9161

Library of Congress Cataloging-in-Publication Data

ISBN: 979-8-88697-198-9

Published by Nova Science Publishers, Inc. † New York

Contents

Preface

Classic methods of Computational Intelligence, such as Artificial Neural Networks, Fuzzy Systems, and Evolutionary Computing have been used by scientists for several decades, and the number of their applications increases every year. These methods have been and are used in virtually all areas of scientific research. There are also new methods of Computational Intelligence.

The book Computational Intelligence for Sustainable Development covers the latest advances in the field of Computational Intelligence with a special focus on Sustainable Development Goals such as promoting sustainable agriculture, good health, clean water and sanitation, affordable and clean energy, sustainable cities and communities, sustainable industrialization and innovation, and sustainable use of terrestrial ecosystems. This book is an excellent resource for data scientists, engineers, students and professors of higher education, computer scientists, researchers, and academicians.

The book is organized into 17 chapters.

Chapter 1 explains the various applications of Artificial Intelligence for Sustainable development.

The use of Artificial Intelligence (AI) models has been bringing efficient solutions to a vast number of real-world clinical problems. With the multifaceted use of AI in many places of the healthcare industry, there are some security problems that arise from time to time on healthcare ethics. But the move towards the application of Computer data and the use of AI in every sector is most welcome. The time and cost of different healthcare facilities have been reduced. Real-time treatment facilities and AI-governed robotic systems not only give treatment in a timely manner but also help reduce stigma. AI-based robotic systems ease the burden on health workers with the growing population and the need for quick treatment facilities. To implement United Nations Sustainable Development Goals (UNSDGs), AI application in the healthcare system with proper ethical consideration should be welcome.

Chapter 2 describes a unique computational intelligence-based approach for detecting and identifying rice leaf diseases using Particle Swarm Optimization (PSO) and Deep Convolutional Neural Networks (DCNN).

The annual rice crop has a direct and indirect impact on India's economic, political, and social stability. Rice production is the sole source of income for hundreds of millions of people. It is vital to detect any illness early and treat the afflicted plants to ensure the healthy and proper growth of rice plants. As a result, the farmer will be able to treat the crop properly and on time. In a broad field, automatic illness detection and identification is particularly helpful since it reduces the work, time, and cost associated with watching and analysing disease symptoms. Thus, automated disease recognition on leaves is critical in the agriculture business. This research describes a unique computational intelligence-based approach for detecting and identifying rice leaf diseases using Particle Swarm Optimization (PSO) and Deep Convolutional Neural Networks (DCNN). The Particle Swarm Optimization technique is used to segment rice leaf pictures, which is critical for disease categorization. Deep convolutional neural networks are employed for the recognition of rice leaf disease classification. Our proposed Rice leaf detection recognition method attained an average accuracy of 98.43%. In the field of rice leaf disease prediction, our technique has shown outstanding results.

Chapter 3 investigates the deployment of IoT, Artificial Intelligence and Machine Learning for Sustainable Smart Cities and Communities.

In Chapter 4, the authors have intended to build a prototype of a web-based platform with the implementation of machine learning techniques. Various aspects in which machine learning can help in building a web-based platform and the ways it is being benefitted are also discussed. Modules are elaborated with the dataflow diagram and use case diagram. The authors intend to implement the prototype with a high-level programming language.

Chapter 5 discusses the recent advances in Data Science and its applications. Big Data is a phrase that refers to a collection of methods that elicit a kind of integration that is used to discover enormous unknown values. These are very complicated values with a massive scope. The phrase Big Data was coined in 1997 by NASA scientists in a study report. Big Data is a phrase that refers to data that has a high velocity, diversity, and volume. Big Data is a term used to describe a huge quantity of information. It is possible for this data to be both organized and unstructured. It's also tough to practice using conventional tools and methods. There is a large quantity of big data in the IT industry that is common to many departments, i.e., there is a vast amount of data in the store of industries and no tool exists to manage it before big data comes into play. Big Data Analytics is concerned with the storage and analysis of various, difficult, complicated, and massive datasets.

Chapter 6 introduces a novel Intuitionistic Fuzzy Stream Cipher, where two LFSR are used which are initialised with a secret key. Then fuzzification is done using and from the outputs of these LFSRs; the fuzzy function used here is Intuitionistic Fuzzy (IF). This IF introduces the non-linearity in the system and encryption done by XORing the output bit with the plaintext. The decryption is the repetition of the same process using the same secret key. The application domain is the blockchain platform, which is now a part of much sustainable devolvement. Here, we give security in the lightning network. It is now seen that for huge nodes in bitcoin, one transaction takes almost five minutes, so, the lightning network is the second layer network introduced before cryptocurrencies to make the transactions faster. This lightning network enables transaction in various crypto-apps in a very short time and a commit protocol is executed over actual cryptocurrency at regular intervals of time. Finally, we evaluated our stream cipher using mono bit frequency test, entropy test, floating frequency test, histogram, N-Gram, Autocorrelation and periodicity to measure the randomness against the widely used stream cipher, RC4.

Chapter 7 notes that many studies have been carried out in order to conclude that bio-fertilizers may give nutrients and certain amino acids, as well as plant hormones as an exogenous source, which can boost the absorption rate and improve the growth rate.

In Chapter 8, the authors proposed a project which is useful to detect willing and unwilling entry of a person into a restricted area. It has applications in various areas. It is used in door security in banks, locker rooms, jewelry shops, etc. It is also used in the detection of unauthorized entry into confidential areas of a company or a building. It is also used as a laser-based alarming system. ESP-32 with laser can also be used as a rangefinder in a car-parking system to help the driver safely park the car. It can be used in different ways by adding different sensors, for example adding a temperature sensor, a distance sensor, and an EC sensor. We can use it as a hydroponics system. It is useful as a child monitor system in a house by installing it on each door of the house.

Chapter 9 aims at the importance of column tests in Geoenvironmental applications such as the design of soil liners for landfills, slurry cutoff walls for waste containment and soil remediation techniques such as soil washing and immobilization. The determination of contaminant transport parameters from the column test results was demonstrated with different soils and their role in Geoenvironmental applications is elaborated. The analytical methods and the various processes involved in the modelling of contaminant transport were discussed. The procedures involved in the remediation of contaminated industrial soil and landfill soil using soil washing technique and immobilization technique with different chemicals and solutions were demonstrated.

Chapter 10 includes quantitative research that uses primary data (questionnaire) and a quasi-experimental research method with a two group design. Data was collected by using a questionnaire. The number of samples is 80, of which 40 are samples for ultrasonic sensors and 40 are for sensors (PIR). Results: The results of statistical tests with Tsimples t test between ultrasonic arduino variables and passive infrared receiver (PIR) obtained $p = 10{,}575$ ($p>0.000$) on ultrasonic sensors $P = 3.975$ ($p> 01{,}000$) on PIR sensors. Arduino ultrasonic sensors have a higher effectiveness than passive infrared receiver sensor necklaces.

Chapter 11 considers an orifice as a flow measuring device for the various applications on the variation of shapes and sizes with classification, design and operating parameters for better understanding and designing. The study also presents the flow characteristics such as Δp and Cd of the orifice on the effects of area ratio, space ratio, and Reynolds number along with the upstream and downstream of the flow. The study also carried out a numerical simulation based on the Finite volume method (FVM) and Computational Fluid Dynamics (CFD) based on ANSYS fluent commercial code software to determine the flow parameter. The article gives an overview regarding the numerical analysis of the orifice for a better understanding of the flow characteristics. On the basis of design sustainability, the virtual platform has been chosen for the work and the novelty focuses on the analytical modeling of the orifice meter for the required application.

Chapter – 12 summarises the research hotspots in quantum error correction using AI techniques which reveal that AI may be an exhaustive solution for quantum error correction. These research hotspots require the immediate attention of the scientific community involved in this regime for making quantum computers practical feasible and viable.

In Chapter 13, a proposed model classifies into four classes i.e., Healthy leaf, Bacterial Leaf Blight, Brown Spot, and Hispa among which some are diseases or pests. The recommended model has achieved an overall accuracy of 98.81% whereas the performance of leaf disease in paddy class like bacterial leaf blight and brown spot has achieved an overall 100% accuracy. The objective of this work is to develop a system that is capable to detect and identify the type of disease based on Image Processing and Machine Learning techniques.

In Chapter 14, the authors discussed about their development of a model for an automatic face detection attendance system for which we had developed complete hardware and we had used the VGG face deep learning model to predict faces accurately and deployed the same in hardware. Then the setup was tested and it was found the setup was able to give accurate results. Faces were detected with more than 95% accuracy and attendance was marked and stored in the database effectively.

Chapter 15 focuses on how the fuzzy concept is helpful to improve the quality-of-service parameter in the field of vehicular ad hoc network and image processing. At the end of the chapter, the authors discuss the fuzzy C-means Clustering algorithm which provides better results as compared to other clustering algorithms in real time systems such as disease prediction in leaf and efficient data transmission in VANET.

Chapter 16 elaborates on how cognitive IoT has been beneficial for human beings.

The complete process and implementation steps are explained in detail in Chapter 17, which can be considered by a developer to understand the development of a Blockchain application.

Brojo Kishore Mishra
GIET University, India

Chapter 1

Sustainable Healthcare and Artificial Intelligence: Some Facts

Sandeep Poddar*
Lincoln University College, Selangor D. E., Malaysia

Abstract

Artificial Intelligence (AI) models have been shown to solve a variety of real-world healthcare problems. With the multifaceted use of AI in many places in the healthcare industry, some security problems arise time to time in healthcare ethics. But the use and move towards application of computer data oriented and use of AI in every sector is most welcome. The time and cost of different healthcare facilities have been reduced. Real time treatment facilities and AI governed robotic systems not only give the treatment in a timely manner, it also help the population stay away from different stigmas. The AI-based robotic systems ease the burden on health workers with the growing population and the need for quick treatment facilities. To implement the United Nations Sustainable Development Goals (UNSDGs), AI applications in the healthcare system should be implemented with proper ethical consideration.

Keywords: AI, deep learning, healthcare, drug discovery, COVID 19, mobile technologies

* Corresponding Author's Email: sandeeppoddar@lincoln.edu.m.

In: Computational Intelligence for Sustainable Development
Editor: Brojo Kishore Mishra
ISBN: 979-8-88697-198-9

Introduction

Artificial intelligence (AI) is quickly being applied in clinical medicine, with developments in sectors such as image analysis, lesion detection, healthcare management assisted by AI, etc. This article elucidates AI-related medicine and AI-assisted visualised medicine's emerging applications, including novel diagnostic approaches, metadata analytical methods, and versatile AI-assisted treatment applications in preclinical and clinical settings, as well as the future prospects of AI-assisted disease prediction. The basics of deep learning frameworks, as well as certain prevalent disorders and a synopsis of deep learning prediction methodologies for specific diseases, are discussed. AI can identify a lot of faults in present disease prediction and provide a future prognosis.

In recent years, artificial intelligence (AI) has progressed fast in terms of software algorithms, hardware implementation, and applications in a wide range of fields. In this article, some of the most recent advances in AI applications in biomedicine, including disease diagnosis, living assistance, biomedical information processing, and biomedical research, have been discussed. The purpose of this review is to keep track of current scientific achievements, comprehend the availability of technology, recognise the enormous potential of AI in biomedicine, and provide inspiration to researchers in related fields. The application of AI in biomedicine, like AI itself, is still in its early stages. New advancements and discoveries will continue to push the AI frontier and broaden the breadth of its applications, with rapid advancements expected in the near future (Rong et al. 2020).

Paranjape, et al. 2021 survey denotes the "organisational value" and anticipated improvements in "quality of care" AI will be useful to them in the near future, but critical prerequisites such as support systems, strategic plans, and budgets must be considered. AI education among healthcare personnel is the need of the hour.

AI applications being data-intensive, a new problem has arisen: the need for more and more precise data to improve AI algorithms, which conflicts with the need for more transparent data handling and protection (Ramirez et al. 2014). Panch et al. (2019) state that although the vast amount of personal healthcare data could be used to develop extremely powerful tools for diagnosis and treatment, there are numerous issues concerning data ownership and privacy that necessitate careful policy intervention. The negative consequences and sustainability of AI applications in healthcare are the opening for novel research arenas.

Computer advancements have provided a potential instrument to aid in the management of the medical information explosion. Medical diagnosis is a mentally demanding profession that has piqued the interest of computer scientists who want to construct expert systems utilising artificial intelligence approaches. Banks G. (1986) developed INTERNIST-1/CADUCEUS programmes to examine some of the issues surrounding medical diagnosis and illustrate several approaches to overcoming these issues.

Miller and Brown (2018) described how AI algorithms are performing on par or better than humans in various tasks. For instance, in analysing medical images or correlating symptoms and biomarkers from electronic medical records (EMRs) with the characterization and prognosis of the disease.

Artificial intelligence (AI) is becoming more widely used in different sectors of society, including the pharmaceutical industry. The use of AI in various sectors of the pharmaceutical industry, such as drug discovery and development, drug repurposing, improving pharmaceutical productivity, and clinical trials, is highlighted in this review; such use reduces human workload while also achieving targets in a short period of time (Paul et al., 2021). Paul et al. (2021) describe how AI tools and methodologies interact, as well as current issues and solutions, and the future of AI in the pharmaceutical business.

AI in Different Healthcare Segments

Laboratory Medicine and AI

Laboratory medicine is important in the healthcare sector because it aids in diagnosis, decision-making, and patient safety (Greaves et al. 2019). The use of artificial intelligence (AI) models has provided efficient answers to a large variety of sustainable healthcare issues. At present, increased workload among support staff, high cost, and demand for accuracy in laboratory processes are the main reasons (Naugler and Church 2019). This is equally important in both healthcare and laboratory medicine (Tizhoosh and Pantanowitz 2018). The use of big data analytics and AI is the present need in healthcare industry. AI uses software based complex algorithms to enhance human cognition for the analysis of healthcare data generated from diagnostics, medical records, claims, clinical trials, etc. (Cabitza and Banfi 2018).

Laboratory medicine continues to undergo digitalization and automation. Clinical laboratories will undoubtedly encounter challenges in analysing, implementing, and verifying artificial intelligence (AI) algorithms both within and beyond the laboratory.

Current research into laboratory-related diagnostic errors highlights the following.

- *Pre- and post-analytical phases are more vulnerable to errors.*
- *Analytical quality is a persistent issue.*
- *Analytical interferences must be focused on.*
- *Errors do not only concern clinical chemistry tests.*

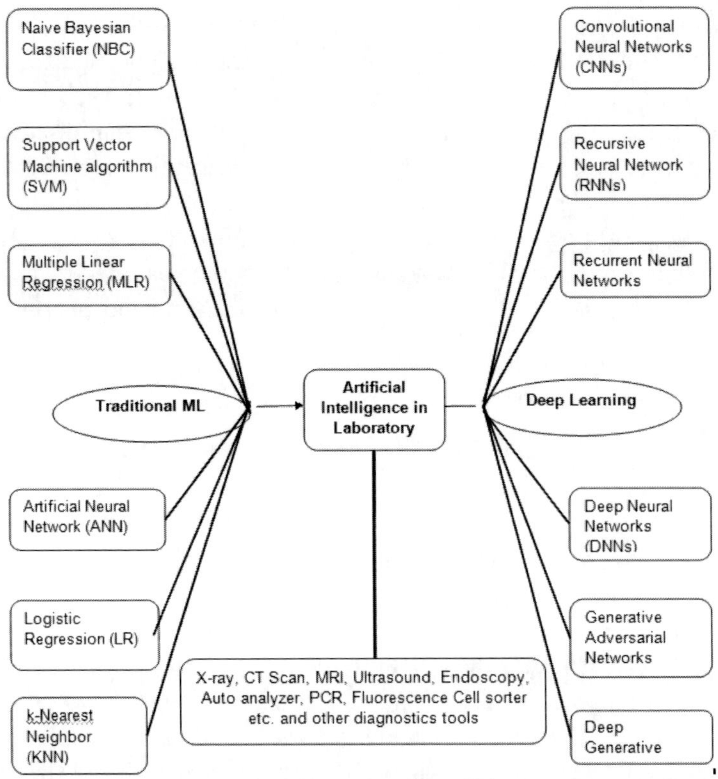

Figure 1. AI algorithms in the laboratory.

The database of images uploaded to clouds are categorised by analysing the occurrence of different diseases using AI algorithms. This helps to generate personalised recommendation based on laboratory reports. Providing recommendations that are precisely tailored to our needs. This includes the use of validated equipment and computer systems; the secure retention of electronic health records to enable instant reconstruction of analyses; user-independent, computer-generated, time-stamped audit data; system and data security; data integrity and confidentiality through limited authorised system access; and secure electronic signatures for closed (individuals with restricted access) and open (everyone else, including data transmitters) systems (Huber L 1999).

AI algorithms are used in many steps of the current laboratory process. Traditional Machine Learning (ML) algorithms include decision trees (Domingos and Pazzani 1997), the nave Bayesian classifier (NBC) (Cortes and Vapnik 1995), the support vector machine algorithm (SVM) (Cao et al. 2010), random forests, multiple linear regression (MLR), the artificial neural network (ANN) algorithm, boosting and bagging algorithms, logistic regression (LR), and k-nearest neighbour (KNN). The domain and connections of AI, ML, and medicine are shown in Figure 1. ML's learning process simulates the human capacity for acquisition.

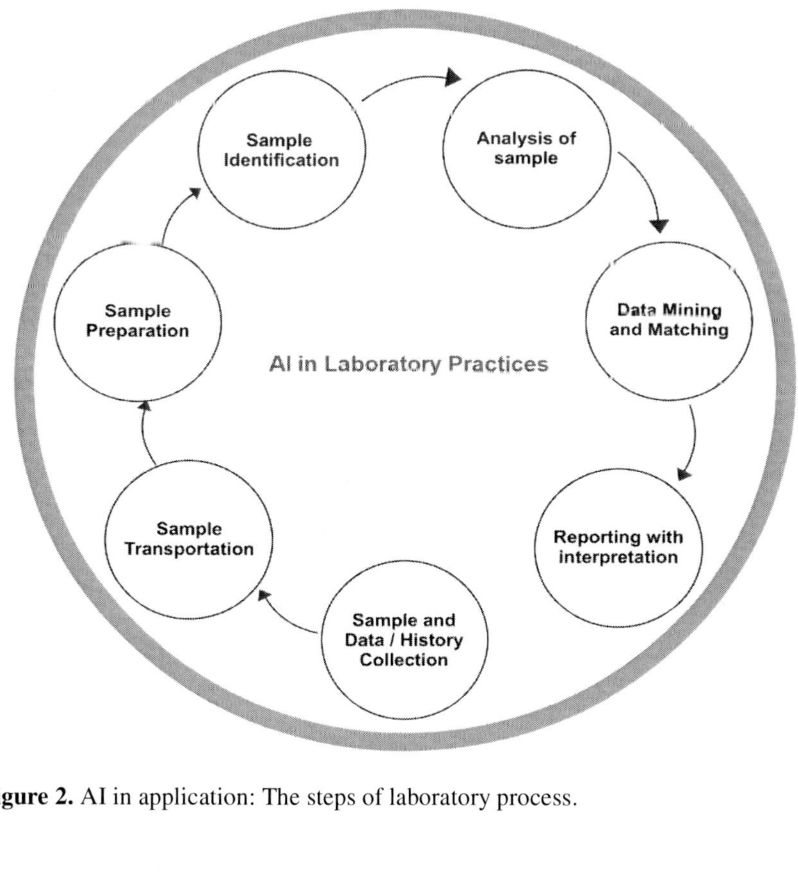

Figure 2. AI in application: The steps of laboratory process.

Test Selection

There was very little laboratory test prediction research reported using AI. Wright et al. (2010) suggested an association and data mining technique to advise laboratory tests. Islam et al. (2020) developed a Deep Learning (DL)-based automated model to recommend laboratory tests based on individual patient clinical history based on cardiology department patent. Cheng et al. (2019) show retrospective analysis of examining lab tests from an ICU and identifying those factors that resulted in an actual change in patient care.

ML models, such as regularised logistic regression, regress and round, naive Bayes, neural network multilayer perceptions, decision trees, and random forest algorithms, were used to identify possibly unneeded laboratory tests. (Xu et al. 2019).

AI Consequences of Error

An error made by any health expert may result in harm to a single patient, whereas an error made by an AI system assisting in medical decision making may result in harm to a far greater number of individuals. To prevent loss of precision, technology must be subjected to stringent regulatory scrutiny and meet legal standards such as FDA 21 CFR Part 11 (USDHH 2002).

Big Data Analysis and AI

Analog data such as radiographs, medical reports, and handwritten notes have traditionally been used in the medical sector to make decisions based on the best available scientific information and the experience of specialists. That is because new technologies have made huge volumes of digital data available to academics and doctors in formats that may be used to supplement or aid in difficult tasks like the aforementioned decision-making, which is altering medicine thanks to the "Big Data" phenomenon. In the medical profession, decision-making has typically been based on the best available scientific information and the experience of specialists using data contained in analogue formats such as radiographies, medical reports, and handwritten notes, among other things. Medical research and practise are being transformed by the Big Data phenomenon, which has enabled researchers and doctors to easily access vast amounts of digital data that may be utilised to assist with difficult tasks like the aforementioned decision-making process (Diez-Sanmartin et al., 2021).

AI in Modern Biosciences Research

Artificial intelligence (AI) based on computer technology aims to mimic intelligent human behaviour and critical thinking. Researchers are under pressure to employ novel approaches to fulfil human wants and requirements for improved healthcare as the world's population grows. In the realm of human health, artificial intelligence is becoming more significant in the development of diverse applications for human welfare and society. Using computers, AI can be used to solve and understand problems in health care for long-term illnesses like Alzheimer's, diabetes, heart disease, and some types of cancer. As a consequence, the severity of illness and the economic burden will be minimised. In recent years, AI-based research has grown in popularity in a range of biological domains in recent years, covering not just human health but also other areas (Kumar et al., 2021).

Detection of Different Diseases

Melanoma Detection

The discipline of dermatology is another medical subspecialty where AI solutions are advancing quickly. Esteva et al. (2017) built a single CNN for the identification of skin cancer and reported an AUC of 0.96 for carcinoma detection and 0.96 for melanoma detection, which was comparable to expert judgments. There are also smartphone applications with a sensitivity of 80% and a specificity of 78% for the diagnosis of malignant or premalignant lesions; nonetheless, their accuracy is inferior to that of expert advice. (Freeman et al. 2020). Several advances in science and technology have led to the availability of non-invasive imaging techniques for the early detection of melanoma, which can be effectively treated with surgical treatments (Davis et al. 2019).

Diabetic Retinopathy Detection

Machine learning is used in clinical care software to detect diabetic retinopathy from diagnostic imagery (Gershgorn D. 2018, FDA 2018).

AI in Alzheimer's Disease

Alzheimer's disease is a category of neurodegenerative diseases marked by a progressive and frequently severe deterioration in cognitive function over time. The elderly are disproportionately affected since age is the key risk factor for brain diseases. Due to the severity of the disease's worldwide growth, organisations and researchers are continuing to invest in the early identification and prevention of such disorders, with a focus on correct treatment and therapy. There is a need for cost-effective and scalable approaches for detecting dementia from the earliest stages, such as Subjective Memory Loss (SML), to more severe stages, such as Mild Cognitive Impairment (MCI) and Alzheimer's Dementia (AD). Using the OASIS, ADNI, and WMH challenge datasets, the purpose of this study is to construct a reliable deep learning system for identifying cognitive impairment (CI). This study looked at the many methodologies and algorithms used in different frameworks to effectively and automatically segment WMH (White Matter Hyperintensities) in the brain in order to find lesions and regions linked to various abnormalities, one of which being Alzheimer's disease (Pious and Sridevi, 2021).

AI in Cardiovascular Disease

Cardiovascular disease is a silent killer that is one of the main causes of death worldwide each year. Premature death affects 7% of people in high-income countries and 43% of people in low-income countries (Balakumar et al. 2016). Obesity, diabetes, and other lifestyle-altering variables are the major risk factors for cardiovascular disease. Early detection of cardiac disease plays a critical role in reducing premature mortality. Early detection of cardiac disease requires combining clinical and pathological data. The goal of this research is to develop a medical prediction support system for heart disease prediction. A computational model based on a deep learning approach has been developed for diagnosis. Sivaranjani and Yuvaraj (2019) described three primary steps in their proposed system. First, a dataset of 13 attributes (13 clinical features) is gathered from the website. Second, the datasets are trained with an artificial neural network with a backpropagation approach algorithm. In order to improve accuracy, it can include one or more hidden layers. Finally, a Cardiac Function Prediction System (CFPS) is constructed, which is an interactive GUI where the user may enter clinical features and learn about a patient's current health status. This system improves medical care while also lowering treatment costs. This approach will be a useful tool for medical practitioners in making accurate diagnoses (Sivaranjani and Yuvaraj, 2019).

AI in Diabetic Retinopathy

Artificial intelligence in medicine is a developing technology that has the potential to help with mass screening and possibly even accurate diagnosis. Complicated computing's capacity to perform pattern recognition by constructing complex associations from input data and then comparing it to performance standards is a significant step (Thrall et al. 2018). Diabetic retinopathy is becoming more common. The burden of sight-threatening retinopathy can be reduced by early detection and treatment (Knickelbein et al. 2016). Any tool that might aid in the rapid screening of this illness while reducing the need for trained human resources would be a boon to both patients and ophthalmologists. Padhy et al. (2019) looked at how AI is being used to treat diabetic retinopathy and a few other common diseases of the retina.

AI in Lung Cancer

In recent years, breath analysis has emerged as a promising approach for detecting respiratory illnesses. The development of an artificial intelligence (AI) based e-nose system for distinguishing lung cancer patients from healthy controls has been discussed by Liu et al. (2021). An electronic nose (e-nose) made up of several gas sensor arrays may easily evaluate volatile organic compounds (VOC). E-noses based on metal oxide semiconductors (MOS) are becoming increasingly common in the analysis of VOCs in human exhaled breath (Wilson, A. D. 2015). Liu et al. (2021) created an e-nose system with a sensor array system comprising five MOS gas sensors, as well as a controller and machine learning algorithms for data processing. The sensor array was created with Figaro USA MOS sensors, and data was collected using the Arduino Uno developer board (Sun et al. 2017). This technique can also be used to diagnose a variety of other lung illnesses (Binson & Subramoniam, 2021).

AI in Kidney Disease

Artificial intelligence (AI) has been widely employed in medical fields to advance medical science, with applications mostly in disease detection, diagnosis, and management. Due to the enormous number of sufferers, kidney disease is a global health issue. Diagnosis and treatment are still difficult to come by. Artificial intelligence has the capacity to consider individual situations, make appropriate decisions, and make significant advancements in renal disease treatment. Yuan et al. (2020) reviewed the most recent research on artificial intelligence (AI) applications in renal disease, including alerting systems, diagnostic assistance, therapy guidance, and prognosis evaluation. Despite the small amount of research associated with AI applications in kidney illness, physicians recognise the promise of AI in the management of renal disease. AI will significantly increase clinicians' capabilities in their clinical practise in the future (Yuan et al. 2020).

AI in Fatty Liver Disease

Our daily lives are increasingly associated with AI, which now includes healthcare applications. Patients with non-alcoholic fatty liver disease and liver fibrosis can now be better cared for by a novel approach to diagnosis supported by AI. Artificial intelligence (AI), can be used in conjunction with traditional methods, such as liver biopsy and noninvasive tests (such as transient elastography and blood biomarkers). The topic of this review study is using AI to analyse electronic health data, liver biopsies, and liver imaging. Among the many Artificial Intelligence (AI) methods that are commonly used include logistic regression, decision trees, random forests, XGBoost, and recurrent neural networks (Wong et al., 2021).

Weight loss and the prevention of non-alcoholic fatty liver disease (NAFLD) can be achieved with a good diet and regular exercise. Dietary composition has not been confirmed, however. Saturated fatty acids (SFA), trans-fats, simple carbohydrates, and animal proteins all have a detrimental effect on the liver when consumed in large amounts. Other than the liver's health benefits, MUFAs, PUFAs, omega-3s, plant protein, and dietary fibre are thought to be good for the liver's function. Specific micronutrient effects are less well understood. The food we eat contains nutrients. Our dietary patterns are formed by the foods we eat. Diets high in soda, frozen junk food, juice, processed meats, whole fat dairy, fatty snacks and take-away foods, cakes and biscuits, low in cereals, whole grains, fruit and vegetables, and extra virgin olive oil (EVOO), are common among individuals with non-alcoholic fatty liver disease. The Mediterranean Diet's (MD), on the other hand, is advantageous for NAFLD even if it is iso-caloric or there are no changes in body weight. The Mediterranean Diet 'Nutritional Geometry' is a new technique to integrate nutrition, animals, and the environment. Nutrients and foods are combined in a model so that the features of diets impacting health and disease can be better understood. NAFLD can be prevented and treated more effectively with the use of AI algorithms that generate a patient-specific diet plan (Berná and Romero-Gomez, 2020).

AI in Inflammatory Bowel Disease (IBD)

Crohn's disease and ulcerative colitis are two kinds of inflammatory bowel disease (IBD), which is a chronic ailment that affects millions of people worldwide. It is possible to enhance the clinical outcome while reducing the associated risks and costs by precisely defining the disease activity and using the appropriate medications. Medical conditions and scenarios have utilised artificial intelligence (AI). An overview of AI for endoscopy in IBD is presented here. Takenaka et al. (2022) describe how artificial intelligence (AI) can improve clinical practise and how some components have already begun to influence understanding. It is imperative that we keep up-to-date on the latest developments in IBD health care as AI algorithms help us make more precise diagnoses and treatments for human illness (Takenaka et al. 2022).

AI in Myeloproliferative Neoplasms (MPNs)

The integration of clinical, morphological, and genetic information is necessary for the accurate diagnosis and categorization of myeloproliferative neoplasms (MPNs). The morphological assessment of bone marrow trephines (BMT) is crucial in distinguishing between distinct subtypes of MPNs and their reactive mimics, despite considerable breakthroughs in our understanding of the molecular and genetic foundation of MPNs. Morphological evaluation is severely restricted by its reliance on subjective, qualitative, and only weakly reproducible criteria. Automated identification of MPNs has already been developed using a machine learning approach. To identify and delineate significant histological features from routinely generated BMTs using an automated method. There were clear signs that MPN subtypes could be distinguished from each other using subsequent analyses, which had a good prediction accuracy (AUC = 0.95). Abstracted megakaryocyte properties in the context of assessed patient cohorts are shown visually in a way that goes beyond typical techniques for sample interpretation and monitoring. MPN patients could benefit from the automated BMT phenotyping approach presented here, either as part of the usual diagnosis process or as a way to monitor disease progression and response to treatment (Sirinukunwattana et al. 2020).

AI in Parkinson's Disease (PD)

In Parkinson's disease, which affects both the motor and non-motor elements of movement, patients lose control over all three phases of movement: planning, initiating, and executing. Tremor, stiffness, and difficulty initiating are among the symptoms that may be seen in the early stages of Parkinson's disease progression. Patients' quality of life (QoL), social functions, and family connections are significantly impacted by Parkinson's disease (PD). PD also imposes significant economic consequences on individuals and society as a whole. Traditionally, motor signs have been used to make the diagnosis of Parkinson's disease (PD). Despite the fact that clinical examinations have established cardinal symptoms of PD, the majority of the rating scales used to determine the severity of the illness have not been well examined and validated. Patients with Parkinson's disease often have non-motor symptoms prior to the beginning of the disease, but these symptoms lack specificity, are difficult to measure, and/or vary from patient to patient. Although various non-motor symptoms have been used as supporting criteria for the diagnosis of PD, they do not now allow for an independent diagnosis of PD. Handwriting patterns, movement, neuroimaging, cerebrospinal fluid (CSF), cardiac scintigraphy, serum, and optical coherence tomography (OCT) have all been used to identify Parkinson's disease (PD) (OCT). In the diagnosis of Parkinson's disease, machine learning also enables the combination of several imaging modalities, such as MRI and SPECT data. New biomarkers that may be used to identify patients at an earlier stage with Parkinson's disease (PD) might lead to a more systematic clinical decision-making system. Consequently, machine learning technologies have the potential to give new tools to doctors for the screening, detection, or diagnosis of PD. Data collection, preprocessing techniques, model deployment, and research findings need to be more transparent in order to allow for greater repeatability of studies on machine learning-based diagnosis and evaluation of Parkinson's disease (PD) (Mei et al. 2021).

Others

An automated method was used to identify and characterise important histological characteristics in regularly produced tissues. A growing body of research indicates that medical images include several new elements that might enhance tumour classification, therapy prediction, and prognostication beyond what is currently possible with positron emission tomography or single-photon emission computed tomography. There have been some promising results from early studies on radiomics' ability to enhance patient care and outcomes. We do not know how these new measurements link to disease-causing molecular pathways. It is also vital for future clinical practise to be able to cope with growing volumes of data from medical imaging and beyond in a fast, repeatable, open and transparent way. Artificial intelligence (AI) may have an impact on language processing, knowledge representation, problem solving, and planning, which are just a few of the many 'intelligent' operations that computers can do. In contrast to rule-based algorithms, which have been used in medical imaging since the 1990s, the revived interest in AI is a result of increased processing capacity and advancements made in machine learning (ML). Cook and Goh (2019) showed non-small-cell lung cancer is utilised as an example and a focus since it is the most prevalent tumour type in which AI and ML techniques have been evaluated (Cook and Goh, 2019).

For the prevention of bone fractures caused by illness and the ageing of the population, 3D bone microarchitecture must be studied and defined. Bone microarchitecture and mechanical stiffness have been studied using a variety of image processing and modelling approaches. In certain cases, artificial intelligence algorithms were used to analyse the morphological, topological, and mechanical characteristics of bone. In a clinical trial, two groups of arthritic and osteoporotic bone samples are studied in a clinical trial. The performances of the Adaptive Neuro Fuzzy Inference System (ANFIS), Support Vector Machines (SVM) and Genetic Algorithms (GA) in identifying the varied samples have been compared. As a result of the results, the Genetic Algorithm has the highest success rate in separating (100 percent) (Akgundogdu et al. 2010).

COVID-19 and AI

Pawar et al. 2022 vividly described the application of AI to combat the COVID 19 outbreak. The applications are mainly alerts and early warnings; tracking and prediction; dashboards with info, diagnosis and prognosis, medication and cures, social distancing. Agarwal et al. 2022 listed 39 AI and computer vision technology based smart technologies to combat COVID 19 in different aspects; starting from early diagnosis to treatment.

A systemic review of different interventions of AI with its main features approach is an example of the multiple ways application of AI in controlling disease has been displayed in Table 1.

Other Uses

Armitage H. (2018) has developed an algorithm capable of interpreting chest X-rays for 14 unique diseases within a fraction of a minute. AI technology will have a significant impact on radiation oncology, organ allocation, robotic surgery, and other healthcare disciplines. (Thompson et al. 2018, Canadian et al. 2019, Rabbani et al. 2018, O'Sullivan et al. 2019, Hashimoto et al. 2018.)

Table 1. A comparison of intervention target and main features

Intervention	Main features	References
Tracking app	Risk assessment and advice	Kodali et al. (2020)
Education and information	Information for patients	Timmers et al. (2020)
Education and information	Information for health care professionals	Ros and Neuwirth (2020) and Zamberg et al. (2020)
Hospital information system	Adaptation to pandemic	Bourdon et al. (2020)
Health assessment	General	Yamamoto et al. (2020)
Health assessment	Self-monitoring	Bae et al. (2020), Medina et al. (2020), Timmers et al. (2020), and Yamamoto et al. (2020)
Home monitoring	Telephone outreach	Medina et al. (2020)
Home monitoring	Vital signs with smart devices	Bourdon et al. (2020)
Home monitoring	Internet of Things	Drew et al. (2020)
Interactive map	Demographics and health status	Timmers et al. (2020)
Data sharing	Patient email to specific recipients	Yamamoto et al. (2020)
Data sharing	Nurse outreach	Medina et al. (2020)
Student mental health	Behavior during the pandemic	Ben Hassen et al. (2020)
Prediction model	Progress of disease	Menni et al. (2020)
Teleophtalmology	Emergency eye care	Huckins et al. (2020)

Face Beatification

Bose et al. (2022) developed an image and face enhancement software. Different tools and filters have been developed to enhance the naturalness and drama of the picture. This application is entirely up to the individual.

- The incorporation of laboratory experts with new abilities and skills into interdisciplinary teams, with the participation of all stakeholders.
- Knowledge of the effects of laboratory mistakes on patient care.
- Adopting effective quality assurance systems to find and stop any mistakes that might happen in the lab.

AI in Pharmaceutical Industry

AI in Drug Discovery

In the last decade, artificial intelligence (AI) has revolutionised the field of drug research. Many drug discovery applications, such as virtual screening and drug design, have made use of AI approaches. In this survey, we first provide an overview of drug discovery and associated applications, which may be boiled down to two main tasks: predicting chemical properties and creating molecules. The common data resources, molecular representations, and benchmark platforms are then presented. AI techniques are divided into model architectures and learning paradigms as a major elements. Different types of work have been carried out, including a machine learning approach in the early 2000s. (Ma, R., and Luo, T. 2020. Deng et al. 2022). Machine learning models, such as random forest (RF), have been exploited for VS and QSAR (Ma et al. 2015, Lavecchia A. 2015). In 2012 the advent of the Deep learning era was marked by AlexNet (Krizhevsky et al. 2012, Alom et al. 2018). Merck Kaggle competition, deep neural networks (DNN) outperformed the standard RF model in predicting molecular activities (Ma et al. 2015). Recent advances in computer vision and natural language processing using AI techniques have a significant impact on drug discovery (Chen et al. 2018, Vamathevan et al. 2019, Ozt urk et al. 2020, Jim enez-Luna et al. 2021), leading to a deep learning approach in chemistry (Mater and Coote 2019). There are several other works related to AI application in drug discovery has been reviewed by Deng et al. (2022)

AI in Drug Repurposing

In the area of healthcare, artificial intelligence (AI) is proving its worth. In the aftermath of the COVID-19 pandemic, technology appears to be useful in limiting the spread of new diseases and assisting medical personnel in their treatment. As a result, the goal of this article is to look at how artificial intelligence can be used to combat the COVID-19 outbreak. Researchers discovered the following findings using a multiple case study approach. AI has been used in medication discovery and knowledge exchange, as well as tracking and prediction, clinical decision making and diagnosis, social distancing, and medical chatbots. AI is also used in contact tracking and social distance applications in detail. AI technologies could have a cross-cutting impact, with a new corporate social responsibility vein focusing on prevention methods. Medical professionals and policymakers get a tool to better comprehend artificial intelligence and make investment decisions based on the practical needs of the hour (Calandra D. and Favareto M., 2020).

AI in Clinical Trials

In the age of big data, artificial intelligence has advanced substantially, and it is increasingly having a profound impact on many parts of life. Its influence on the medical profession is inescapable (Balkin J. M. 2017). Voice input, medical image analysis, customised health data analysis, and independent drug research are only a few examples of the many medical applications of artificial intelligence. The positive effects of artificial intelligence in accelerating the work efficiency of the medical industry, advancing the medical cause, a series of medical ethical issues arising from the use of artificial intelligence in the medical industry, and a series of problems for medical industry employees. The negative consequences of artificial intelligence in the medical profession are addressed in three ways. It is hoped that it will make it easier to look at the pros and cons of artificial intelligence in an unbiased way (Li H. 2020).

Data Security and Encryption

Medical algorithms must comply with restrictions such as general data protection requirements. (GDPR, 2016), the Medical Devices Regulation (MDR) or the *in vitro*-Diagnostics Regulation (IVDR). No matter what the laws are in a country, the challenge of trying to use machine learning or even deep learning algorithms in laboratory diagnostics remains the same..

Eric Topol's books: "*Your ATM card works in outer Mongolia, but your electronic health record can't be used in a different hospital across the street*" (Topol E. 2019). Liang et al. (2019) tried for the first time to use deep learning algorithms to sort through patient records in an EHR and pull out clinically relevant data in order to find common diseases.

Even though some of these violations of patient's privacy may have occurred despite existing privacy laws, regulations, and policies, the Deep Mind case makes it clear that appropriate safeguards must be in place to protect patient privacy and patient autonomy in the context of these public–private partnerships. In addition to the danger of general abuses of power, AI poses a unique challenge since algorithms often need access to vast volumes of patient data and may utilise the data in a variety of ways over time (He et al. 2019).

Reidentification requires innovative and enhanced data security and anonymization methods. This requires innovation, along with regulatory components to ensure cutting-edge technology and safe methods to protect patient privacy (Murdoch B. 2021).

Social Stigma and AI

The harmful impact of societal stigma on the emotional, mental, and physical well-being of COVID-19 survivors and their families has been documented. (Chen et al. 2020, Chew et al. 2021, Hasinuddin et al. 2021). By using AI based mobile apps with a helpline facility in some countries, e.g., in Malaysia, the necessary food and other items have been delivered to the doorstep. It not only prevents the spread of disease, it also keeps the person's identity confidential. Implementation of artificial intelligence technologies into the healthcare industry is challenging, but it is unavoidable. Abdulaal et al. (2021) use an application based on ANN to predict what will happen to COVID-19 patients.

Discussion

Artificial Intelligence (AI) models have been shown to solve a variety of real-world healthcare problems. With the different uses of AI in many places of the healthcare industry, some security problems arise from time to time in healthcare ethics. But the use and move towards application of computer data oriented and use of AI in every sector is most welcome. The time and cost of different healthcare facilities have been reduced. Real time treatment facilities and AI governed robotic systems not only give the treatment in a timely manner, they also help the population stay away from different stigmas. To emphasise, proper implementation of the United Nations Sustainable Development Goals (UNSDGs) AI application in the healthcare system of society with proper ethical consideration should be welcomed.

Conclusion

Artificial Intelligence (AI) based medical diagnosis as discussed in X-ray, CT scan, MRI, ultrasonography, endoscopy and other diagnostics, including cancer research on diseases like melanoma, breast cancer, opthalmic diseases etc., is unavoidable. Although medical report analysis presents many challenges, AI can improve the working efficiency of health workers. The end users need to focus on interpreting the information with the patient's history and records to facilitate proper support and treatment. The confidentiality and data security and access option of data during a sudden disaster also needs to be kept in mind, which opens a new segment of the research arena for sustainable healthcare.

References

Abdulaal, A., Patel, A., Al-Hindawi, A., Charani, E., Alqahtani, S. A., Davies, G. W., ... & Moore, L. S. P. (2021). Clinical Utility and Functionality of an Artificial Intelligence–Based App to Predict Mortality in COVID-19: Mixed Methods Analysis. *JMIR Formative Research*, *5*(7), e27992.

Agarwal, P., Swami, S. and Malhotra, S. K. (2022), "Artificial Intelligence Adoption in the Post COVID-19 New-Normal and Role of Smart Technologies in Transforming Business: a Review", *Journal of Science and Technology Policy Management*, Vol. ahead-of-print No. ahead-of-print. https://doi.org/10.1108/JSTPM-08-2021-0122.

Akgundogdu, A., Jennane, R., Aufort, G., Benhamou, C. L., & Ucan, O. N. (2010). 3D image analysis and artificial intelligence for bone disease classification. *Journal of medical systems*, *34*(5), 815-828.

Alom, M. Z., Taha, T. M., Yakopcic, C., Westberg, S., Sidike, P., Nasrin, M. S., ... & Asari, V. K. (2018). The history began from alexnet: A comprehensive survey on deep learning approaches. *arXiv preprint arXiv:*1803.01164.Armitage, H. (2018). Artificial intelligence rivals radiologists in screening X-rays for certain diseases.Stanford Medicine News Center. 2018. https://med.stanford.edu/news/all-news/2018/11/ai-outperformed-radiologists-in-screening-x-rays-for-certain-diseases.html.

Bae, Y. S., Kim, K. H., Choi, S. W., Ko, T., Jeong, C. W., Cho, B., ... & Kang, E. (2020). Information technology–based management of clinically healthy COVID-19 patients: lessons from a living and treatment support center operated by Seoul National University Hospital. *Journal of medical Internet research,* 22(6), e19938. https://doi.org/10.2196/19938.

Balakumar, P., Maung-U, K., & Jagadeesh, G. (2016). Prevalence and prevention of cardiovascular disease and diabetes mellitus. *Pharmacological research,* 113, 600-609.

Balkin, J. M. (2017). *2016 Sidley Austin Distinguished Lecture on Big Data Law and Policy: The Three Laws of Robotics in the Age of Big Data.* Ohio St. LJ, 78, 1217.

Banks, G. (1986). Artificial intelligence in medical diagnosis: the INTERNIST/ CADUCEUS approach. *Critical reviews in medical informatics,* 1(1), 23-54.

Ben Hassen H., Ayari N., Hamdi B. A home hospitalization system based on the Internet of Things, Fog computing and cloud computing. *Inform Med Unlocked.* 2020;20:100368. https://doi.org/10.1016/j.imu.2020.100368.

Berná, G., & Romero-Gomez, M. (2020). The role of nutrition in non-alcoholic fatty liver disease: pathophysiology and management. *Liver International*, *40*, 102-108.

Binson, V. A., & Subramoniam, M. (2021, August). Artificial Intelligence Based Breath Analysis System for the Diagnosis of lung cancer. In *Journal of Physics: Conference Series* (Vol. 1950, No. 1, p. 012065). IOP Publishing.

Bose, P., Bandyopadhyay, S. K., Bhaumik, A., & Poddar, S. (2022). An Application for Facial Image Beautification. In *Ambient Communications and Computer Systems* (pp. 117-128). Springer, Singapore. https://doi.org/10.1007/978-981-16-7952-0_12

Bourdon, H., Jaillant, R., Ballino, A., El Kaim, P., Debillon, L., Bodin, S., & N'Kosi, L. (2020). Teleconsultation in primary ophthalmic emergencies during the COVID-19 lockdown in Paris: experience with 500 patients in March and April 2020. *Journal francais d'ophtalmologie,* 43(7), 577-585. https://doi.org/10.1016/j.jfo.2020.05.005.

Cabitza, F., & Banfi, G. (2018). Machine learning in laboratory medicine: waiting for the flood?. *Clinical Chemistry and Laboratory Medicine (CCLM),* 56(4), 516-524. Calandra, D., & Favareto, M. (2020). Artificial Intelligence to fight COVID-19 outbreak impact: an overview. *European Journal of Social Impact and Circular Economy,* 1(3), 84-104.

Canadian Blood Services. *Kidney Paired Donation (KPD) Program.* 2019. https://profedu.blood.ca/en/organs-and-tissues/programs-and-services/kidney-paired-donation-kpd-program. Accessed 15 Mar 2021.

Cao, D. S., Xu, Q. S., Liang, Y. Z., Chen, X., & Li, H. D. (2010). Automatic feature subset selection for decision tree-based ensemble methods in the prediction of bioactivity. *Chemometrics and Intelligent Laboratory Systems,* 103(2), 129-136. https://doi.org/10.1016/j.chemolab.2010.06.008.

Chen, H., Engkvist, O., Wang, Y., Olivecrona, M., & Blaschke, T. (2018). The rise of deep learning in drug discovery. Drug discovery today, 23(6), 1241-1250.Chen, Z. D., Zhao, L., Chen, H. Y., Gong, J. N., Chen, X., & Chen, C. Y. C. (2020). A novel artificial intelligence protocol to investigate potential leads for Parkinson's disease. *RSC Advances*, *10*(39), 22939-22958.

Cheng H.-Y., Jian S.-W., Liu D.-P., Ng T.-C., Huang W.-T., and Lin H.-H. (2020) Contact tracing assessment of COVID-19 transmission dynamics in Taiwan and risk at different exposure periods before and after symptom onset. *JAMA Intern Med*. (2020) 180:1156–63. https://doi.org/10.1001/jamainternmed.2020.

Cheng, L., Prasad, N., Engelhardt, B. E. (2019) An Optimal Policy for Patient Laboratory Tests in Intensive Care Units. In *Proceedings of the Pacific Symposium on Biocomputing*, Puako, HI, USA,

Chew, C. C., Lim, X. J., Chang, C. T. Rajan, Ph., Nasir, N, Low, W.-Y., (2021) Experiences of social stigma among patients tested positive for COVID-19 and their family members: a qualitative study. *BMC Public Health* 21, 1623. https://doi.org/10.1186/s12889-021-11679-8.

Cook, G. J., & Goh, V. (2019). What can artificial intelligence teach us about the molecular mechanisms underlying disease?. *European journal of nuclear medicine and molecular imaging*, *46*(13), 2715-2721.

Cortes C., Vapnik V. (1995) Support-vector networks. *Mach Learn.* 20:273-97. https://doi.org/10.1007/BF00994018.

Davis, L. E., Shalin, S. C., & Tackett, A. J. (2019). Current state of melanoma diagnosis and treatment. *Cancer biology & therapy,* 20(11), 1366-1379.Deng, J., Yang, Z., Ojima, I., Samaras, D., & Wang, F. (2022). Artificial intelligence in drug discovery: applications and techniques. *Briefings in Bioinformatics,* 23(1), bbab430.

Diez-Sanmartin, C., Sarasa-Cabezuelo, A., & Belmonte, A. A. (2021). The impact of artificial intelligence and big data on end-stage kidney disease treatments. *Expert Systems with Applications*, *180*, 115076.

Domingos P., Pazzani M. (1997) On the optimality of the simple Bayesian classifier under zero-one loss. *Mach Learn* 29:103-30. https://doi.org/10.1023/A:1007413511361.

Drew, D. A., Nguyen, L. H., Steves, C. J., Menni, C., Freydin, M., Varsavsky, T., ... & COPE Consortium § (2020). Rapid implementation of mobile technology for real-time epidemiology of COVID-19. *Science,* 368(6497), 1362-1367.https://doi.org/10.1126/science.abc0473.

Esteva, A., Kuprel, B., Novoa, R. A., Ko, J., Swetter, S. M., Blau, H. M., & Thrun, S. (2017). Dermatologist-level classification of skin cancer with deep neural networks. *Nature*, 542(7639), 115-118. *European In-Vitro-Diagnostica Regulation (IVDR).* Available online: https://eur-lex.europa.eu/legal-content/DE/TXT/?uri=CELEX:02017R0746-20170505

European Medical Devices Regulation (MDR). Available online: https://eur-lex.europa.eu/legal-content/DE/TXT/?uri=CELEX:02017R0745-20200424

FDA. *FDA permits marketing of artificial intelligence-based device to detect certain diabetes-related eye problems.* 2018. https://www.fda.gov/news-events/press-announcements/fda-permits-marketing-artificial-intelligence-based-device-detect-certain-diabetes-related-eye.

Freeman, K., Dinnes, J., Chuchu, N., Takwoingi, Y., Bayliss, S. E., Matin, R. N., ... & Deeks, J. J. (2020). Algorithm based smartphone apps to assess risk of skin cancer in adults: systematic review of diagnostic accuracy studies. *BMJ,* 368. *General Data Protection Regulation (GDPR).* Available online: https://eur-lex.europa.eu/eli/reg/2016/679/oj

Gershgorn D. The FDA just opened the door to let AI make medical decisions on its own. *Quartz.* 2018. https://qz.com/1251502/the-fda-just-opened-the-door-to-let-ai-make-medical-decisions-on-its-own/.

Greaves, R. F., Bernardini, S., Ferrari, M., Fortina, P., Gouget, B., Gruson, D., Lang, T., Loh, T. P., Morris, H. A., Park, J. Y., Roessler, M., Yin, P., Kricka, L. J. (2019) Key questions about the future of laboratory medicine in the next decade of the 21st century: a report from the IFCC-Emerging Technologies Division. *Clin Chim Acta*; 495:570–89. https://doi.org/10.1016/j.cca.2019.05.021.

Hashimoto D. A., Rosman G., Rus D., Meireles O. R. (2018) Artificial intelligence in surgery: promises and perils. *Ann Surg.*; 268(1):70–6.

Hasinuddin, M., Rusana, R., Noviana, U., Ekawati, H., Aini, N., & Poddar, S. (2021). Parents' Psychological Aspect in Caring for the Child During COVID-19 Pandemic. *Jurnal Ners,* 16(2), 193-197.

He J., Baxter S. L., Xu J., Xu J., Zhou X., Zhang K. (2019) The practical implementation of artificial intelligence technologies in medicine. *Nat Med.* 25(1):30–6.

Huber, L. (1999) Implementing 21 CFR Part 11 in Analytical Laboratories—Part 1: Overview and Requirements. *BioPharm*, *12*, 28–34.

Huckins J. F., daSilva A. W., Wang W., Hedlund E., Rogers C., Nepal S. K., Wu J., Obuchi M., Murphy E. I., Meyer M. L., Wagner D. D., Holtzheimer P. E., Campbell A. T.

(2020) Mental health and behavior of college students during the early phases of the COVID-19 pandemic: longitudinal smartphone and ecological momentary assessment study. *J Med Internet Res.* 17;22(6):e20185. https://doi.org/10.2196/20185.

Islam, M. M., Yang, H. C., Poly, T. N., Li, Y. J. (2020) Development of an Artificial Intelligence-Based Automated Recommendation System for Clinical Laboratory Tests: Retrospective Analysis of the National Health Insurance Database. *JMIR Med. Inform.* 8, e24163.

Jiménez-Luna, J., Grisoni, F., Weskamp, N., & Schneider, G. (2021). Artificial intelligence in drug discovery: Recent advances and future perspectives. *Expert opinion on drug discovery*, 16(9), 949-959.Knickelbein, J. E., Abbott, A. B., & Chew, E. Y. (2016). Fenofibrate and diabetic retinopathy. *Current diabetes reports,* 16(10), 1-6.

Kodali, P. B., Hense, S., Kopparty, S., Kalapala, G. R., & Haloi, B. (2020). How Indians responded to the Arogya Setu app?. *Indian Journal of Public Health,* 64(6), 228. https://doi.org/10.4103/ijph.IJPH_499_20. Krizhevsky, A., Sutskever, I., & Hinton, G. E. (2012). Imagenet classification with deep convolutional neural networks. Advances in neural information processing systems, 25.1097–1105.

Kumar, P., Yadav, A. K. S., & Singh, A. (2021). Prospective Of Artificial Intelligence: Emerging Trends In Modern Biosciences Research. In *IOP Conference Series: Materials Science and Engineering* (Vol. 1020, No. 1, p. 012008). IOP Publishing.

Lavecchia, A. (2015). Machine-learning approaches in drug discovery: methods and applications. *Drug Discovery Today*, 20(3), 318-331.Li, H. (2020). Impact of artificial intelligence based on big data on medical care. *Journal of Physics: Conference Series,* 1533(3): p. 032077. IOP Publishing.

Liang, H., Tsui, B. Y., Ni, H., Valentim, C. C. S., Baxter, S. L., Liu, G., Cai, W., Kermany, D. S., Sun, X., Chen, J., He, L., Zhu, J., Tian, P., Shao, H., Zheng, L., Hou, R., Hewett, S., Li, G., Liang, P., Zang, X., Zhang, Z., Pan, L., Cai, H., Ling, R., Li, S., Cui, Y., Tang, S., Ye, H., Huang, X., He, W., Liang, W., Zhang, Q., Jiang, J., Yu, W., Gao, J., Ou, W., Deng, Y., Hou, Q., Wang, B., Yao, C., Liang, Y., Zhang, S., Duan, Y., Zhang, R., Gibson, S., Zhang, C. L., Li, O., D. Zhang, E. D., Karin, G., Nguyen, N., Wu, X., Wen, C., Xu, J., Xu, W., Wang, B., Wang, W., Li, J., Pizzato, B., Bao, C., Xiang, D., He, W., He, S., Zhou, Y., Haw, W., Goldbaum, M., Tremoulet, A., Hsu, C., Carter, H., Zhu, L., Zhang, K., Xia, H. (2019) Evaluation and accurate diagnoses of pediatric diseases using artificial intelligence. *Nat. Med.*, *25*, 433–438.

Liu, L., Li, W., He, Z., Chen, W., Liu, H., Chen, K., & Pi, X. (2021). Detection of lung cancer with electronic nose using a novel ensemble learning framework. *Journal of Breath Research,* 15(2), 026014.

Ma, J., Sheridan, R. P., Liaw, A., Dahl, G. E., Svetnik, V. (2015). Deep neural nets as a method for quantitative structure–activity relationships. *J. Chem. Inf. Model*, 55, 263–274.

Ma, R., & Luo, T. (2020). PI1M: a benchmark database for polymer informatics. *Journal of Chemical Information and Modeling*, 60(10), 4684-4690.

Mater, A. C., Coote, M. L. Deep learning in chemistry. *J. Chem. Inf. Model* 2019, 59, 2545–2559.

Medina M., Babiuch C., Card M., Gavrilescu R., Zafirau W., Boose E., Giuliano K., Kim A., Jones R., Boissy A. Home monitoring for COVID-19. *Cleve Clin J Med.* 2020 Jun 11;1. https://doi.org/10.3949/ccjm.87a.ccc028.

Mei, J., Desrosiers, C., & Frasnelli, J. (2021). Machine learning for the diagnosis of Parkinson's disease: A review of literature. *Frontiers in aging neuroscience*, *13*, 184.

Menni C., Valdes A., Freidin M., Sudre C. H., Nguyen L. H., Drew D. A., Ganesh S., Varsavsky T., Cardoso M. J., El-Sayed Moustafa J. S., Visconti A., Hysi P., Bowyer R. C. E., Mangino M., Falchi M., Wolf J., Ourselin S., Chan A. T., Steves C. J., Spector T. D. (2020) Real-time tracking of self-reported symptoms to predict potential COVID-19. *Nat Med.* 26(7):1037–1040. https://doi.org/10.1038/s41591-020-0916-2.

Miller D. D., Brown E. W. (2018) Artificial intelligence in medical practice: the question to the answer? *Am J Med.* 131(2):129–133.

Murdoch, B. (2021) Privacy and artificial intelligence: challenges for protecting health information in a new era. *BMC Med Ethics* 22, 122. https://doi.org/10.1186/s12910-021-00687-3.

Naugler C., Church D. L. (2019) Automation and artificial intelligence in the clinical laboratory. *Crit Rev Clin Lab Sci.* 56:98-110.

O'Sullivan S., Nevejans N., Allen C., Blyth A., Leonard S., Pagallo U., Holzinger K., Holzinger A., Sajid M. I., Ashrafian H. (2019) Legal, regulatory, and ethical frameworks for development of standards in artificial intelligence (AI) and autonomous robotic surgery. *Int J Med Robot Comput Assist Surg*.;15(1):e1968.

Öztürk, H., Özgür, A., Schwaller, P., Laino, T., Ozkirimli, E. (2020). Exploring chemical space using natural language processing methodologies for drug discovery. *Drug Discov. Today*, 25, 689–705.

Padhy, S. K., Takkar, B., Chawla, R., & Kumar, A. (2019). Artificial intelligence in diabetic retinopathy: A natural step to the future. *Indian journal of ophthalmology,* 67(7), 1004.

Panch, T., Mattie, H., & Celi, L. A. (2019). The "inconvenient truth" about AI in healthcare. *NPJ digital medicine*, 2(1), 1-3.

Paranjape, K., Schinkel, M., Hammer, R. D., Schouten, B., Nannan Panday, R. S., Elbers, P. W., ... & Nanayakkara, P. (2021). The value of artificial intelligence in laboratory medicine: current opinions and barriers to implementation. *American Journal of Clinical Pathology*, 155(6), 823-831. https://doi.org/10.1093/ajcp/aqaa170.

Paul, D., Sanap, G., Shenoy, S., Kalyane, D., Kalia, K., & Tekade, R. K. (2021). Artificial intelligence in drug discovery and development. *Drug Discovery Today,* 26(1), 80.

Pawar, M. V., Pawar, A. M., Bhapkar, H., Anuradha, J., Bachate, R., Sharma, A., ... & Shardoor, N. (2022). Artificial intelligence-based solutions for COVID-19. In *Data Science for COVID-19* (pp. 167-189). Academic Press. https://doi.org/10.1016/B978-0-323-90769-9.00004-9

Pious, A. E., & Sridevi, U. K. (2021). A study of automatic segmentation of White Matter Hyperintensity for detection of Alzheimer's disease. *Journal of Physics: Conference Series*, . 1831(1):p. 012003. IOP Publishing.

Rabbani, M., Kanevsky, J., Kafi, K., Chandelier, F., & Giles, F. J. (2018). Role of artificial intelligence in the care of patients with nonsmall cell lung cancer. *European journal of clinical investigation*, 48(4), e12901.Ramirez, E., Brill, J., Maureen, K., Wright, J.

D. & McSweeny, T. Data Brokers: A Call for Transparency and Accountability (Federal Trade Commission, 2014).

Rong, G., Mendez, A., Assi, E. B., Zhao, B., & Sawan, M. (2020). Artificial intelligence in healthcare: review and prediction case studies. *Engineering,* 6(3), 291-301.

Ros M, Neuwirth L. (2020) Increasing global awareness of timely COVID-19 healthcare guidelines through FPV training tutorials: portable public health crises teaching method. *Nurse Educ Today.* 21;91:104479. https://doi.org/10.1016/j.nedt.2020.104479.

Sirinukunwattana, K., Aberdeen, A., Theissen, H., Sousos, N., Psaila, B., Mead, A. J., ... & Royston, D. (2020). Artificial intelligence–based morphological fingerprinting of megakaryocytes: a new tool for assessing disease in MPN patients. *Blood advances, 4*(14), 3284-3294.

Sivaranjani, R., & Yuvaraj, N. (2019). Artificial intelligence model for earlier prediction of cardiac functionalities using multilayer perceptron. *Journal of Physics: Conference Series..* 1362(1) :. 012062. IOP Publishing.

Sun, H., Tian, F., Liang, Z., Sun, T., Yu, B., Yang, S. X., ... & Liu, X. (2017). Sensor array optimization of electronic nose for detection of bacteria in wound infection. *IEEE Transactions on Industrial Electronics,* 64(9), 7350-7358.

Takenaka, K., Kawamoto, A., Okamoto, R., Watanabe, M., & Ohtsuka, K. (2022). Artificial intelligence for endoscopy in inflammatory bowel disease. *Intestinal Research*, *20*(2), 165.

Thompson R. F., Valdes G., Fuller C. D., Carpenter C. M., Morin O., Aneja S., Lindsay W. D., Aerts H. J., Agrimson B., Deville C. Jr, Rosenthal S. A. (2018) Artificial intelligence in radiation oncology: a specialty-wide disruptive transformation? *Radiother Oncol.*;129(3):421–6.

Thrall, J. H., Li, X., Li, Q., Cruz, C., Do, S., Dreyer, K., & Brink, J. (2018). Artificial intelligence and machine learning in radiology: opportunities, challenges, pitfalls, and criteria for success. *Journal of the American College of Radiology,* 15(3), 504-508.

Timmers, T., Janssen, L., Stohr, J., Murk, J. L., & Berrevoets, M. A. H. (2020). Using eHealth to support COVID-19 education, self-assessment, and symptom monitoring in the Netherlands: observational study. *JMIR mHealth and uHealth*, 8(6), e19822.. https://doi.org/10.2196/19822.

Tizhoosh, H. R., & Pantanowitz, L. (2018). Artificial intelligence and digital pathology: challenges and opportunities. *Journal of pathology informatics,* 9(1), 38.Topol, E. *Deep Medicine: How Artificial Intelligence Can Make Healthcare Human Again*; Hachette Book Group: New York, NY, USA, 2019.

U.S. Department of Health and Human Services Food and Drug Administration. *Guidance for Industry Part 11, Electronic Records; Electronic Signatures—Scope and Application*; U.S. Department of Health and Human Services Food and Drug Administration: Silver Spring, MD, USA, 2002.

Vamathevan, J., Clark, D., Czodrowski, P., Dunham, I., Ferran, E., Lee, G., Li, B., Madabhushi, A., Shah, P., Spitzer, M., Zhao, S. (2019) Applications of machine learning in drug discovery and development. *Nat. Rev. Drug Discov.* 18, 463–477.

Wilson, A. D. (2015). Advances in electronic-nose technologies for the detection of volatile biomarker metabolites in the human breath. *Metabolites*, 5(1), 140-163.

Wong, G. L. H., Yuen, P. C., Ma, A. J., Chan, A. W. H., Leung, H. H. W., & Wong, V. W. S. (2021). Artificial intelligence in prediction of non-alcoholic fatty liver disease and fibrosis. *Journal of gastroenterology and hepatology*, *36*(3), 543-550.

Wright A., Chen E. S., Maloney F. L. (2010) An automated technique for identifying associations between medications, laboratory results and problems. *J Biomed Inform.*; 43(6):891–901. https://doi.org/10.1016/j.jbi.2010.09.009. https://linkinghub.elsevier.com/retrieve/pii/S1532-0464(10)00141-3.

Xu, S., Hom, J., Balasubramanian, S., Schroeder, L. F., Najafi, N., Roy, S., Chen, J. H. (2019) Prevalence and Predictability of Low-Yield Inpatient Laboratory Diagnostic Tests. *JAMA Netw. Open*, *2*, e1910967.

Yamamoto K., Takahashi T., Urasaki M., Nagayasu Y., Shimamoto T., Tateyama Y., Matsuzaki K., Kobayashi D., Kubo S., Mito S., Abe T., Matsuura H., Iwami T. (2020) Health observation app for COVID-19 symptom tracking integrated with personal health records: proof of concept and practical use study. *JMIR mHealth uHealth.* 06;8(7):e19902. https://doi.org/10.2196/19902.

Yuan, Q., Zhang, H., Deng, T., Tang, S., Yuan, X., Tang, W., ... & Xiao, X. (2020). Role of artificial intelligence in kidney disease. *International Journal of Medical Sciences*, *17*(7), 970.

Zamberg I., Manzano S., Posfay-Barbe K., Windisch O., Agoritsas T., Schiffer E. (2020) A mobile health platform to disseminate validated institutional measurements during the COVID-19 outbreak: utilization-focused evaluation study. *JMIR Public Health Surveill.* 14;6(2):e18668. https://doi.org/10.2196/18668.

Chapter 2

Rice Leaf Diseases Prediction using PSO Segmentation and Deep CNN Classification

D. Akila[1,*], Kusum Yadav[2,†], Ranbir Singh Batth[3,‡], Souvik Pal[4,§] and Jayakarthik[5]

[1] Saveetha College of Liberal Arts and Sciences, SIMATS Deemed to be University, Chennai, India
[2]College of Computer Engineering and Science, University of Hail, Kingdom of Saudi Arabia
[3]Lovely Professional University, Punjab, India
[4]Sister Nivedita University, Kolkata, India
[5]Saveetha College of Liberal Arts and Science, SIMATS Deemed University, Chennai, India

Abstract

The annual rice crop has a direct and indirect impact on India's economic, political, and social stability. Rice production is the sole source of income for hundreds of millions of people. It is vital to detect any illness early and treat the afflicted plants to ensure the healthy and proper growth of rice plants. As a result, the farmer will be able to treat the crop properly and on time. In a broad field, automatic illness detection and identification is particularly helpful since it reduces the work, time, and cost associated with watching and analysing disease symptoms. Thus, automated disease recognition on leaves is critical in the agriculture business. This research describes a unique computational intelligence-

* Corresponding Author's Email: akiindia@yahoo.com,
† Corresponding Author's Email: Y.kusum@uoh.edu.sa.
‡ Corresponding Author's Email: ranbir.21123@lpu.co.in.
§ Corresponding Author's Email: souvikpal22@gmail.com.

In: Computational Intelligence for Sustainable Development
Editor: Brojo Kishore Mishra
ISBN: 979-8-88697-198-9

based approach for detecting and identifying rice leaf diseases using Particle Swarm Optimization (PSO) and Deep Convolutional Neural Networks (DCNN). The Particle Swarm Optimization technique is used to segment rice leaf pictures, which is critical for disease categorization. Deep convolutional neural networks are employed for the recognition of rice leaf disease classification. Our proposed rice leaf detection recognition method attained an average accuracy of 98.43%. In the field of rice leaf disease prediction, our technique has shown outstanding results.

Keywords: rice, leaf disease detection, image segmentation, image classification, PSO, deep convolutional neural networks

Introduction

Rice is now regarded as the best source of food all across the world, particularly in India. During the cultivation process, the paddy crop typically encounters a variety of illnesses at various phases of development. Early detection and treatment of such diseases can save a lot of money and result in a profitable harvest that is both plentiful and of high quality. Because the time required to get to such locations is insufficient, the best agricultural specialist advice is not available in remote areas. Rice is a staple diet for the majority of people in the country, accounting for two-thirds of per capita daily calorie intake. Rice agriculture that is disease-free would play a key role in sustaining economic growth and maintaining target levels [1, 2]. Figure 1 depicts the distinct patterns and shapes of these diseases.

The spread of several plant diseases has increased in recent years. Plant diseases, such as viral, bacterial, and fungal infections, can affect both above and below-ground plant sections. It's difficult to recognise the proper disease signs and know when to treat them. In the agricultural industry, the detection of numerous diseases using pictures is an attractive research subject. As a result, disease detection in different plant components has become increasingly significant and relevant in contemporary culture. For most of the processes in the workflow sequence to evaluate photos that can be used to identify illness in plants, image processing and machine learning provide a variety of techniques [3].

Infectious diseases are wreaking havoc on rice plants in the agriculture business, reducing crop quality and quantity. Automatic plant disease detection and identification are necessary to enhance production. Rice cultivation is considered the key agricultural sector in our country, as it is one of the most important cereal crops. Apart from pests and environmental variables, the primary issues faced by farmers during rice growth are illnesses that occur in the plant, which cause massive losses. Although numerous technologies for detecting agricultural diseases are available, such as image processing tools and remote sensing, the accuracy of such systems is poor [4]. If pests on crops and leaves are not properly evaluated and treated in a timely manner, food production quality and quantity may suffer, resulting in increased poverty, food insecurity, and mortality. This large effect can cause economic disruption in any country, especially in countries where 70% of the population depends on agricultural products for their livelihood and survival. Reducing or eliminating the proliferation of pests that affect agricultural productivity is one of the most difficult issues for agriculturists. Pests are organisms that carry disease, cause property damage, or cause annoyance. Plant pests include aphids, fungus, gnats, flies, thrips, slugs, snails, mites, and caterpillars. Pests create disease epidemics on occasion, resulting in famine and food scarcity [5].

Rice is farmed in over 100 countries in a variety of agroecological and socioeconomic settings. Rice is one of the most significant human food crops, accounting for one-tenth of all arable land on the planet. However, there are some plant diseases that can cause catastrophic economic and ecological losses. As a result, accurate and prompt disease diagnosis is critical. In the last four decades, rice insect outbreaks have risen, and the insect pest complex has altered. 42 of the 266 bug species discovered in rice habitats are considered pests. Identifying pests and illnesses, as well as the percentage of pest or disease occurrence, is critical for successful crop production [6]. Farmers frequently confront challenges in paddy production, harm to arable land, population growth, climate change, pests and diseases, and so on. Farmers are losing interest in paddy cultivation as a result of these different issues. This work has solely addressed pests and illnesses as they relate to rice farming difficulties [7].

Crop production is critical in locations where there is a food shortage. Crop diseases or pests can cause the grains to become loose. As a result, early diagnosis of these diseases can prevent crop loss and boost crop productivity. Early detection and identification of rice crop illnesses are essential for good yield, high quality, and high efficiency in rice production, which feeds around half of the world's population. Professionals must have a tremendous degree of experience and knowledge in order to manually identify rice illness [8].

Figure 1. (a) Leaf Smut, (b) Bacterial leaf blight, (c) Brown Spot disease [1].

Deep learning approaches to image classification have shown a lot of promise. Tea, apple, tomato, grapevine, peach, and pear diseases have all been investigated using similar procedures in recent years. Some have suggested building a feed-forward back propagation neural network from the ground up to identify plant species from leaf images. The neural network ensemble was used to identify five different tea plant diseases from the leaves (NNE). They used weather data, including temperature, humidity, rainfall, and wind speed, to build a neural network that could predict rice blast illness [9].

Indian farmers grow a wide variety of crops, including maize, rice, wheat, and cotton. Indian food production relies heavily on maize, the country's most important food crop. In order for it to grow, it requires a larger space than normal crops. Plant leaf disease has been detected using a number of image processing techniques in recent years. Yuxia Zhao et al. (2007) used threshold, regional labelling, and freeman chain coding methodologies to segment the corn leaf disease image and construct a recognition system based on the features of maize leaf diseases. A diagnostic expert system model was developed using an image analyzer by M EI-Helly et al. (2003). Abnormal characteristics on cucumber leaves might be identified by the machine. Using CCM textural analysis and the Mahalanobis statistical classifier, R. Pydipati (2007) classified leaf samples with an accuracy of above 95% using hue and saturation texture data. However, it has not yet been reported in the field of diagnosing rice leaf disease [11].

Researchers have recently become interested in deep learning algorithms due to their greater performance in image classification. Deep learning, unlike traditional machine learning techniques, does not require the extraction of complicated handcrafted features and allows for end-to-end learning. The most widely used deep learning technology for image categorization is the deep CNN. The CNN model connects image layers with spatial information, making it suitable for image classification. There have been few research utilising CNN to classify rice plant diseases [12]. Lighting, humidity, nutrients, fertiliser, water management, and farming conditions are all factors to consider. Manual diagnosis and identification of these disorders can take a long time, and recognition accuracy can be low. It could lead to incorrect pesticide diagnosis and application [13]. The capacity of deep CNN approach for the categorization of various rice diseases is investigated in this study.

As computer and internet technology advances, new image processing and pattern recognition techniques are constantly being developed to help identify individuals. Recent years have seen a rise in the use of digital image processing technologies, which has fueled new research and development in these areas. The most typical applications in agriculture are the identification and diagnosis of agricultural diseases, pests, and weeds. In the field of crop disease diagnosis, the classification and inspection of crop seed quality have generated some results, as has the quality detection and classification of agricultural products. For the objective of recognising vegetables and fruits, deep learning and convolution neural networks (CNN) were used [14].

Farmers will be able to use a simple digital camera to detect early signs of rice illness using the deep convolutional networks model. It's difficult to get good feature learning features for recognising rice diseases from raw inputs, and CNNs are generally assumed to automate this process [15]. The learnt features are viewed by the deep architecture as a more abstract representation of low-level raw images of rice sickness by the architecture. The best classification for improving diagnostic findings in pattern recognition tasks is considered to be CNNs. In this paper, we offer a model for identifying rice illnesses using deep convolutional neural networks. Using Particle Swarm Optimization (PSO) and Deep Convolutional Neural Networks, this study describes a novel computational intelligence-based approach for detecting and classifying rice leaf illnesses (DCNN). Rice leaf images are segmented using the Particle Swarm Optimization approach, which is important for disease classification. The recognition of rice leaf disease classification is done using deep convolutional neural networks. Our method has produced exceptional results in the field of rice leaf disease prediction.

Related Works

As a result of this, crop quality control has long been concerned about the detection of leaf illnesses and the extent of the disease. Computer vision and image processing techniques have been used in a variety of research to build various leaf disease detection systems. Small areas can only be examined for illness because of the difficulty of identifying disease on leaves in large areas. Automated disease detection in plants is an important research topic because it might be used to monitor large amounts of agricultural land more effectively. The following sections explain some of the existing systems in the problem area:

Shampa Sengupta et al. [16] an association rule mining-based incremental classification method based on the Particle Swarm Optimization algorithm was proposed by the researchers. Changes in categorization criteria are automatically applied to new data. The algorithm can construct dynamic categorization rules for such data without considering the current data, instead relying on previously extracted information from the existing data, if the data is not totally available at any given time, i.e., it is constantly incoming or rising. As a result, the system under consideration is limited to small to medium-sized data sets and must be updated on a regular basis. Not only does the method reduce computing time, but it also improves accuracy and other statistical measurements. Another advantage of the method is that it may be used to create categorization rules in both static and dynamic situations, where data arrives in stages. The study's purpose was to close the gap between empirical rice disease prediction and contemporary agricultural machine learning algorithms. Other crop diseases can also be predicted using this method.

Moumita Chanda et al. [17] particle swarm optimization is a computationally efficient evolutionary algorithm with theoretical and practical applications. When compared to the other approaches, PSO with BPNN offers better training performance, higher accuracy, and a faster convergence rate. The only issue in our suggested strategy is determining the initial values of the PSO parameters. In the future, the work could be expanded by devising a mechanism to ease the process of selecting the initial values of PSO parameters and applying it to a big dataset.

Taranjeet Singh et al. [18] digital images are more reliable for disease identification than human eyes because many diseases have similar traits. As a result, recognition is completely reliant on the vision of a human expert. A lot of progress is still possible in the field of Machine Learning and Image Processing, so that manual disease detection methods can be substituted for everyone's benefit. Large and high-quality image samples can be used in future study to propose a robust and dependable technique that can overcome the limitations of present techniques.

Kanthan Muthukannan et al. [19] for disease-affected portion segmentation, PSO-based image segmentation was examined. The hybrid features were then extracted from the diseased area using co-occurrence matrices of various diseases. Preprocessing was carried out, and the results were assessed using PSNR and Maxerror before segmentation and feature extraction. Different diseased plant leaves were studied and assessed for textural features such as contrast, correlation, homogeneity, and energy. Using a hybrid feature extraction technique, researchers found that it enhances classification accuracy and minimises misclassification in the categorization of plant leaf diseases. Finally, the proposed hybrid feature extraction approach could be helpful in better categorising various plant leaf diseases.

Ashraf Darwish et al. [20] using an ensemble of pre-trained CNNs, the proposed approach in their work could categorise maize photos into four distinct categories: three indicating sick leaves and one representing healthy leaves. The OLPSO optimization approach is used to optimise some of the hyperparameters of each individual model in the ensemble model. Pre-trained CNNs were created with EDLR and then applied to the data in this investigation. There are around 174,000 images of diseased and healthy leaves from a variety of plants in a publicly accessible database that was used in this investigation. In this experiment, only photographs of maize plants were used. It is possible to expand the size of an image in a given dataset by using the optimal model provided in this research, which takes into consideration the natural variability of each image by dividing it up. For plant image categorization, the suggested diagnostic model is both competitive and accurate, according to the results.

Lawrence C. Ngugi et al. [21] they gave a thorough overview of recent IPT-based plant disease recognition studies. We discussed the results of numerous designs that have been published recently. Shallow classifiers trained using handwritten features have been surpassed by deep learning algorithms. Deep learning algorithms may be able to identify pests and diseases with high accuracy if adequate data is given for training. Big Data with high variability, augmentation, transfer learning, and visualisation of CNN activation maps have all been highlighted as ways to improve accuracy.

Daniya T. et al. [22] the rice plant disease recognition system is primarily concerned with disease prediction in crop disease management that is quick and accurate. Rice plant disease can be detected early, allowing paddy researchers and farmers to take appropriate action to safeguard the plant. The fundamental purpose of this research is to survey and summarise how image processing and machine learning technologies can be used to detect rice plant illnesses. Various segmentation algorithms were used to retrieve the diseased leaf picture of a rice plant. The strategies discussed in this study assist researchers in overcoming the many difficulties that directly or indirectly influence society.

Moumita Chanda et al. [23] principal component analysis picks and selects the uncorrelated and independent features that are utilised for classification, and particle swarm optimization is a fast evolutionary approach of theoretical and practical significance. After lowering the amount of features with PCA, PSO with BPNN produces a stable and reliable classification. We expanded our dataset and applied our proposed classification algorithm BPNN with PSO to six plant diseases in this research. On our dataset, we also saw a comparison of our method with the other six selected methods.

Kavitha D. et al. [24] disease identification and categorization using plant leaf photos is an important component in agriculture. This research describes picture segmentation using particle swarm optimization and a minimal distance classifier for identifying illnesses in rice leaves. The suggested algorithm's effectiveness in classifying rice leaf diseases is proved effectively. Only a few samples were misclassified during the analysis testing. The proposed method has an average accuracy of 98.28%.

Nageswara Rao Moparthi et al. [25] they proposed this technique to assist farmers in increasing productivity and improving agricultural systems. This technology can detect diseases in paddy crops and also provides the medicines that may be employed to control that specific disease. We may also begin adopting new and smart technologies in agriculture, which would have a significant impact on the Indian economy. This system is built on Python language libraries and has an accuracy of 94 to 96 percent.

Punam Bedi et al. [26] it's tough and time-consuming to detect disease early in plants. Automated detection of plant diseases has been achieved using Machine Learning and Deep Learning algorithms. Most of these algorithms, on the other hand, need a large number of training parameters and have low classification accuracy. It was developed in this paper using the Convolutional Autoencoder (CAE) network and the Convolutional Neural Network (CNN) (CNN). Use of a CAE encoder network to compress leaf pictures was then employed for CNN classification in the hybrid model. The number of features and, as a result, the number of training parameters decreased significantly in comparison to the existing state-of-the-art systems due to the reduction in dimensionality using CAE. Bacterial Spot disease on peach plants was used to evaluate the model. A training accuracy of 99.55% and a testing accuracy of 98.08% were achieved with only 9,914 training parameters. Using fewer training parameters, it takes less time to train the model for automatic plant disease detection and to identify the illness in plants with the trained model.

Pritam C. Patil et al. [27] the use of Multiclass SVM and PSO Technique to detect and identify leaf diseases will be crucial in agricultural problem solving. These technologies automatically detect leaf diseases. To combat bacterial blight, brown spot, scald and blast on leaves, these methods have been used successfully on many different kinds of plants. As a final point, it has been proved that the best results can be reached with the least amount of effort to demonstrate the productivity of planned processes. By using the PSO technique to optimise the feature as diseases of the leaves, SVM will be shown to be a promising technique for distinguishing and categorising rice leaf diseases.

Divakar Yadav et al. [28] they suggested a new method for detecting disease in apple leaves that combined contrast stretching-based preprocessing with fuzzy c-means segmentation and CNN. Everything from photograph collecting to feature extraction and classification was meticulously documented. On the basis of the results of preprocessing and segmentation, a new deep convolutional neural network model has been developed that can automatically detect different features and properly determine apple leaf pathologies. When compared to the present state-of-the art, the outcomes of the suggested study were impressive. With 98 percent accuracy and a significantly smaller dataset size, the created technique outperforms the competition.

Kapilya Gangadharan et al. [29] a wide variety of plant diseases and insect attacks may be quickly and accurately detected and classified using the new technology. For the purposes of this study, a statistical feature selection technique and an RBPNN algorithm with a GA based PSO optimization method were investigated. To distinguish between healthy and diseased leaf patterns, S-PCA picks the features that the classifier needs, whereas EFFTA extracts the optimum texture feature values. We used a sigma noise filter to improve the image's noise sensitivity for the RBPNN. Image scaling is used to train the dataset from a variety of sources. The pattern classification problem can be handled by optimising the weight and the constant coefficient of activation function of the Neural Network structure. It solves cost estimating variational problems that are expressed analytically as unknown functions and multiclass problems. Over 3000 images were used to train the illness and infectious area. Furthermore, our method distinguishes between healthy and sick leaves. It performs better with the maximum prediction capabilities, with an accuracy of roughly 98.76 percent. This strategy is also effective when dealing with complex backgrounds.

Navneet Kaur et al. [30] the most significant contribution of this research is the development of an efficient ensemble classification system that utilises a variety of feature extraction strategies. Bell peppers and potatoes were both affected by two disease categories; tomatoes were affected by 10 disease categories. Image capture, segmentation, feature extraction, and classification are all involved, despite the fact that the focus is solely on the extraction and classification of features. These include GLCM, LBP, Gabor feature extraction, and SIFT feature extraction. Classifiers like SVM, ANN, KNN, logistic regression, and Nave Bayes were employed to construct an effective ensemble classifier. It has been investigated how well a categorization system with several features performs as a whole. When combined with the proposed research, our ensemble classifier produced the greatest results in terms of accuracy, precision, and recall. Bell pepper accuracy was 92.13 percent (2 categories), potato accuracy was 95.66 percent (3 categories), and tomato accuracy was 90.23 percent (10 categories).

Archana Chaudhary et al. [31] using the EnsPSO method has made a substantial contribution to the current research effort. Vegetable crop illnesses can be identified using the EnsPSO algorithm for multiclass classification. EnsPSO improves disease categorization accuracy by 96 percent when compared to Vote's 84 percent. EnsPSO outperforms Vote in a variety of other metrics. Using a 10-fold cross validation procedure, the EnsPSO technique is further validated for classification accuracy on three standard UCI datasets. Vote is outperformed by EnsPSO, according to these findings. The current study leaves room for future research. The EnsPSO method can also be used to identify diseases in other crops.

Krishnamoorthy N et al. [32] we developed an InceptionResNetV2 deep convolutional neural network with a transfer learning technique for detecting rice leaf diseases. There are three main diseases that target rice plants: leaf blast, bacterium blight, and brown spot. The healthy class is also taken into account in this study. The simple CNN model was fine-tuned using numerous hyper parameters and obtained an accuracy of 84.75 percent after 15 epochs. Using 10 epochs and fine-tuning several hyper parameters, InceptionResNetV2 achieved an optimum accuracy of 95.67 percent.

Rishabh Sharma et al. [33] Crop diseases have become a common part of the agricultural field and with the growth of the agricultural field, these crop diseases are also increasing day by day. Rice crop is one of the main crop and its plantation has spread in almost every region of India and many parts of the globe also. Rice infections are highly prevalent, and different machine learning algorithms have been established in recent decades to detect them. In this report, we did a survey study on eight main rice diseases, including bacterial leaf blight, false smut, rice hispa, blast, stemborer, sheath blight, brown spot, and brown planthopper, and we used CNNs to examine them. The paper is separated into two sections: the survey methodology used to conduct the work and the state of the art used for rice disease detection (RDD) utilising CNNs approach.

Summary:

- Instead of relying on human leaf recognition, agricultural specialists can use these new technology to rapidly identify the proper pesticide and dosage.
- Some quick recognition measures are necessary for paddy leaf disease diagnosis.
- The Artificial Intelligence Revolution has made maintaining a standard of living easier. Plant disease is one of several issues in agriculture that technology has made much easier to deal with. Machine learning and deep learning are currently being used to diagnose a wide range of diseases.

Proposed Method

Because of its considerable impact on the economy and society, the majority of countries now consider agriculture to be one of the most important components of sustainable development. As a result, continual monitoring of plants is critical in detecting plant illnesses. Typically, traditional crop inspection methods rely on visual approaches with the assistance of expertise and training to detect plant illnesses. Traditional approaches concentrate on the mental and cognitive processes that might lead to dishonesty and mistakes. As a result, we suggested a rice leaf disease detection method based on computational intelligence.

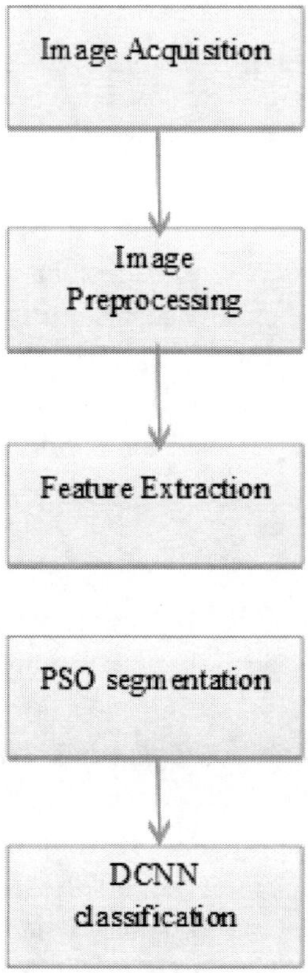

Figure 2. Rice leaf disease recognition system.

This study looked at how deep learning technology could be used to detect rice disease automatically. This paper employs Particle Swarm Optimization (PSO) and Deep Convolutional Neural Networks to detect and identify rice leaf diseases (CNN). Rice leaf images are segmented using the Particle Swarm Optimization approach, which is important for disease classification. The Deep Convolutional Neural Network (DCNN) architecture was used to categorise the images. Figure 2 shows the general framework of a Rice Leaf disease recognition system.

The proposed model's algorithm is presented in the following steps:

1. Gathering the database and categorising it based on the plant's characteristics.
2. Using an image as the test/input image.
3. Array conversion and data storage for the image.

4. The data is then Preprocessed and stored for further extraction.
5. The stored image data then features extracted.
6. The extracted image data is used for segmentation using PSO algorithm
7. The segmented image data we got from datasets are then used to classify the diseases using DCNN classification.
8. Comparison of these both test and train images.
9. Diseases in the plant are found if there are any, remedies and the accuracy is given.

Image Acquisition

The researchers interviewed Department of Agriculture workers, specifically the staff of the Regional Crop Protection Center, and also examined and assessed documents relevant to rice pests and diseases to get the necessary data on the numerous rice pests and diseases. Image capturing tools such as a digital camera and a smart phone were used to capture images of different rice pests and illnesses. All photographs were pre-processed and encoded after collection to become part of the dataset.

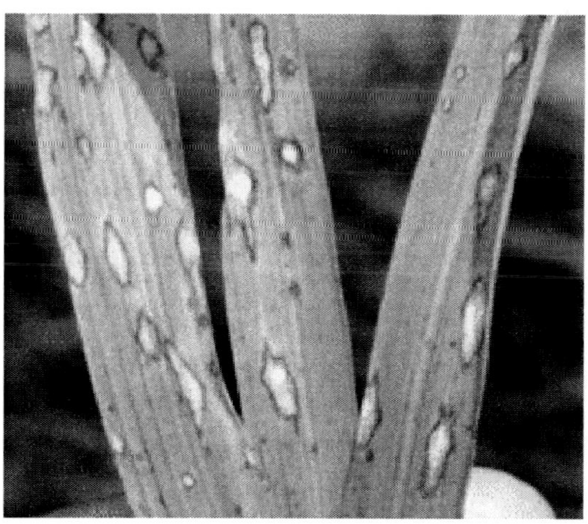

Figure 3. Blast disease affected leaf.

The rice fields of the Indira Gandhi Agricultural University in Raipur, Chhattisgarh, India, provided the setting for the study's photographs. These photos were taken between 6.30 a.m. and 9.30 a.m., as well as 4 p.m. and 5.30 p.m. When the sun is shining, the image received at this time is clearer than at other times of the day. The photos were taken with a Canon Powershot SX530HS digital camera, as well as a Gionee and LYF mobile phone. To reduce background complexity and hence processing expense, the image backdrop is prepared to be black. There are 979 images in four categories, including Brown Spot (RB), Leaf Blight, Leaf Blast, Leaf Smut, and Healthy Leaves (HL). First time a huge dataset of rice illnesses from the Indian farming industry has been gathered.

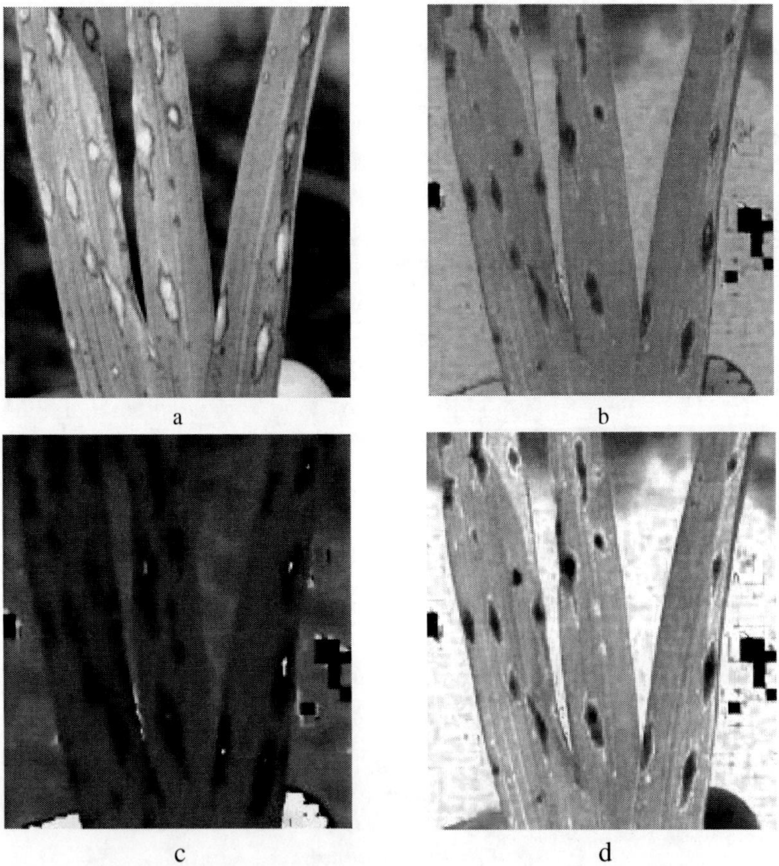

Figure 4. (a) Rgb Image, (b) Rgb to HSV, (c) Hue part, (d) Saturation part.

Image Preprocessing

Image samples may contain noise and are not suited for direct processing, thus preprocessing processes are utilised instead. In this category are operations such as cropping, enhancing, filtering, smoothing, and changing the colour of an image. This stage enhances the optical evaluation of the picture samples. To sharpen the photos, Auzi et al. (2013) used Laplacian filters after converting the images to HSV colour space, extracting the S component, and then using histogram equalisation. Pictures of a diseased area were removed, after which the photos were changed to their suitable grey levels and the image was enhanced with the Laplacian filter. The collected picture data was changed to HIS colour space since some infected portions of plant leaves have higher intensity values than others. Cropped photos were used to increase CPU efficiency, decrease processing time, and increase disc storage. Consequently, 30 percent of storage space was conserved, CPU speed was boosted by 1.4 times, and no regions of interest were lost. The backgrounds of the collected photographs were separated after converting RGB image samples to YCbCr colour space and extracting the Y, Cb, and Cr components.

Feature Extraction

In image processing, extraction of image features is a trait that extracts the basic attributes that must be evaluated. This is the most important stage in image processing after segmentation.

In plant disease detection, feature extraction is utilised to extract the contaminated part of the leaf image. Every leaf disease has distinguishing traits that aid in its classification. Color, shape, and texture are characteristics that are used in the best plant disease detection. These features are the foundation of illness detection methods, and their effectiveness is dependent on their right selection. In this article, the area, perimeter, eccentricity, mean, standard deviation, and vein features are all analysed. These sections of the leaf are critical to understanding what is going on. Thus, these characteristics were chosen to represent this idea.

In photography, an area refers to a section of the image that is distinct from the rest. The number of pixels with power 1 from a data leaf image represents the number of real pixels in the area.

$$Area = \sum_{s,d \in S} 1 \quad (1)$$

where s,d reflect the region's size. S

The perimeter of an oval leaf image is defined as the space surrounding the locale's limit.

Eccentricity: The eccentricity of the circle encloses the precise second minute as the region in a picture of a maize plant leaf. An oval's focal point and its vast rotational length are displayed as the distance between them. It is a measure of how far two points are separated from one another. In most cases, the least bouncing square shape approach or the boss hatchets procedure will be employed.

The mean is frequently used to illustrate the brilliance of an image via the lens of motivation. Bright images have a high mean value across the board, whereas faint images have a low mean value Following is a breakdown of how the average is calculated:

$$F = \frac{1}{O} \sum_{i=1}^{O} Qij \quad (2)$$

Here Qji represented the value of the j colour channel at the i image pixel is Qji.

When discussing standard deviation, the square root of change is usually used. Low contrast images are less likely to fluctuate, whereas high contrast images are more likely. It is laid out as depicted in the diagram below:

$$\sigma_j = \sqrt{(\frac{1}{O} \sum_{i=1}^{O} (Qij - fj)^2)} \quad (3)$$

In a modified plant collection and identification technique, the leaf vein, one of the largest and most complex components of the leaf, is used as a confirmation for customised plant recognition. It is possible to identify fundamental plant types from leaf photographs using these characteristics. On the other hand, it's here that one of the "leaf veins" becomes apparent. This process is essential for determining plant parts. The following are details on the vein memories we intend to use in our method:

$$X(l) = \frac{X*(l)}{\sqrt{X*(l) Dx*(l)}} \quad (4)$$

The formula for our proposed leaf venation can be found in Equation (4). X l is the complex conjugate of X l, which is symbolised by the letter l. Based solely on the colour and shade of the leaf photos, this method is only applicable to the best examples.

PSO Segmentation

Because of the complexity and variety of images, image segmentation is a difficult procedure. Lighting, contrast, interference, and other factors all have an impact on segmentation results. Using segmentation, you can pinpoint trouble spots and narrow down the condition.

PSO is a rapidly increasing field in all areas, as well as a very powerful approach that may be used effectively in a variety of image processing fields. Image processing techniques such as segmentation and thresholding, enhancement and classification can greatly boost the performance of PSO.

Advantages of the PSO algorithm:

- A continuous optimization strategy, PSO effectively optimises data points.
- It is possible to control a large number of variables simultaneously.
- The ability to search across a broad testing group of price surfaces.
- Using highly composite cost surfaces, the ability to modify data points.
- Provides a variety of ideal options. As a result, multiple image dissection results may be acquired simultaneously.

As a nature-inspired algorithm, PSO is capable of mimicking the behaviour of a natural swarm. It disseminates essential information and imitates group affiliate communication. It has been used to solve a number of problems and can be combined with a variety of machine learning techniques. Different agents, sometimes known as particles, hunt for the best potential solution in this algorithm using fine-tuned pathways via a stochastic and deterministic component.

The 'best' acquired position as well as the 'best' position of the group drive each particle despite its tendency to wander randomly. Particle I's location and velocity vectors are denoted by xi and vi, respectively. As shown below, each particle in each cycle changes its position in response to the new velocity:

$$v[\,] = v[\,] + c1 * rand(\,) * (pbest[\,] - present[\,]) + c2 * rand(\,) * (gbest[\,] - present[\,]) \quad (5)$$

$$present\,[\,] = present\,[\,] + v\,[\,] \quad (6)$$

A random number between (0,1), rand (), pbest[] and gbest[] are the best local and global solutions for each particle, while c1 and c2 are constants that have been pre-specified.

$$c1 = (c11 - c12) * (it\,/\,nPop * 10) + c12 \quad (7)$$

$$c2 = (c22 - c21) * (it\,/\,nPop * 10) + c21 \quad (8)$$

According to empirical investigations, the values of parameters (w, c1 and c2) for typical PSO are as follows: w should be around 0.7 to 0.8, and c1 and c2 should be around 1.5 to 1.7. The experimental results of our study reveal that the values of w and (c1 and c2) for picture segmentation should be 0.9 and 1.2, respectively.

Algorithm 1. PSO segmentation Algorithm

For each particle:
Get the particle going.
END
Perform
Identify each particle
Calculate your fitness level.
If the fitness value is higher than any other time in history.
Make pbest the current value.
END
Create an initial swarm of size swarm size using Eq. (1). (2).
For each particle, create a velocity array.
END for.
If (epsum _ pbest)
upgrade pbest value point
upgrade pbest direction

End if
Repeat
There is a particle in the swarm for every particle in the swarm.
Calculate the fitness value using Eq (5).
pbest = the particle's best position at the moment. gbest denotes the particle's best global position.
Update the particle's velocity with pbest and gbest.
Substitute a new particle for an old one.
end-for
Until the requirements for termination are met
In R, type gbest.
Return
end-for 2
end-for (R)
End

Minimum error or maximum number of iterations are not met, so long as these conditions exist. The method uses a high number of clusters to lessen the complexity of the initial circumstances. In order to determine the optimal number of clusters, the binary PSO is used, and k-means clustering refines the centres of the selected cluster. Clustering is a method of unsupervised image segmentation. Using the PSO, each pixel is assigned to a cluster. Using the proposed technique, the user can select any validity index depending on the available data. Functions like this one's fitness assessment are based on quantitative data.

DCNN

CNN is a type of deep, feedforward ANN that has been effectively applied to computer vision applications in machine learning. In contrast to ANN, whose training requirements in terms of time are unfeasible in some large-scale issues, CNN can learn complicated problems extremely quickly due to weight sharing and more complex models utilised, which enable huge parallelization.

Deep Convolutional neural networks can improve their chances of proper classification if big enough data sets (hundreds to thousands of measurements, depending on the complexity of the subject under study) are available to describe the problem. Convolutional, pooling, and/or completely linked layers make up these layers. The convolutional layers extract features from the input images, which the pooling layers then reduce in dimensionality, while the fully connected layers act as classifiers. The completely linked layers at the final layer typically use the learned high-level properties to classify input images into specified classes.

Because of its highly hierarchical structure and huge learning capacity, DCNN models excel at classification and prediction while remaining versatile and adaptive to a wide range of problems.

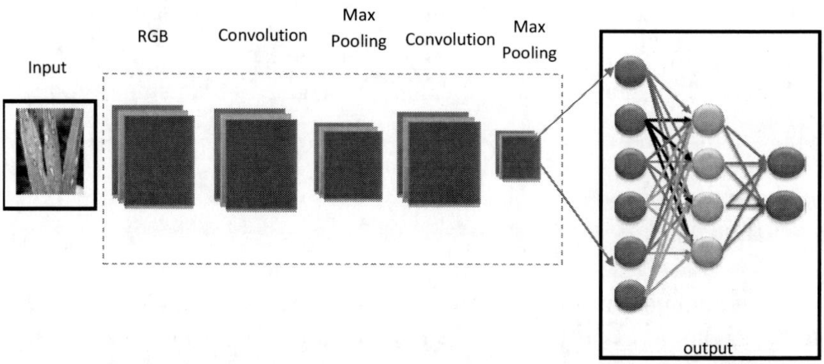

Figure 5. DCNN leaf disease classification.

A DCNN's construction is depicted in Figure 5. An Input Layer, an Output Layer, and a number of hidden layers are all part of it. Hidden layers include convolution, pooling, and totally linked layers. A convolutional layer is added after the input layer, and it is responsible for calculating what happens when certain input-connected neurons are stimulated. The data will be downsampled along the spatial dimensions by the pooling layer. The class scores will be computed by the fully linked layer.

Rice Disease Type and Dataset Descriptions

Data Types and Description

The rice fields of the Indira Gandhi Agricultural University in Raipur, Chhattisgarh, India, provided the setting for the study's images. Photos were taken between 6.30 and 9.30 am and 4.30 to 5.30 p.m on two separate occasions: one at 4:20 and one at 5.30 pm on two separate days. It is clearer while the sun is shining than at other times of the day. The photos were taken with a Canon Powershot SX530HS digital camera, as well as a Gionee and LYF mobile phone. To reduce background complexity and hence processing expense, the image backdrop is prepared to be black. There are 979 images in four categories, including Brown Spot (RB), Leaf Blight, Leaf Blast, Leaf Smut, and Healthy Leaves (HL). First time a huge dataset of rice illnesses from the Indian farming industry has been gathered.

The rice plant's leaf, stem, and grain show signs of several diseases and pests in Figure 6. Toxins found on the rice leaf include: bacterial leaf blight, brown spot, brown planthopper pest, and hispa pest.

Figure 6. A sample image of each detected class [9].

a. *Brown Spot (BS)*

Early signs of a fungal disease that affects the entire crop are brown oval or circular spots that appear on seedling leaves. As shown in

Figure 6 (c), the culprit is Bipolaris Oryzae, a fungus that not only affects yield but also degrades grain quality.

b. *Leaf Blast (LB)*
It's caused by Magnaporthe oryzae, which causes lesions on the plant's leaves and other parts of the plant, such as stems, roots, and seeds, to be affected. Like the brown spot in this example, the patches on the leaves are boat-shaped with a grey centre and a thick brown edge.

c. *Bacterial Leaf Blight (BLB)*
Bacterial leaf blight is caused by a bacterium that enters the seedling through hyadathodes and cuts lesions in the tip of the leaf, resulting in seedling death (Figure 4 (a)). Within a few days, the incisions enlarge with a wave-like edge, turning the straw yellow. As the disease develops, the lesions turn the entire leaf white or straw-colored. Dew droplets containing bacterial colonies can be found on fresh wounds in the early morning hours.

d. *Leaf Smut (LS)*
Leaf smut is distinguished by vertical scratches on leaf blades that are slightly black in colour; The leaf tips may turn grey and dry as a result of LS.

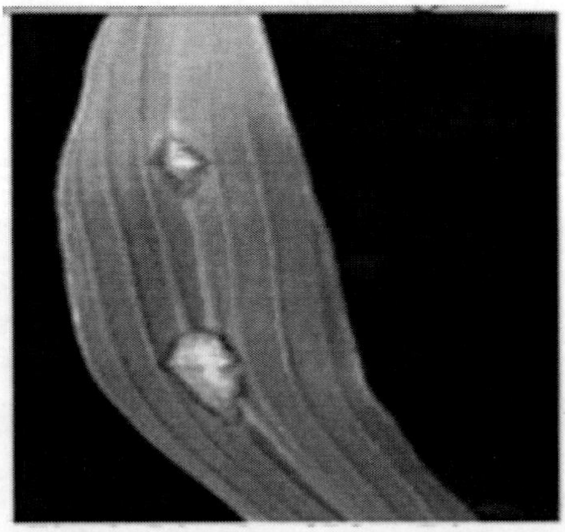

Figure 7. Rice leaf blast disease [2].

It is not a serious sickness, but it can create an atmosphere that fosters and stimulates the growth of fungi, which can lead to the spread of other diseases. As seen in Figure 8, LS wounds on leaves can be oval, round, or irregular in shape, with a rough surface.

e. *Healthy Leaf*
   All illnesses will be absent from a healthy paddy leaf. There should be no symptoms of disease on the leaves, and they should be bright green.

Figure 8. Leaf smut disease [2].

Data Preprocessing

The Keras ImageDataGenerator function was used for data preprocessing. All 979 images in the collection have three colour channels and varying pixel values, making each unique. Take each photo and shrink it to fit on the screen. Changing the point of view on the visual item is made possible by rotating the training images in a 15-degree range continually. Unlike the zoom and shear ranges, the width and height shift ranges are always 0.1. The only common preprocessing approach used in both training and testing datasets is rescaling pictures. There are eight batches in the training dataset, but only one in the test dataset.

Table 1. Each disease's number of training and test images

Rice plant disease name	**Images**	**Number of Images Used for Train and Validation**	**Number of Images used for Test**
Brown Spot	161	137	29
Leaf Blast	158	130	29
Leaf Blight	215	187	29
Leaf Smut	215	187	29
Healthy Leaf	226	198	29
Total	979	839	145

The suggested system's efficiency was evaluated using various assessment metrics. A summary of these metrics is shown below [28].

Accuracy(A): The proportion of correctly classified examples to all classified examples is the accuracy measure.

$$\text{Accuracy(A)} = (\text{TP} + \text{FN})/(\text{TP} + \text{FP} + \text{TN} + \text{FN}) \tag{9}$$

It's the percentage of positive patterns in a class predicted properly that is known as precision (P).

$$\text{Precision(P)} = \text{TP}/(\text{TP} + \text{FP}) \tag{10}$$

In order to calculate the percentage of positive events that are accurately detected, recall (R) is used.

$$\text{Recall(R)} = \text{TP}/(\text{TP} + \text{TN}) \tag{11}$$

For example, the F1-score metric is the harmonic mean of recall and precision values, which is called the F1-score.

$$\text{F1score(F)} = 2 * \text{P} * \text{R}/(\text{P} + \text{R}) \tag{12}$$

where, TP = True Positive, FP = False Positive, TN = True Negative and FN = False Negative.

Table 2. Precision of the proposed system

Rice plant disease name	TP	FP	FN	TN
Brown Spot	148	9	13	92
Leaf Blast	129	12	29	80
Leaf Blight	198	17	17	117
Leaf Smut	203	2	12	94
Healthy Leaf	218	4	8	87

Rice plant disease name	Precision
Brown Spot	94.27
Leaf Blast	91.49
Leaf Blight	92.09
Leaf Smut	98.95
Healthy Leaf	98.32

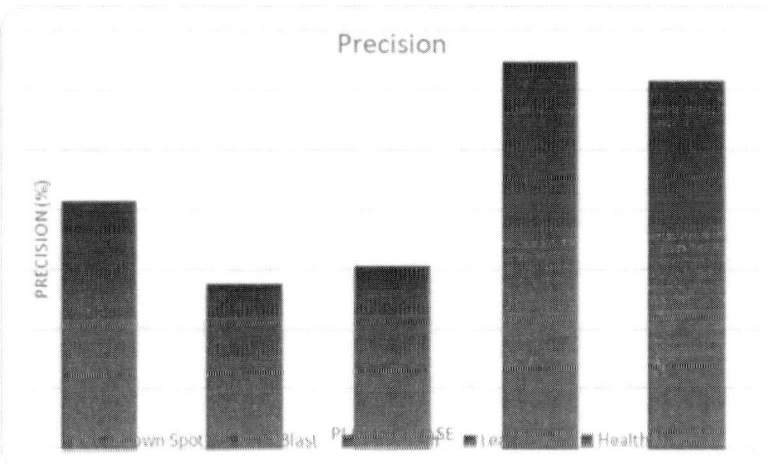

Figure 9. Precision of the proposed system.

Table 2 and Figure 9 shows the Precision of the proposed system. The average Precision is 0.986 for our proposed Rice leaf detection recognition system. The healthy leaf accuracy is 1.0.

Table 3 and figure 10 shows the Recall of the proposed system. The average Recall is 0.981 for our proposed Rice leaf detection recognition system. The healthy leaf accuracy is 1.0.

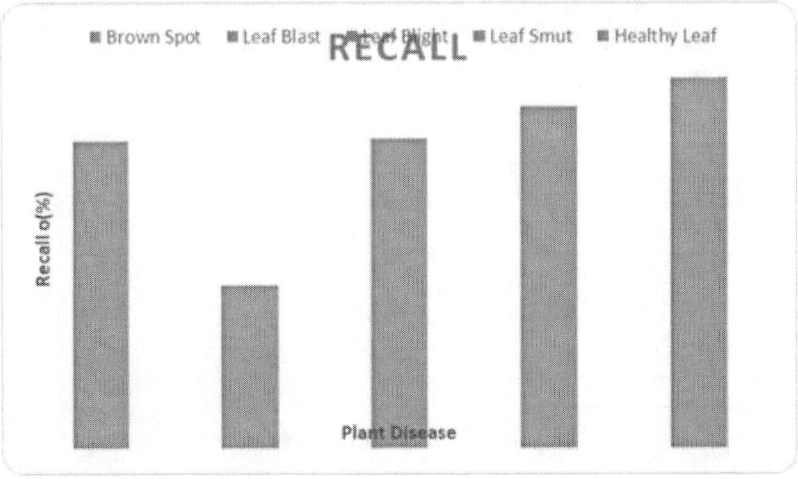

Figure 10. Recall of the proposed system.

Table 3. Recall of the proposed system

Rice plant disease name	**Recall**
Brown Spot	91.93
Leaf Blast	81.65
Leaf Blight	92.09
Leaf Smut	94.42
Healthy Leaf	96.46

Table 4. Accuracy of the proposed system

Rice plant disease name	**Accuracy**
Brown Spot	91.60
Leaf Blast	83.60
Leaf Blight	90.26
Leaf Smut	95.45
Healthy Leaf	96.30

Table 4 and Figure 11 shows the accuracy of the proposed system. The average accuracy is 98.43% for our proposed Rice leaf detection recognition system. The healthy leaf accuracy is 100%.

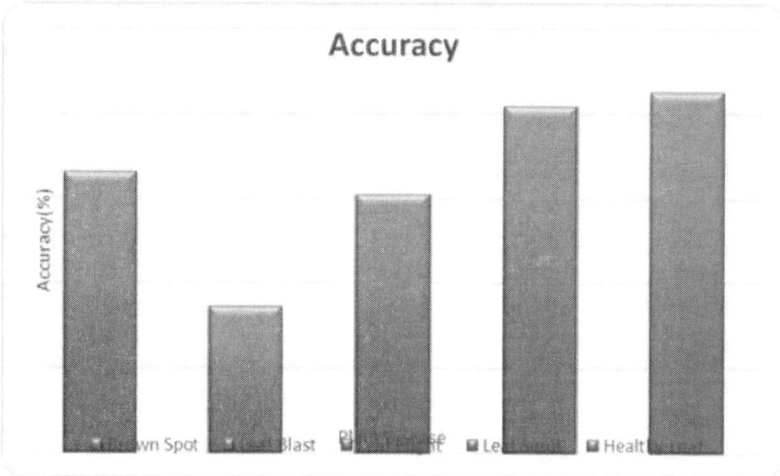

Figure 11. Accuracy of the proposed system.

The segmental results obtained from Deep Convolutional Neural Network (DCNN) based plant disease classification and Particle Swarm Optimization (PSO) based plant disease segmentation are presented in the tables and figures above, and the effectiveness of our approach is evaluated using accuracy measures. Our proposed Rice leaf detection recognition method has an average accuracy of 98.43 percent. The accuracy of healthy leaves is 100 percent. When compared to other methods, the proposed method produces superior outcomes than existing methods. The average Recall for our suggested Rice leaf detection recognition system is 0.981. Our proposed Rice leaf detection recognition system has an average Precision of 0.986. In this case, too, our technique yielded superior outcomes. This is because of the Computational Intelligence technique, which is based on Particle Swarm Optimization (PSO) and Deep Convolutional Neural Networks (DCNN).

Conclusion

In the agricultural realm, illness detection and recognition using deep learning and images of rice crops is a signal. Machine learning and deep learning techniques and technologies have opened up new prospects for academics because they can reveal several previously unknown areas. The best classification for improving diagnostic findings in pattern recognition tasks is considered to be CNNs. In this paper, we offer a model for identifying rice illnesses using deep convolutional neural networks. New computational intelligence-based methods for detecting rice leaf illnesses are described in this paper using PSO and Deep Convolutional Neural Networks (DCNN). Rice leaf images are segmented using the Particle Swarm Optimization approach, which is important for disease classification. Rice leaf disease categorization is recognised using deep convolutional neural networks. In the field of rice leaf disease prediction, our technique has shown outstanding results. Our proposed Rice leaf detection recognition method attained an average accuracy of 98.43 percent. When compared to the old system, we achieved better accuracy.

In the future, for the study challenge at hand, a plan of action will be devised, along with an investigation into machine learning and segmentation techniques that could help in the identification of diseased crops.

References

[1] Ahmed, K., Shahidi, T. R., Alam, S. M. I., & Momen, S. (2019, December). Rice leaf disease detection using machine learning techniques. In *2019 International Conference on Sustainable Technologies for Industry 4.0 (STI)* (pp. 1-5). IEEE.

[2] Vasantha, S. V., Kiranmai, B., & Krishna, S. R. (2021). Techniques for Rice Leaf Disease Detection using Machine Learning Algorithms. *Int. J. Eng. Res. Technol*, *9*(8), 162-166.

[3] Pothen, M. E., & Pai, M. L. (2020, March). Detection of rice leaf diseases using image processing. In *2020 Fourth International Conference on Computing Methodologies and Communication (ICCMC)* (pp. 424-430). IEEE.

[4] Ramesh, S., & Vydeki, D. (2018, September). Rice blast disease detection and classification using machine learning algorithm. In *2018 2nd International Conference on Micro-Electronics and Telecommunication Engineering (ICMETE)* (pp. 255-259). IEEE.

[5] Rastogi, A., Arora, R., & Sharma, S. (2015, February). Leaf disease detection and grading using computer vision technology & fuzzy logic. In *2015 2nd international conference on signal processing and integrated networks (SPIN)* (pp. 500-505). IEEE.

[6] Mique Jr, E. L., & Palaoag, T. D. (2018, April). Rice pest and disease detection using convolutional neural network. In *Proceedings of the 2018 international conference on information science and system* (pp. 147-151).

[7] Islam, M. A., Shuvo, N. R., Shamsojjaman, M., Hasan, S., Hossain, S., & Khatun, T. (2021). An automated convolutional neural network based approach for paddy leaf disease detection. *Int. J. Adv. Comput. Sci. Appl*, *12*(1), 280-288.

[8] Rathore, N. P. S., & Prasad, L. (2020). Automatic rice plant disease recognition and identification using convolutional neural network. *Journal of Critical Reviews, 7*(15), 6076-6086.

[9] Rahman, C. R., Arko, P. S., Ali, M. E., Khan, M. A. I., Apon, S. H., Nowrin, F., & Wasif, A. (2020). Identification and recognition of rice diseases and pests using convolutional neural networks. *Biosystems Engineering, 194*, 112-120.

[10] Akanksha, E., Sharma, N., & Gulati, K. (2021, January). OPNN: Optimized probabilistic neural network based automatic detection of maize plant disease detection. In *2021 6th International Conference on Inventive Computation Technologies (ICICT)* (pp. 1322-1328). IEEE.

[11] Liu, L., & Zhou, G. (2009, December). Extraction of the rice leaf disease image based on BP neural network. In *2009 International Conference on Computational Intelligence and Software Engineering* (pp. 1-3). IEEE.

[12] Shrivastava, V. K., Pradhan, M. K., Minz, S., & Thakur, M. P. (2019). Rice plant disease classification using transfer learning of deep convolution neural network. *International Archives of the Photogrammetry, Remote Sensing & Spatial Information Sciences, 42*(3), W6.

[13] Li, D., Wang, R., Xie, C., Liu, L., Zhang, J., Li, R., & Liu, W. (2020). A recognition method for rice plant diseases and pests video detection based on deep convolutional neural network. *Sensors, 20*(3), 578.

[14] Jiang, F., Lu, Y., Chen, Y., Cai, D., & Li, G. (2020). Image recognition of four rice leaf diseases based on deep learning and support vector machine. *Computers and Electronics in Agriculture, 179*, 105824.

[15] Lu, Y., Yi, S., Zeng, N., Liu, Y., & Zhang, Y. (2017). Identification of rice diseases using deep convolutional neural networks. *Neurocomputing, 267*, 378-384.

[16] Sengupta, S., & Das, A. K. (2017). Particle Swarm Optimization based incremental classifier design for rice disease prediction. *Computers and Electronics in Agriculture, 140*, 443-451.

[17] Chanda, M., & Biswas, M. (2019, April). Plant disease identification and classification using back-propagation neural network with particle swarm optimization. In *2019 3rd International Conference on Trends in Electronics and Informatics (ICOEI)* (pp. 1029-1036). IEEE.

[18] Singh, T., Kumar, K., & Bedi, S. S. (2021). A Review on Artificial Intelligence Techniques for Disease Recognition in Plants. In *IOP Conference Series: Materials Science and Engineering* (Vol. 1022, No. 1, p. 012032). IOP Publishing.

[19] Muthukannan, K., & Latha, P. (2015). A PSO model for disease pattern detection on leaf surfaces. *Image Analysis & Stereology, 34*(3), 209-216.

[20] Darwish, A., Ezzat, D., & Hassanien, A. E. (2020). An optimized model based on convolutional neural networks and orthogonal learning particle swarm optimization algorithm for plant diseases diagnosis. *Swarm and evolutionary computation, 52*, 100616.

[21] Ngugi, L. C., Abelwahab, M., & Abo-Zahhad, M. (2021). Recent advances in image processing techniques for automated leaf pest and disease recognition–A review. *Information processing in agriculture*, *8*(1), 27-51.

[22] Daniya, T., & Vigneshwari, S. (2019). A review on machine learning techniques for rice plant disease detection in agricultural research. *system*, *28*(13), 49-62.

[23] Chanda, M. (2019). Plant diseases classification using feature reduction, BPNN and PSO. *International Journal of Software Engineering and Computer Systems*, *5*(2), 90-103.

[24] Kavitha, D., & Sirisha, J. Particle Swarm Optimization Based Recognition And Classification of Rice Leaf Diseases. *(UGC Care Group I Listed Journal)* Vol-11 Issue-01 2021.

[25] Moparthi, N. R., Bhattacharyya, D., Balakrishna, G., & Prashanth, J. S. (2021). Paddy Leaf Disease Detection using CNN. *NVEO-Natural Volatiles & Essential Oils Journal| NVEO*, 800-806.

[26] Bedi, P., & Gole, P. (2021). Plant disease detection using hybrid model based on convolutional autoencoder and convolutional neural network. *Artificial Intelligence in Agriculture*, *5*, 90-101.

[27] Patil, M. P. C., Patil, D. D., & Subramanium, P. Review on Identification and Detection of Diseases in Rice Leaves Using Multiclass SVM and Particle Swarm Optimization Technique. *International Research Journal of Modernization in Engineering Technology and Science*. Volume:02/Issue:04/April-2020.

[28] Yadav, D., & Yadav, A. K. (2020). A Novel Convolutional Neural Network Based Model for Recognition and Classification of Apple Leaf Diseases. *Traitement du Signal*, *37*(6).

[29] Gangadharan, K., Kumari, G. R. N., Dhanasekaran, D., & Malathi, K. (2020). Detection and classification of various pest attacks and infection on plants using RBPN with GA based PSO algorithm. *Indonesian Journal of Electrical Engineering and Computer Science (IJEECS)*, *20*(3), 1278-1288.

[30] Kaur, N. (2021). Plant leaf disease detection using ensemble classification and feature extraction. *Turkish Journal of Computer and Mathematics Education (TURCOMAT)*, *12*(11), 2339-2352.

[31] Chaudhary, A., Thakur, R., Kolhe, S., & Kamal, R. (2020). A particle swarm optimization based ensemble for vegetable crop disease recognition. *Computers and Electronics in Agriculture*, *178*, 105747.

[32] Krishnamoorthy, N., Prasad, L. N., Kumar, C. P., Subedi, B., Abraha, H. B., & Sathishkumar, V. E. (2021). Rice leaf diseases prediction using deep neural networks with transfer learning. *Environmental Research*, *198*, 111275.

[33] Sharma, R., & Kukreja, V. (2021, March). Rice diseases detection using Convolutional Neural Networks: A Survey. In *2021 International Conference on Advance Computing and Innovative Technologies in Engineering (ICACITE)* (pp. 995-1001). IEEE.

[34] Kanimozhi E., Akila D. (2020) An Empirical Study on Neuroevolutional Algorithm Based on Machine Learning for Crop Yield Prediction. In: Peng SL., Son L., Suseendran G., Balaganesh D. (eds) *Intelligent Computing and Innovation on Data Science. Lecture Notes in Networks and Systems*, Vol 118. Pages 109-116.

[35] V. Vanitha and D. Akila, "Detection and Diagnosis of Hepatitis Virus Infection Based on Human Blood Smear Data in Machine Learning Segmentation Technique," *2021 9th International Conference on Reliability, Infocom Technologies and Optimization (Trends and Future Directions)* (ICRITO), 2021, pp. 1-5, doi: 10.1109/ICRITO51393.2021.9596482. ISBN No : 978-1-6654-1703-7.

Chapter 3

Computational Intelligence for Sustainable Smart Cities and Communities

S. Meenakshi Sundaram[1*], Tejaswini R. Murgod[2] and M. Sowmya[1]

[1]Department of CSE, GSSSIETW, Mysuru, India

[2]Department of ISE, Nitte Meenakshi Institute of Technology, Yelahanka, Bangalore, India

Abstract

Investing in smart cities leads to improving regional and global competitiveness that will attract new residents to improve business. Since the entire city will have access to an open data platform and relevant information, companies will flourish. They can make informed decisions with the available technologies which lead to economic development. Sustainable urban development is an approach for cities to mitigate climate change.

With the widespread deployment of IoT technology, the cities of tomorrow will be able to meet the demands of their residents in an efficient manner. Connected technologies and big data will create smart solutions. These solutions can solve problems, increase the quality of life of residents in the city and lower the consumption of resources. For a truly smart city to function at its full potential, the Internet of Things is a vital ingredient. IoT is integrated into the urban environment and our daily lives, the more data on productivity, challenges, and opportunities for infrastructure and environmental development will be available. IoT technology can provide the strongest impact on the development of the cities of the future.

---

* Corresponding Author's Email: 1965sms@gmail.com.

In: Computational Intelligence for Sustainable Development
Editor: Brojo Kishore Mishra
ISBN: 979-8-88697-198-9

Adopting ICT, Cloud and big data solutions will help address many issues such as providing the storage and analysis tools. As per the United Nations Department of Economic and Social Affairs, currently, 55% of the world population lives in urban areas and is expected to rise to 68% before 2050. In the case of India, cities accommodate nearly 31% of India's current population and contribute 63% of GDP. The cities will be accommodating 40% of India's population with a contribution of 75% to India's GDP before 2030. With such growing urbanization, it is essential to upscale the physical, institutional, social and economic infrastructure of the cities. This is achieved using Artificial Intelligence (AI) and Machine Learning (ML) technologies.

As per the international organization reports, globally more than fifty percent of the population live in urban areas and by 2050 more than two-thirds population will live in cities, providing big investment opportunities of tech development companies. As per the International Data Corporation report, since 2021, the spending on smart cities is likely to reach more than USD 130 billion to make the urban areas and cities more livable with advanced infrastructural facilities and society management system.

Keywords: sustainable urban development, IoT technology, fifth-generation mobile communication, big data, artificial intelligence, machine learning

Introduction

Urbanization is a non-ending phenomenon. Today, 54% of the people worldwide live in cities, a proportion that is expected to reach 66% before 2050. With the overall population growth, urbanization will add another 2.5 billion people to cities over the next three decades. Environmental, societal, and economic sustainability is a must to keep pace with this rapid expansion taxing the city resources. 193 countries agreed upon the Sustainable Development Goals (SDGs). Citizens and local authorities are certainly more agile to launch swift initiatives and smart city technology is paramount to success and meeting these goals.

A smart city uses sensors and connected devices to collect and analyze data. This data is used to optimize city operations, manage resources and improve the daily life of citizens. Smart cities utilize technology to improve access to public transport, manage traffic, optimize water, power supply and to improve law enforcement services, schools, hospitals to name a few. At the technological layer, a large number of sensors and connected devices are used to provide a wide range of services. For dedicated applications, information systems are used by city officials and citizens to improve the city operations. A smart city has the potential to drastically improve the current level of transportation throughout a city. It will have a better traffic management system with the ability to track public transportation and to serve its citizens better with constant information and cheaper cost. Technological advances and partnerships with the private sector benefits the society, since there will be lesser criminal activities. Smart cities will have technologies and the necessary tools to cut down on usage of natural resources and decrease water waste, electricity etc. Energy-efficient buildings can improve the air quality, use renewable energy sources and decrease the dependence on non-renewable energy sources. These will help to decrease the ecological impact we have on the environment. Similarly access to high-speed internet services at affordable costs and devices need to be made available. If they have access to public Wi-Fi in local areas of the city, all residents will have equal opportunities.

Investing in smart cities will lead to improving regional and global competitiveness and will attract new residents and improve business. Since the entire city will have access to open data and relevant information, companies will flourish. They can make informed decisions with the available technologies and lead to economic development. Smart cities will have the ability to analytically predict and identify the areas that can cause infrastructure failures before it occurs. Businesses and job opportunities will enhance, since people will get equal access to basic resources such as transportation, internet connection etc. Since the authorities can monitor the dealings of people closely with the help of technologies, there will be reduced crime rates. Crimes increase with availability of fewer jobs and more unemployed people. If job opportunities increase, this will simultaneously lead to decrease in crime rates.

Since the authorities or the government will have access to security cameras and intelligent systems connected through many different spaces, the citizens will have difficulty in maintaining their anonymity. Facial recognition and related things will drastically change the concept of privacy or personal space. Government or private agencies can track data of security cameras. They will have the power to control a citizen's data and can easily manipulate public opinions. As the citizen of these smart cities will rely almost entirely on electronics and networks they will lose autonomy in their decision-making and could become incompetent. They would not be able to react appropriately in a scenario where these tools are not usable. Though money will be available, the smart technologies will be still in their pre-commercial stages. These cities will lack the technology related skills and capacity. If the people of the city do not know about technology, then they will not be able to use it. Without training, they will find it irrelevant to their daily lives and will find it difficult to utilize it.

Sustainable Urbanism and Smart Urbanism Paradigms

Sustainable urbanism is both the study of cities and the practices to build them. These focus on promoting their long term viability by reducing consumption, waste and harmful impacts on people and place. This will enhance the overall well-being of both people and place. Sustainable urbanism is a recent term prevalent in urban design and planning. Within the contemporary metropolitan environment, it is rooted in study of sustainability and urban design in a rapidly urbanizing world.

Sustainable urban development is the way forward for cities to mitigate climate change. Integrated urban places are designed to bring people, activities, buildings, and public spaces together. This means inclusive access to all with local and citywide opportunities. The inclusive development is an essential foundation for long-term sustainability, equity, shared prosperity and civil society in cities. The classification of cities depends on the functionalities of cities as given below:

Sustainable Cities

Despite the fact that the discourse of sustainable cities is now mature and powerful. Precise conceptualizations are still rare and often contested. Notwithstanding the near universal recognition of sustainable cities being a desirable vision or goal of policy, there is less certainty about what this might mean in practice. Sustainable cities are so complex and intangible that the notion of what the concept means is constructed in a variety of ways within different city-related disciplines. Consequently, there are multiple views on what a sustainable city should be or look like and thus various ways of conceptualizing it. A sustainable city can be understood as an approach to practically applying the knowledge about sustainability to the planning and design of existing and new cities. It represents an approach to sustainable urban development, which is a strategic process to achieve the long term goals of urban sustainability. It is required to balance between the environmental, economic, and social dimensions of sustainability. As an integrated process of change, a sustainable city strives to maximize the efficiency of energy and material use, minimizes waste generation, supports renewable energy production and consumption. It also promotes carbon–neutrality, reduces pollution, provides efficient and sustainable transport. It emphasizes compactness, supports design scalability, spatial proximity, preserves ecosystems, green space to promote livability and community oriented human environments.

In a nutshell, a sustainable city must meet the needs of the population without undermining the ability of future generations to meet their own needs. There are different approaches to sustainable cities, which tend to be identified as models of sustainable urban form. These include compact cities, eco cities, new urbanism, urban containment, landscape ecological urbanism etc. Compact cities and eco-cities are the central paradigms of sustainable urbanism and the most prevalent and advocated models of sustainable urban form. Compact cities and eco-cities are the central paradigms of sustainable urbanism and the most prevalent and advocated models of sustainable urban form.

Compact Cities

A compact city is taken to mean a relatively high density, mixed use city, based on an efficient public transport system and dimensions that encourage walking and cycling. The compact city is characterized by high density and mixed land use with no sprawl. Every compact city has "unique individuality", own life, and physiognomy. A compact city is the most advocated model of sustainable urban form due to its ability to produce the benefits of sustainability as to its tripartite composition, though to varying degrees. When strategically planned and well–designed, a compact city becomes able to support the balancing of the three goals of sustainability.

Eco-Cities

The idea of the eco-city is widely varied in conceptualization and operations. Consequently, there are multiple definitions of the eco-city, depending on the context where it is embedded in the form of urban projects and initiatives in terms of the practices and strategies adopted to achieve the goals of the eco-city.

The concept of the eco-city has also been used to describe a wide range of urban projects and initiatives, mostly large scale new districts. This encapsulates a diversity of conceptualizations and pulling together an ensemble of normative and prescriptive principles supported by various policies about how to develop and design sustainable urban areas. An eco-city is based on three analytical categories: a development on a substantial scale, occurring across multiple domains and supported by policy processes.

Smart Cities

It is difficult to identify the common trends of smart cities at the global level. The smart city concept is still without a universally agreed definition, as there seems Tobe neither a pre-determined template for nor a one-size-fits-all approach to its definition. As a consequence, there is still a lack of conceptual clarity around the smart city concept, or inconsistent understanding of what it means, despite the popularity of smart cities. The concept of having different connotations and being approached from a variety of perspectives is clearly manifested in various ways. Many governments across the globe set initiatives or implement projects to enable their cities to badge or regenerate themselves as smart, or to plan to become smart. As a result, a large number and variety of definitions have been suggested, with a plethora of scopes, which has led to confusion amongst urban policymakers, working on establishing public policies to enable the transition to smarter cities.

Data-Driven Smart Cities

Data-driven city is one of the recent faces of smart cities. As such, it represents an emerging paradigm of smart urbanism: data-driven smart cities. It is too often associated with "smartness" under which is labeled "data-driven smart cities." This is due to the fact that the big data technology is an advanced form of ICT, which is an enabler of all approaches to smarter cities, such as ambient city, sentient city, ubiquitous city and real-time city. There is no single conceptual unit of a data driven city, nor is there an agreed industry or academic description thereof. In a broader sense, the data-driven city implements datafication for enhancing and optimizing its operations, functions, services, strategies, and policies to some purpose. The concept employs big data technologies to bring about changes to city life, which are for the better. The phenomenon of the data-driven city is materialized as a result of the emergence of big data science and computing. With the wider adoption of the underlying technologies, there is an extensive growth of urban data that has transformed of urban landscape in the light of urbanization. These developments can be used in a range of proposals for a conceptual framework for the data-driven city (Bibri et al. 2021).

Artificial Intelligence for Smart and Sustainable Cities

Smart cities yields many benefits to the society, including enhanced opportunities for education, improved job prospects, better access to healthcare and cleaner water. Yet it is also associated with immense societal and environmental challenges. Inefficient planning and management practices lead to unsustainable settlements that do not enable people to advance personally, socially or economically. Smart and innovative technologies, including artificial intelligence, are revolutionizing the way cities address the challenges associated with smart growth.

AI in Smart Cities

Such technologies help cities to utilize existing assets more effectively, allocate resources more efficiently and improve how data and information are managed and shared across systems. Increasingly, satellite data is becoming a fundamental component of smart cities and an essential tool for city management and governance. From understanding connectivity between cities to measuring economic growth, detecting power outages or identifying where resources should be allocated after disasters, the increasing availability of satellite data is transforming how cities are managed and helping to improve their functionality.

From the point of view of urban governance, machine learning and artificial intelligence (AI) provide near-real-time information on how cities change in practice, e.g., through the conversion of green spaces into built-up structures. By 'teaching' computers what to look for in satellite images, rapidly expanding sources of satellite data are leveraged in combination with machine learning algorithms to quickly reveal how actual city development aligns with planning and zoning or which communities are most prone to flooding.

Today, AI is almost a buzzword, a kind of magic formula, based on some 'intelligent agents' and sophisticated algorithms that make decisions and take action for humans. But AI will never replace human validation or effective governance on the ground. 'Smart' technologies that collect data on the ground or from space must be leveraged to monitor and manage urban systems and to provide guidance and recommendations for better decision-making, which will in turn make cities more sustainable (Voda 2018, Inclezan 2017).

How AI Can Be Used within Smart Cities

Cities are continuously innovating with the help of artificial intelligence. Due to the AI technology used within smart cities, these cities are then able to control the traffic, waste and maintenance as well as predicting the energy consumption, pollution risks, and the effects on the environment. Smart cities cause lesser effects on climate change, smarter decisions and improved quality of life, in this list we take a look at ten ways artificial intelligence helps make this possible.

Security

Security camera footage is typically reviewed when a crime has been reported, this doesn't prevent or stop crime. Security cameras that use artificial intelligence have the ability to analyze footage in real time and detect criminal behavior which can then be instantly reported and dealt with. These cameras can also detect people from their clothes, allowing the technology to find suspects quicker than ever.

Maintenance

A company called RoadBotics has developed a technology using artificial intelligence that has the ability to analyze road imagery to then assess issues and produce cost effective solutions. This allows cities to know when and where repairs need to take place and deal with them while saving money. This type of technology also improves safety within cities as problems will not go unnoticed.

Predicting Future Needs

Smart cities that are innovating with AI technology are able to better predict future needs of the people. Using energy tracking technology allows cities to know when new energy sources are needed or when more sustainable methods can be implemented. Some AI technology can also predict and help plan on property developments, this means houses are put onto the market during periods they are needed.

Controlling Pollution

Scientists have developed technology which uses AI and machine learning to analyze the current pollutants and predict the pollution levels for the next 2 hours. This type of technology allows authorities to make decisions in advance to reduce their effect on the environment.

Parking Systems
Using license plate recognition technology, car parks are able to detect cars that have outstayed hours; this can also enforce payments and tickets. When AI systems are integrated throughout car parking areas space availability is able to be presented to awaiting users, some more advanced technology has the ability to recommend spaces depending on the car.

Public Transportation
Public transport has been innovated with the use of AI to be used within smart cities. This technology allows public transport users to receive and access live updates and tracking, which improves timing and customer satisfaction. Automated busses are also planned to be used within cities, these can reduce emissions, improve routes, and increase the frequency.

Waste Management
Smart cities are the beginning to use artificial intelligence within their waste management. This type of technology allows cities to track recycling, and identify what can be recycled in the area. Some cities in Sydney take this a step further and use AI powered robots to sort rubbish, as well as clean areas such as lakes and rivers.

Traffic Management
AI technology is being implemented within the transportation industry to reduce traffic and accidents. A traffic management technology known as CIRCLES has the ability to predict and reduce traffic, using deep learning algorithms, this can then reduce the pollution created by traffic too.AI can also be used throughout traffic camera systems to detect road crimes in real time, making them easier to deal with.

Energy Tracking
Artificial intelligence can be used within smart cities to analyze and track business and citizen energy usage, with this data it can then be decided where to implicate renewable sources of energy. This can also show cities where energy is being wasted and how it can be saved.

The Environment
Smart cities can use artificial intelligence to see their effect on the local environment, global warming, as well as the pollution level. Using AI and

machine learning enables pollution control and energy consumption within limits. This allows authorities and cities to make well informed decisions that are best for the environment. Smart cities also use AI to detect carbon dioxide which can then lead to decisions around transportation.

How AI and Machine Learning Accelerates the Development of Smart Cities

A smart city is defined as an amalgamation of data and digital technology to address arising city problems, from healthcare, road infrastructure, public transportation, security, and education to waste management. The smart city is a way of empowering individuals, organizations, or systems to be self-sufficient and futuristic with the use of the latest technologies such as AI and Machine Learning. These technologies acquire a massive amount of data which is used to upgrade the systems and components that run the cities and makes them capable of making intelligent decisions without much human intervention.

As per the United Nations Department of Economic and Social Affairs, currently, 55% of the world population lives in urban areas and is expected to rise to 68% by 2050. In case of India, cities accommodate nearly 31% of India's current population and contribute 63% of GDP. The cities will be accommodating 40% of India's population with a contribution of 75% to India's GDP by 2030. With such growing urbanization, it is essential to upscale the physical, institutional, social and economic infrastructure of the cities. This is achieved through heuristic devices using Artificial Intelligence (AI) technologies and Machine Learning (ML).

Artificial intelligence and Machine Learning algorithms are now fundamental parts of industries and hence are crucial in building smart cities. The deployment of AI-enabled intelligent machines powered by Machine Learning creates cyber-physical space which includes traffic sensors, medical monitoring, industrial control systems, video cameras, environment sensors, smart meters, etc. The data obtained through all these intelligent machines when analyzed, help in predictive analysis and decision making for smart city planning. Furthermore, AI paves a smooth way in creating a dialogue between citizens, systems, and cities. It is the most efficient way to address thousands of questions coming from citizens accurately and consistently. The chatbots integrated with AI and ML algorithms are 24*7 available to citizens, who can anticipate citizens' needs and cater to them.

As per the international organization reports, globally more than fifty percent population is living in urban areas, and by 2050, more than two-third population will live in cities providing big investment opportunities of tech development companies. As per the International Data Corporation report, in 2021 the spending on smart cities has almost reached to more than USD 130 billion to make the urban areas and cities more livable with advance infrastructural facilities and society management system.

AI in smart cities is going to play big role in making urbanization smarter, aiming to achieve sustainable growth making the cities equipped with advanced features to live, walk, shop, and enjoy a safe and more convenient life in such environments. While developing cities to make them smart, several challenges like administration, sanitation, traffic congestion, security surveillance, parking management and many more are overcome by AI as it can help to provide a sustainable solution to habitants (Shafiq 2020).

Role of Internet of Things in Smart Cities

The term IoT (Internet of Things) refers to the rapidly growing number of digital devices the quantity is now billions – these devices can communicate and interact with others over the network/internet worldwide and they can be remotely monitored and controlled. The IoT includes only smart sensors and other devices. On the operational level of IoT, for example weather data is collected. IoT offers new opportunities for cities to use data to manage traffic, cut pollution, make better use of infrastructure and keep citizens safe and clean.

Companies use IoT for innovative management and for monitoring widely dispersed processes. As a result, they even can control the latter from distant places as information is continuously fed into applications and data storage. IoT provides an advantage of knowing things in advance. Due to the low cost of IoT, it is now possible to monitor and manage activities that were previously unreachable. The financial aspect is the best advantage, because this new technology could replace humans who are in charge of monitoring and maintaining supplies. Consequently, costs can be significantly reduced and optimized. IoT also makes it possible to gain completely new insights.

IoT is important for every city. Currently, the world's largest cities are Tokyo, Delhi, Shanghai, and Sao Paolo, with populations of 38 million, 29 million, 26 million, and 21 million respectively. Today, these megacities are notable because of their huge populations. In the future, there will be many more like them, with even denser populations.

It's predicted that more than 60% of the planet's population will live in cities by the year 2030. It's a bold prediction and one that could spell disaster if the appropriate measures are not taken. Large populations demand large resources. Residents will need access to water, efficient and environmentally-friendly transportation, clean air, practical sanitation and waste management. With the clever use of smart city practices and widespread deployment of IoT technology, the cities of tomorrow will be able to meet the demands of their residents in an effective and efficient way. Connected technologies and big data can create smart solutions. These solutions can solve problems, increase the quality of life for a city's residents, and lower the consumption of resources. For a truly smart city to function at its full potential, the Internet of Things is a vital ingredient (Rajab 2018).

IoT Technology and Implementation in Smart Cities

Secure wireless connectivity and IoT technology are transforming traditional elements of city life - like streetlights - into next-generation intelligent lighting platforms with expanded capabilities. The scope includes integrating solar power and connecting to a cloud-based central control system that connects to other ecosystem assets. These solutions shine far beyond simple lighting needs.

1. High-power embedded LEDs alert commuters about traffic issues, provide severe weather warnings and provide heads up when fires arise, for example.
2. Streetlights can also detect free parking spaces and EV charging docks and alert drivers where to find an open spot via a mobile app. Charging might even be able from the lamppost itself in some locations.

The following factors enable the sustained development of smart cities that are implemented using IoT technologies.

1. Influenced by information technology, the world is changing rapidly. More and more active connectible devices contribute to the transformation of both separate industries and our familiar urban

environment. Information technology and IoT facilities are integrating with urban infrastructures, resulting in the appearance of the "smart city" concept and its development.

2. The smart city term has not got the only and exact definition. It is interpreted quite broadly. However, in general, we can provide the following functionalities:

A smart city functions based on data collection, processing, and exchanges processes which are then used to make decisions (by both the city government and individual residents at different levels). Connected sensors and IoT devices are used to collect the necessary data.

If a city develops based on the smart city model actively using IoT and other information technologies, such a city and its residents will get several important benefits like:

I. improved city utilities efficiency
II. city lighting systems efficient management
III. improves the city's traffic flow quality
IV. increases public transport convenience and efficiency
V. general increase in city residents' living standards
VI. all city services operations with centralized monitoring
VII. increased road safety and timely transport services
VIII. resource use reduction and environmental improvement
IX. collecting a large amount of data on different aspects of city functioning.

The number of active sensors is already huge. The most interesting thing is the possibility to combine them with each other and further expand the potential for the implementation of new solutions. Given below are some examples of IoT technologies used as part of the smart city concept.

1. Sensors installed at various points of the engineering infrastructure allow detecting emergencies.
2. Energy resource meters (water, gas, electricity) send consumption data to the relevant service providers by themselves.
3. Smart street lighting adapts its luminous power to changing lighting environments (for example, the rapid darkness onset due to weather changes).

4. Smart garbage bin sensors detecting fill levels transmit data to waste disposal services to optimize collection routes.
5. Self-driving vehicles are a prerequisite for a smart city. Driverless public transportation will reduce traffic congestion, carbon emissions and accidents due to human errors.
6. ITS (Intelligent Transport Systems) are used to control traffic lights, road info boards and signs, car navigation, car number plate recognition and data flow integration.
7. As for the security area, automatic face recognition systems are widely used. They analyze data from cameras installed throughout the city. Such systems allow locating criminals or assist in searching for missing adults and children.
8. Intelligent parking solutions register when drivers leave certain parking lots and inform other drivers that free lots are available.
9. Electric power, water and other resources consumption optimization directly influences the city's environmental condition.
10. Residents get informed electronically and receive common services, for example, from government agencies, exclusively online.

The future of IoT in smart cities can be summarized as follows:

1. IoT is integrated into the urban environment and our daily lives, the more data on productivity, challenges and opportunities for infrastructure and environmental development will be available. Out of all the technologies that have recently appeared, IoT can provide the strongest impact on the development of the cities of the future (Chatterjee 2018).
2. If we compare the past and the present, we can see that most of the cities have formed whole IoT ecosystems, surrounding their residents with multiple benefits of mobility, security, health care, and higher efficiency. The world is changing, moving confidently towards the next industrial revolution.
3. The Fifth Generation (5G) Mobile Communication implementation will give a new impetus to the Internet of Things. The technological development will lead to the emergence of new solutions making modern cities more sustainable, more comfortable and safer for their citizens.

Big Data for Smart Cities

Many governments are considering adopting the smart city concept in their cities and implementing big data applications that support smart city components to reach the required level of sustainability and improve the living standards. Smart cities utilize multiple technologies to improve the performance of health, transportation, energy, education and water services leading to higher levels of comfort of their citizens. This involves reducing costs and resource consumption in addition to more effectively and actively engaging with their citizens.

One of the recent technologies that have a huge potential to enhance smart city services is big data analytics. As digitization has become an integral part of everyday life, data collection has resulted in the accumulation of huge amounts of data that can be used in various beneficial application domains. Effective analysis and utilization of big data is a key factor for success in many business and service domains, including the smart city domain.

Undoubtedly, the main strength of the big data concept is the high influence it will have on numerous aspects of a smart city and consequently on people's lives. Big Data is growing rapidly; currently at a projected rate of 40% growth in the amount of global data generated per year versus only 5% growth in global IT spending. Around 90% of the world's digitized data was captured recently in the past two years. As a result, many governments have started to utilize big data to support the development and sustainability of smart cities around the world. This has allowed cities to maintain standards, principles and requirements of the applications of smart city through realizing the main smart city characteristics. These characteristics include sustainability, resilience, governance, enhanced quality of life and intelligent management of natural resources and city facilities.

There are well-defined components of the smart city, such as mobility, governance, environment and people as well as its applications and services such as healthcare, transportation, smart education and energy. To facilitate such applications and services large computational and storage facilities are needed. One way to provide such platforms is to rely on Cloud Computing and utilize the many advantages of using cloud services to support smart city big data management and applications.

Big Data applications have the potential to serve many sectors in a smart city. It helps to provide better customer experiences and services, which help businesses, achieve better performance (e.g., higher profits or increased market shares). Transportation systems can greatly benefit from big data to optimize routes, schedules, to accommodate varying demands for being more environment friendly (Hashem 2016, Karimi 2021).

Deploying big data applications require the support of good Information and Communication Technology (ICT) infrastructure. ICT supports smart cities, as it provides useful solutions and also unique solutions that may not be possible without it. For example, it enables efficient transport planning by providing easy ways to handle their services from different fields/locations to reduce the transportation costs. Other examples includes: providing better water management and improved waste management by applying innovations to effectively manage these services.

For example, waste management includes waste collection, disposal, recycling and recovery, all of which can be efficiently managed using ICT solutions. More examples include new construction and structural methods for the health of buildings and better environment; risk management; safety and security; air quality and pollution; public health; urban sprawl; bio-diversity loss; and energy efficiency. In general, a smart city can be made smarter when utilizing ICT and big data for many of its applications and services.

Adopting ICT, Cloud and big data solutions will help address many issues such as providing the storage and analysis tools. In addition this will help to reach the innovation stage and encourage collaboration and communication between the different entities of a smart city. This can be done by building big data communities to work as one entity to foster collaborative and creative solutions addressing applications for areas like education, health, energy, law, manufacturing, environment, and safety. This also helps in real-time solutions to challenges in agriculture, transportation, and crowd management, as applications and systems are integrated while information flows easily across applications and entities. There are many examples of big data applications serving smart cities like:

Smart Education

ICT provides a solution to enhance the education processes efficiency, effectiveness and productivity using education smart services that are flexible and intelligent to provide better use of information, enhanced control and assessment, higher support for life-long learning for all people (citizens and stakeholders). Smart education applications will engage people in active learning environments that allow them to adapt to the rapid changes of society and the environment. In addition, by relying on big data collected in the field and correctly processed to generate the required information, we will have a positive effect on the knowledge levels and teaching / learning tools to deliver or acquire knowledge.

Furthermore, technology can make such opportunities available everywhere including remote or rural areas where commuting to schools may not be possible or the economic status of people is low and they cannot afford other more expensive models. Using ICT and big data will also help create a knowledge-based society, which will enhance the nation's capability in competitiveness. Big Data in education is generated mainly by collecting data on people (e.g., students, teachers, parents, administrators and other support personnel), infrastructures (e.g., schools, libraries, computing facilities, educational locations, museums, universities and other related entities)and information (e.g., courses, books, exams, grades, economic surveys, assessments, reports and much more).

This data can create a useful resource for analysis and extracting useful trends, models and using them to offer better and more enhanced education. Big Data in education can be also utilized to observe educational shortages to enhance study curriculums.

Smart Traffic Lights

One of the main aspects of smart cities is a good control of the traffic flow within the city, which will enhance the transportation systems and improve the commutation aspects and overall traffic patterns in cities. When the population increases, traffic problems, pollution and economic problems happen. Due to this, the use of smart traffic lights and signals is one of the most important techniques that smart cities use to deal with high volumes of traffic and congestions. Smart traffic lights and signals should be interconnected across the traffic grids to offer more information about traffic patterns. Each sensor detects different parameter of the traffic flow (e.g., the speeds of cars, traffic density, waiting time at the lights, traffic jams, etc.).

The system makes decisions according to the values of these parameters and gives the appropriate instructions to the lights and signals. Thus, the more data available to this system, the more informed decisions it will be able to make. As a result, to offer the best possible services in smart traffic lights, it will be best to collect data from all traffic lights across the city and build intelligent decision systems using this data. This requires the use of real-time big data analytics. As an example, implementing smart traffic lights and signals designed by the Traffic 21 project in Pittsburgh, Pennsylvania, USA obtained significant results, which reduced traffic jams and waiting times resulting in reduced emissions by over 20%.

Smart Grid

The smart grid is an important component of a smart city. It is a renovated electrical grid system that uses information and communication technology to collect and act on available data, such as information about the behaviors of suppliers and consumers, in an automated fashion to add some values. It improves the efficiency, reliability, economics and sustainability of the production and distribution of electric power. A smart grid uses computer-based remote controls with two-way communication technology between power producers and consumers to increase grid efficiency and reliability through system self-monitoring and feedback.

This involves placing smart sensors and meters on production, transmission and distribution systems in addition to consumer's access points to get granular near real-time data about the current power production, consumption and faults. It implements dynamic pricing models for power usage to smooth out peaks by applying high charges during peak times and lower charges during other periods. This helps to avoid potential power outages due to high consumer demands. It can provide consumers with near real-time information about their energy use and allows them to manage their usage based on both their needs and their affordable prices.

Consumer devices such as washing machines and water heaters can be more cost-effective by controlling them automatically to operate during lower pricing periods. Although the smart grid has many potential benefits, it requires the collection of huge amount of data from power procedures, transmissions, distributors and consumers. In addition, it requires processing of the collected data, which is considered as big data analytics in real-time, to avail control information for the improvement of overall performance of the electric power system (Pal 2018).

Data Storage in Smart Cities

Smart cities need a huge amount of archived and real-time data to function properly:

1. Autonomous and smart cars are becoming more integrated with mobile systems. As a result, cars can communicate with each other and with cities. This communication can reduce congestion, prevent road incidents and improve navigation.
2. Camera surveillance can help with traffic enforcement, improve public safety, build intelligent lighting systems and serve as a crime detector.

3. Air quality sensors enable cities to find and take action against polluters, locate green areas with low air quality and provide quality air status alerts to citizens. Smart cities can store their data in three main locations namely the Cloud, Edges and hybrid mode of both Edges and Cloud.

Cloud Storage
Smart cities require large amounts of data for analytical purposes. Cloud data systems use solid-state drives in their data centers, remove redundant data and encrypt the transmission of data. Cloud-based solutions usually have more flexible payment options than on-premise data centers.

Edge Computing
Edge computing enables you to process data close to the source. Edge computing can be cheaper than streaming data to a remote storage location and then to relevant city authorities. Edge computing functionalities like artificial intelligence (AI) traffic management is already under development in some cities. AI traffic management makes use of intelligent automation to detect traffic congestion and accidents and provide faster response to different conditions.

Hybrid Data Storage
Hybrid data storage systems combine the benefits of cloud and edge storage. This enables cities to make new decisions based on real-time alerts on conditions as well as rich data stores.

Challenges of Big Data in Smart Cities

We face a variety of challenges in the design, development and deployment of big data applications for smart cities. These are considered very dynamic and evolving environments, thus it is important to avoid or at least reduce the challenges involved in smart applications design and development for smart cities.

There are also some controversies related to the definition, use and benefits of big data for smart cities. These relate to available big data tools, real-time analytics, accuracy, representation, cost and accessibility. Such issues can affect the performance of smart city applications and services relying on big data. We will address some of the key challenges in using big data in smart cities.

Data Sources and Characteristics

Data is generated from many different sources in many different formats. There are a lot of new data formats many of which are unstructured (e.g., images, audio, tweets, video, server logs, etc.). This data need to be managed and classified into a structured format using some form of advanced database systems. Many identified different Vs of big data the most agreed upon are the 3Vs: Velocity, Volume and Variety. Several more were added such as Validity, Veracity, Volatility, Variability and Value. Just trying to encompass these different attributes of big data generates very complex models and approaches and makes it hard to manage. This is simply because the current methodologies or data mining software tools cannot handle the large size and complexity.

In addition, there are some challenges that may be faced in the future, such as analytics architecture, evaluation, distributed mining, time evolving data, compression, visualization and hidden big data. When considering smart city applications utilizing big data difficulties arise in various areas. For one, collecting the data by itself is complicated by the existence of multiple sources with different formats and types and different usage and access policies. In addition, the unstructured nature of the data makes it hard to categorize, organize and an easily accessible way for applications to use.

Data and Information Sharing

Sharing data and information among different city departments is another challenge. Each government and city agency or department typically has its own warehouse or silo of confidential or public information. Most of which are often reluctant to share what might be considered proprietary data. In addition, some data may be governed by certain privacy conditions that make them hard to share across different entities.

The challenge here is to make sure not to cross the fine line between collecting and using big data and ensuring citizens' rights of privacy. This is applicable within any smart city since there are many sectors and industries involved. Smart city applications will need to find ways to prevent or reduce the barriers to achieve seamless information sharing and exchange among different entities. Furthermore, with multiple diverse data sources distributed among related departments, some data types such as spatio-temporal data can be updated quickly.

Therefore, it is difficult to create a unified understanding of data semantics and extract new knowledge based on specific cycle data and real-time data. As result, it will be difficult to create a knowledge base for a smart city.

Data Quality

Looking at more fundamental aspects of big data, there are a number of challenges that are associated with the quality of the data. The data captured by different people under special regimes are stored in distinctive databases and is rarely stored in any of the standard formats. Relying on crowd sourcing and collaboration of multiple providers this results in data that suffers from a lack of structure and consequently consistency, heterogeneity and disparity issues that will have a greater chance to occur. Accordingly, "there is no universal way to retrieve and transform the data automatically and universally into a unified data source for useful analysis." This will cause more challenges like data uncertainty and trustworthiness. For example, sensor data collected through a third party without a centralized control could have been produced by sensors that are faulty, wrongly calibrated or beyond their lifetime.

The challenge may also extend to the outputs of analyzing existing data (given the possibility of errors) and reporting the results for use by others, who may not be aware of such issues. Therefore, continuously updating data gathering and usage policies, sharing and discussing them among all entities in a smart city, ensuring that the citizens understand and apply the policies correctly is vital and challenging at the same time.

Security and Privacy

Another major challenge in a smart city and with using big data is the security and related privacy issues. In basic terms this means that databases may include confidential information related to the government and people, so they need high levels of security policies and mechanisms to protect this data against unauthorized use and malicious attacks. In addition, smart applications integrated together across agencies also require high security since the data will move over various types of networks, some of which may be unsecure.

What makes such an issue more complex is that most big data technologies today, including Cassandra and Hadoop, suffer from a lack of sufficient security. In addition to the need to secure data as it travels and as it is being used by the different components of smart city applications, there is also the need to clearly identify and protect privacy rights of organizations and individuals this data represents. Although specific smart city entities can claim ownership of most big data, a lot of it includes personal and private information about individuals.

Health and medical records, financial and bank records, retail history etc. provide intimate views of the people they represent. Many view access to this type of data as a violation of a person's legal rights for privacy. Making sure that stringent privacy policies are put in place and properly enforced represents a major challenge for big data smart city applications developers and users.

Cost

Cost is a sensitive subject that involves the ways public authorities may affect people when they use ICT solutions. For example, using an energy usage reduction system, that forces the government to use new systems, components or features to monitor consumption and record information. This leads to creating a smart energy management system; however, it is also a very expensive to implement. In addition if such a project is not implemented correctly from the beginning, it might cause a big problem, result in very high costs and the city may be negatively affected.

For example, the testing of a smart traffic light and signal system has a very high cost. This test produce not only high costs in resources but also brings traffic problems while physically deploying and testing the system. Hence this will require replacing expensive hardware and software for further development and monitoring of smart city infrastructure and applications.

Smart City Population

People affect and are also affected by the smart applications. Particularly the city's population sizes have a great effect on the size of big data. As the population grows, the size of generated data also rapidly grows and can become massive. This is one of the main challenges because of the rapid growth traffic congestion, pollution and increasing social inequality besides increased urbanization. This raises a variety of technical, social, economic and organizational problems that tend to risks the economic and environmental sustainability of cities (Osman 2019).

As a result, smart city applications need to evolve quickly and extend efficiently to handle the growing volume and variety of big data to help avoiding such problems. Ultimately, the goal is to develop and deploy smart city applications that are smart enough to evolve and intelligently handle the rapid growth of big data to generate better result. The above challenges have varying effects and implications on such applications and pose varying levels of difficulty and complexity. Furthermore, different applications have different requirements for data usage.

For example, traffic control requires immediate responses from the application to control traffic in real-time; while environmental sustainability applications may be able to handle more delayed responses as decisions are generally made over longer periods of time. Therefore, real-time transfer, discovery, analysis, decision-making, and responses are an issue; however, the degrees of its importance varies with the application.

Conclusion

Smart Cities use sensors and connected devices to collect and analyze data. This data is used to optimize city operations, manage resources, and improve the everyday life of citizens. Smart cities utilize technology to improve access to public transport, manage traffic, optimize water and power supply, and improve law enforcement services, schools, hospitals and more. Investing in smart cities leads to improving regional and global competitiveness and will attract new residents to improve business. Since the entire city will have access to an open data companies will flourish. They can make informed decisions with the available technologies and lead to economic development. Sustainable urban development is the way forward for cities to mitigate climate change. Sustainable urban development provides inclusive access for all from local to citywide opportunities. Inclusive development is an essential foundation for long-term sustainability, equity, shared prosperity and civil society in cities.

Cities can be classified depending on the functionalities as sustainable cities, compact cities, eco-cities, smart cities and data driven smart cities. Smart cities yields many benefits for society, including enhanced opportunities for education, improved job prospects, better access to healthcare and cleaner water. Yet it is also associated with immense societal and environmental challenges. Smart and innovative technologies, including artificial intelligence, are revolutionizing the way cities address the challenges associated with smart growth. 'Smart' technologies collect data on the ground or from space needs to be leveraged to monitor and manage urban systems to provide guidance and recommendations for better decision-making. This will in turn make cities more sustainable. Smart cities cause lesser effects on climate change, smarter decisions and improved quality of life using artificial intelligence.

Artificial intelligence and Machine Learning algorithms are now fundamental parts of industries and hence are crucial to building smart cities. The deployment of AI-enabled intelligent machines powered by Machine Learning creates cyber-physical space which includes traffic sensors, medical monitoring, industrial control systems, video cameras, environment sensors, smart meters, etc. The data obtained through all these intelligent machines when analyzed, help in predictive analysis and decision making for smart city planning. While developing cities to make them smart, several challenges like administration, sanitation, traffic congestion, security surveillance, parking management and many more are overcome by AI as it can help to provide a sustainable solution to habitants. With the clever use of smart city practices and widespread deployment of IoT technology, the cities of tomorrow will be able to meet the demands of their residents in an efficient manner. Connected technologies and big data can create smart solutions. These solutions can solve problems, increase the quality of life for a city's residents and lower the consumption of resources.

A smart city functions based on data collection, processing and exchange processes which are then used to make decisions (by both the city government and individual residents at different levels). Connected sensors and IoT devices are used to collect the necessary data. One of the recent technologies that have a huge potential to enhance smart city services is big data analytics. As digitization has become an integral part of everyday life, data collection has resulted in the accumulation of huge amounts of data that can be used in various beneficial application domains. Effective analysis and utilization of big data is a key factor for success in many business and service domains, including the smart city domain.

Adopting ICT, Cloud and big data solutions will help address many issues such as providing the storage and analysis tools. In addition this will help to reach the innovation stage and encourage collaboration and communication between the different entities of a smart city. Big Data applications include smart education, smart traffic lights and smart grids. There are also some controversies related to the definition, use and benefits of big data for smart cities. These relate to available big data tools, real-time analytics, accuracy, representation, cost, and accessibility. Such issues can affect the performance of smart city applications and services relying on big data. As per the international organization reports, globally more than fifty percent population are living in urban areas. By 2050, more than two-third population will live in cities providing big investment opportunities of tech development companies.

References

Bibri, Simon Elias. "Data-driven smart sustain able cities of the future: An evidence synthesis approach to a comprehensive state-of-the-art literature review." *Sustainable Futures* 3 (2021): 100047.

Chatterjee, Sheshadri, Arpan Kumar Kar, and M. P. Gupta. "Success of IoT in smart cities of India: An empirical analysis." *Government Information Quarterly* 35, no. 3 (2018): 349-361.

Hashem, Ibrahim Abaker Targio, Victor Chang, NorBadrul Anuar, Kayode Adewole, Ibrar Yaqoob, Abdullah Gani, Ejaz Ahmed, and Haruna Chiroma. "The role of big data in smart city." *International Journal of information management* 36, no. 5 (2016): 748-758.

Inclezan, Daniela, and Luis I. Pradanos. "A criticalview on smart cities and AI." *Journal of Artificial Intelligence Research*60 (2017): 681-686.

Karimi, Yaghoob, Mostafa Haghi Kashani, Mohammad Akbari, and Ebrahim Mahdipour. "Leveraging big data in smart cities: A systematicreview." *Concurrency and Computation: Practice and Experience*33, no. 21 (2021): e6379.

Osman, Ahmed M. Shahat. "A novelbig data analytics framework for smart cities." *Future Generation Computer Systems* 91 (2019): 620-633.

Pal, Debajyoti, Tuul Triyason, and Praisan Padungweang. "Big Data in smart-cities: Current research and challenges." *Indonesian Journal of Electrical Engineering and Informatics (IJEEI)* 6, no. 4 (2018): 351-360.

Rajab, Husam, and Tibor Cinkelr. "IoT based smart cities." In 2018 *international symposium on networks, computers and communications (ISNCC),* pp. 1-4. *IEEE,* 2018.

Shafiq, Muhammad, Zhihong Tian, Ali Kashif Bashir, Alireza Jolfaei, and Xiangzhan Yu. "Data mining and machine learning methods for sustainable smart cities traffic classification: A survey." *Sustainable Cities and Society* 60 (2020): 102177.

Voda, Ana Iolanda, and Laura Diana Radu. "Artificial intelligence and the future of smart cities." *BRAIN. Broad Research in Artificial Intelligence and Neuroscience* 9, no. 2 (2018): 110-127.

Biographical Sketches

S. Meenakshi Sundaram, PhD

Affiliation:
GSSS Institute of Engineering and Technology for Women, Mysuru, India

Education:

PhD Degree in Computer Science and Engineering from Anna University Chennai, Tamil Nadu State, India in 2014

M. Tech Degree in VLSI Systems from National Institute of Technology (NIT), Tiruchirappalli, Tamil Nadu State, India in 2006

B.E. Degree in Computer Science & Engineering from Bharathidasan University, Tiruchirappalli, Tamil Nadu State, India in 1989

Business Address:

Professor and Head
Department of Computer Science and Engineering
GSSS Institute of Engineering and Technology for Women,
KRS Road, Mysuru – 570016
Karnataka State, India

Research and Professional Experience:

32+ Years of teaching experience and 15 years of research experience.

Professional Appointments:

Dr. Sundaram is a reviewer of Springer – SN Computer Science, Soft Computing Journal, International Journal of Ah Hoc Network Systems, Journal of Engineering Science and Technology, Taylor's University, Malaysia and International Journal of Computational Science & Engineering, Inderscience Publishers, UK.

Honors:

Best Research Excellence Award, 2021

Publications from the Last 3 Years:

1. Rajashekar MB, Meenakshi Sundaram S (2021),' Enhanced CP-ABE with RSA for Secure and Revocable Data Transmission of Big Data in Cloud', *'International Journal of Intelligent Engineering and Systems,*Vol.15, No.2, pp 47 – 56, doi: 10.22266/ijies 2022.0430.05doi-https://inass.org/wp-content/uploads/2022/02/ 2022043005.pdf.
2. Tejaswini RM, Meenakshi Sundaram S (2021),'A Comparative Study of Different Network Simulation Tools and Experimentation Platforms for Underwater Communication', *'Bulletin of Electrical Engineering and Informatics',* Vol. 10, No. 2, pp. 879 - 885 ISSN:

2302-9285,doi: 10.11591/eei.v10i2.1466 - https://beei.org/index.php/EEI/article/view/1466/2128.
3. Manjula K, Meenakshi Sundaram S (2021),'SVRGGA approach to investigate and predicte dynamic transformation amongst the nodes in the Distributed Systems', *'International Journal of Electrical Engineering and Technology (IJEET)'*, Volume 12, Issue 1, January 2021, pp. 88-97, Article ID: IJEET_12_01_010 - ISSN Print: 0976-6545 and ISSN Online: 0976-6553 -doi: 10.34218/IJEET.12.1.2021.010.

Tejaswini R.Murgod

Affiliation: Nitte Meenakshi Institute of Technology (NMIT), Yelahanka, Bengaluru, India

Education:

PhD Degree in Computer Science & Engineering from Visvesvaraya Technological University, Belagavi, India in 2022

M.Tech Degree in Software Engineering from Visvesvaraya Technological University, Belagavi, India in Jan 2015

B.E. Degree in Computer Science and Engineering from Visvesvaraya Technological University, Belagavi, India on June 2008.

Business Address:

Associate Professor
Department of Information Science and Engineering
Nitte Meenakshi Institute of Technology (NMIT)
Yelahanka, Bengaluru – 560063
Karnataka State, India

Research and Professional Experience:

13 Years of teaching experience and 06 years of research experience.

Honors: Best Researcher Award, 2021

Publications from the Last 3 Years:

1. Tejaswini RM, MeenakshiSundaram S (2021),'A Comparative Study of Different Network Simulation Tools and Experimentation

Platforms for Underwater Communication', *'Bulletin of Electrical Engineering and Informatics',* Vol. 10, No. 2, pp. 879 - 885 ISSN: 2302-9285,doi: 10.11591/eei.v10i2.1466 - https://beei.org/index.php/EEI/article/view/1466/2128.

2. Tejaswini RM, Dr. Meenakshi Sundaram S (2020),' Design of an Optimal Distributed Energy Efficient Hybrid Optical - Acoustic Cluster Based Routing Protocol (EEHRCP) to Minimize the Energy Consumption in Underwater Wireless Sensor Networks, *'International Journal of Recent Technology and Engineering (IJRTE)',* Volume-8 Issue-5, pp. 873 - 880, ISSN: 2277-3878, DOI:10.35940/ijrte.E6124.018520, Scopus Indexed Journal https://www.ijrte.org/wp-content/uploads/papers/v8i5/E6124018520.pdf.
3. Tejaswini RM, Dr. MeenakshiSundaram S (2020),'Design of an Energy Efficient Cluster based Localization Algorithm (EECBLA) for Underwater Wireless Sensor Networks', *'International Journal of Innovative Technology and Exploring Engineering',* Volume-9 Issue-7, 37-42. ISSN: 2278-3075 doi: 10.35940/ijitee.G4861.059720 - *Scopus Indexed Journal* https://www.ijitee.org/wp-content/uploads/papers/v9i7/G4861059720.pdf.

M. Sowmya

Affiliation: GSSS Institute of Engineering and Technology for Women KRS Road, Mysuru, India

Education:

M.Tech Degree in Software Engineering from Visvesvaraya Technological University, Belagavi, India in Jan 2016

B.E. Degree in Computer Science and Engineering from Visvesvaraya Technological University, Belagavi, India on June 2010.

Business Address:

Assistant Professor
Department of Computer Science and Engineering
GSSS Institute of Engineering and Technology for Women,
KRS Road, Mysuru, Karnataka State, India

Research and Professional Experience:
11 Years of teaching experience and 01 years of research experience.

Publications from the Last 3 Years:
Sowmya M, Anandhi G, Mobile Ad-hoc Networks: A survey on multipath routing protocols (2022), *Journal of University of Shanghai for Science and Technology.*

Chapter 4

DairyShop: A Machine Learning-Based Platform for One-Stop Dairy Product Management

Subrata Paul[1,*] and Anirban Mitra[2]

[1]Brainware University, Barasat, India
[2]Amity University, Kolkata, India

Abstract

Sustainable engineering is a method of modeling or operating systems that allows for the efficient use of goods and assets without jeopardizing the natural environment or the ability of future generations to meet their needs. Machine learning has numerous presentations that range from areas that include healthcare, agriculture, and security to daily life. In sustainable computing, machine learning similarly has a significant role. The presence of machine learning and even its mounting large associations in various markets necessitates an assessment of associations within a sustainable novelty accomplishment. Throughout the work, the authors have tried to develop a prototype of a system that would efficiently manage the dairy products. The intended system aims at providing a one stop solution for common users for the purchase of different dairy products. It would also allow merchants in this industry to seek the government schemes that are available to seek grants and enhance their business. It has also been interpreted how machine learning techniques can be incorporated into the development of the portal.

Keywords: sustainable engineering, machine learning, dairy products

* Corresponding Author's Email: subratapaulcse@gmail.com.

In: Computational Intelligence for Sustainable Development
Editor: Brojo Kishore Mishra
ISBN: 979-8-88697-198-9

Introduction

Sustainability refers to meeting our needs today while keeping in mind the usage of resources in the near future. The usability of resources should be determined after considering the upcoming prospect. Sustainability is built on three key pillars: environmental, economic, and social—also called Earth, gain, and human. Fundamentally, sustainability is a castoff in designating the practicality, arrangements, and measures necessary for efficient utilization of resources. Many examples of sustainable development include health-related solutions, Crop Rotation Farming, and a focus on sustainable construction, repetition reprocessing, and water treatment plants.

Over a specific, modest, and detailed goal, sustainability has been so extremely important that one mustn't withstand the position of humans, the variety of life on the planet, or the assortment of Earth until being developed. There are suggestions for the discussion on sustainability as an important aspect that is under discussion in every sphere, from big to small. All the energy resources would have been exhausted, and in fact, masses of species would have been eliminated. If an attempt is made to consume forest harvests outside of recovery, the community will suffer. Therefore, it shouldn't be turned. Possibly, the core of the progress of a community, neighborhood, or habitat around the world is dependent on understandability and its operations on sustainability (Bruno, 2012). Undoubtedly, the elementary requirements such as water, electricity, and public resources are being demoralized and expended beyond what is not right now. The performance of efficient resource utilization is observed in catering to human development objectives while preserving the potential of local environments and providing indispensable amenities and possessions for ecosystems that rely on society and the economy (Portney, 2015). An arrangement of society is to be made where existing circumstances as well as amenities are castoff as per the anticipated performance. Sometimes, human business interests are shadowed by deprived of violations of the truthfulness and steadiness of the natural habitat. Sustainable development is demarcated as an enlargement which caters to some specific necessities deprived of bargain in potential for the alleviation of self-desires (Bhatnagar et al., 2021). The achievement of sustainability is highly dependent on the usage of machine learning techniques.

Machine learning has been an application that forms a broader technological area of artificial intelligence that comprises generating procedures or sequencers that have the capacity of accessing and learning from data without being programmed by a human. It is quite interesting to know about the techniques which these algorithms use for learning, primarily with the application of the concepts of pattern recognition. Training of machine learning algorithms is done with an abundant amount of data. Subsequent analysis of the data is done in order to search for the existence of trends. Eventually, the algorithm is "intelligent" enough to produce new data sets.

Machine learning algorithms are archetypally divided into three expanses:

- Supervised algorithms are basically applied when the learning procedures of the recent past have been applied with new data. Predictions are made for the upcoming proceedings, and the output is compared to the projected results. Such algorithms are self-improved with "practice."
- Unsupervised algorithms are used for the analysis of unbranded and uncategorized documents. There does not exist any precise illustrations with the base expectations. These programs, further extracts implications in addition to ID hidden constructions or patterns inside data.
- Reinforcement algorithms interrelate within the surroundings for the test of outputs. Accurate performances are determined by the trial and error method, and further replies are modified accordingly (Donald et al., 1994).

The foundation of machine learning is much broader than previously thought. This discipline emerged rapidly once researchers established the workings of the neurons within the brain.

The chapter is organized as follows. In the next section, an elaboration is provided on the problem statement. Further, the way AI and machine learning have helped web development platforms will be discussed in the next section. Available ML frameworks that are successfully implemented are illustrated in the next section. The business benefits of the developed system will be discussed further. The authors illustrated a workflow diagram of the proposed system and discussed various aspects related to it. Phase-wise illustration of ML approaches is discussed.

Problem Statement

The government is making efforts to strengthen the infrastructure that would enhance the fabrication of quality milk, obtain, dispensation, and publicize milk along-with various milk products with the introduction of numerous dairy development schemes. For rationalizing the entire data, there is a requirement for a web-based gateway that would generate and reinforce dairy assets or infrastructure by entire dairy processors belonging to cooperatives as well as the private dairy sector. The author has tried to develop an online marketplace for the procurement of dairy products with the aim of covering the entire system's objectives. An extension of this system is aimed at supporting the tagging and monitoring of the capital assets shaped under various dairy development schemes grounded in geographical localities.

The Need to Incorporate Artificial Intelligence and Machine Learning Frameworks in Web Development

AI and machine learning can strengthen our systems in various ways. The way in which they can help is being discussed in the present section.

AI and ML Delight the Customer Support Team

Establishment of an AI chatbot within a website can enhance customer service involvement and assist users in a better way. It interconnects in a natural manner, recognizes sentiments, and provides responses to quite complex queries. Answers are also provided for those queries which are not associated with the concerned product, service or industry. The expansion of businesses and taking businesses to the next level needs the implementation of AI chatbot. Such chatbots can be incorporated into the proposed system, which will help the users in guiding the product according to their necessities. A few examples of existing chatbots are Capital One's "Eno" for assisting customers in the management of queries associated with their transactions, account balances, and further performing an analysis on the expenditure traditions. MedWhat's Virtual Medical Assistant delivers rapid reactions to queries that are grounded on performances studied with user interaction (Sharma, 2020).

AI and ML Help Maximize User Experience (UX)

In addition, machine learning aids in understanding customers' fondness for AI. Websites with AI and ML structures assist in the analysis of clients' favorites, search history, and location. This method of design and improvement of websites is done according to the needs of customers. Better customer experience can be provided with the update of UI from time to time.

Implementation of AI solutions on the proposed online web portal will provide customers with a better personalized shopping experience such that the whole thing is shaped precisely for them. It has also crafted a real-world scenario within the website that will encourage customers to make procurement of the products from the website. Amazon.com uses AI and ML for merchandise commendations. This has assisted the company in the identification of user preferences; everything of them is grounded on earlier followed data given by Amazon's AI solution.

AI and ML Provide Fast Access to Information

With its ever increasing popularity, every business proprietor is currently eyeing the addition of voice-dependent search functions on their websites. Several applications, such as Google Assistant, Siri, and Cortana, provide users with voice-empowered exploration functions for providing users with quick access to information. Voice-enabled search systems can be used by retailers to assist customers in searching for clothing categories according to their choice with complete specifications of the product like value of an article, kind of fabric, size, color, and length.

AI and ML Streamline Marketing Strategy

Both AI and ML technologies have been discarded marketing squads in the line of creating key advertising judgments and market predictions based on demand. The operation of this technology is dependent on the analysis of users' behavior with the discovery of the kinds of products preferred or liked by any user. These records assist marketing teams in making a decision regarding the alterations to be made to increase sales with the enhancement of the transformation rate. The prospect of AI and ML within web development hasn't absconded a consideration of top organizations like Google, Facebook, IBM, and Microsoft Launch, who have used AI toolkits with their APIs for containing natural language in the understandability of machine learning characteristics within contemporary web applications (Sharma, 2020).

Available Machine Learning Frameworks

A few of the algorithms are used by Machine Learning for making computers study without explicit programming. It has been the optimum technique for the analysis of data because it computerises the formation of investigative representations. This has been the reason that machine learning agendas play a significant part in web development. A few major machine learning agendas for web development will be discussed in this section.

Microsoft Cognitive Toolkit

This has been an open-source, deep learning toolkit used for training algorithms that would study similarities to the human brain. With the usage of such a tool, numerous machine learning models similar to convolutional neural networks, feed-forward DNNs, and recurrent neural networks can be exploited. Certainly, the tool is intended for the use of neural networks that would go through enormous unstructured datasets. The earliest possible training periods with an easy-to-use structural design makes it largely customizable, thereby permitting the choice of distinct constraints, networks, and algorithms.

Tensor Flow

It's been the most prevalent machine learning context for an environment for Java. It's been an open-source library that customs data flow graphs that concern numerical calculations. Flexibility in the architecture of Tensor Flow makes it easier for manipulators to implement the computations on single or greater GPUs or CPUs with a single API, irrespective of the type of node. The node within a graph characterizes mathematical procedures, although an edge within graphs signifies a multidimensional data arrangement (tensors) interconnected among them.

Apache Mahout

An alternative and most prevalent open source contribution from Apache has been basically intended in the fields of statistics, data science, and mathematics so that they might rapidly implement their individual algorithms. Furthermore, it has been a distributed linear algebra framework for the creation of machine learning applications with accessible presentations. With a primary focus on collaborative grouping, filtering, and classification, it also offers the ability to develop separate mathematical computations within a communicating environment, which essentially functions within a big data podium. Further, there is movement of an exact similar piece of code into the application for its implementation.

Caffe

It is preferred for its speed, countenance, and modularity, and its expressive structural design encourages specific personalized applications as well as innovation. Furthermore, with the availability of configuration options, customers can adjust the GPU and CPU with the configuration of a single pointer. Its extended code has aided in its initial development, making it highly successful.

Apache Singa

It has emerged as a deep learning platform that is commonly used in big data analytics due to its scalability and flexibility. It can be extended to function on common hardware by providing flexibility in structural design for ascendable distributed training for large volumes of data. (Sarkar, 2018).

Business Benefits of Ecommerce Machine Learning

Aside from augmented reality, machine learning offers a slew of business benefits, particularly for online retailers. The capacity of its algorithms to make sense of immense ribbons of data becomes invaluable. Machine learning has extensive applications in all the areas of ecommerce procedures that range from inventory management to customer experience, ecommerce machine learning.

Improved Transformations

Conversion of browsers into online customers is vital for any e-commerce website, which might unquestionably obsess over the transformation rate of a website. The cause of machine learning being convenient in e-commerce lies in the fact that this might facilitate the boosting of the proportion in numerous methods. Its worth in this respect originates in different areas regarding the way empowerment is done in on-site search engines besides endorsement of any particular product.

Machine learning algorithms might lead to the production of smart search consequences with the use of natural language processing. After gathering the knowledge on the searched item in its bar, a learning is done from earlier searches in order to help users demonstrate the product they want to search.

Machine learning also aids in the recommendation of any particular product, making it a smart tool. Analysis of the behavior of any visitor is also done with it. Recognition of products purchased or browsed alongside the content being interacted with is done first. These items, or some similar items that have been shown interest, are showcased to a buyer when he logs in again to the website; the products a visitor browses or buys; and the content with which they interact. Such techniques are used by leading e-commerce platforms in present days.

Functioning of Marketing Campaigns with Greater Relevance

Marketing strategies within e-commerce websites involve numerous correspondences from the viewpoint of sales. The finest promotions are extremely applicable to the target audiences. Machine learning aids an e-commerce company in maintaining the extent of significance. Within this period of big data, e-commerce has greater accessibility to additional information than earlier. Machine learning aids in making sense of the customer statistics for an enhanced adaptation to the marketing campaigns.

The arrangements identified through machine learning algorithms have become vigorous. They have demonstrated an interest in diverse customers on the website, which permits better segmentation of customers with higher accuracy. Predictions might vary on the basis of varying interests. This allows you to direct a user to relevant content.

Retargeting has been an alternative expanse where machine learning is priceless. Algorithms generally comprehend a customer's behavior and thereby recommend extremely pertinent retargeting campaigns. For an illustration, whenever a user visits a website, browses an item, adds an item to the cart but doesn't purchase it, and logs out by providing a mail-id. Through machine learning, Bliss will see that such a visitor will be the main target for a retargeting campaign. Further, this firm might send an email vending exactly the product that might interest the user.

Improvement the Efficiency of In-House Operations

Every assistances of ecommerce machine learning doesn't respect the procedures of facing customers. Sometimes, algorithms might provide some real-time insights that assist in the production of additional operations with better efficiency. Management of stock levels and creating an account for inventory might be illustrated as an example. Several companies, which struggle with traditional FIFO and LIFO decisions, make a method of selection through the analysis of customer data. Such an analysis is made rapidly and accurately with the incorporation of machine learning methodologies. Alongside, future demands can be predicted. Therefore, upon availability of all the complete information, the most efficient procedure can be adapted.

Well Informative Decision Making

Machine learning has been an outstanding tool for enhanced decision-making. Sometimes there is a requirement to decide on whether drop shipping is correct. Sometimes it might be speculated on whether a customer is interested in a new product. Any of the choices can be assisted by machine learning. Backing every decision made with data is done through machine learning, where the progression of any algorithms or programs and getting intellect out of the high capacities of information is done rapidly. This shall convey illegal insights that are further used in providing information to the available options (Shaw, unknown).

Proposed Framework

Within this section, the authors have presented a framework of the proposed system in the form of a flowchart in Figure 1. The details of the modules and data flows are illustrated in the next section.

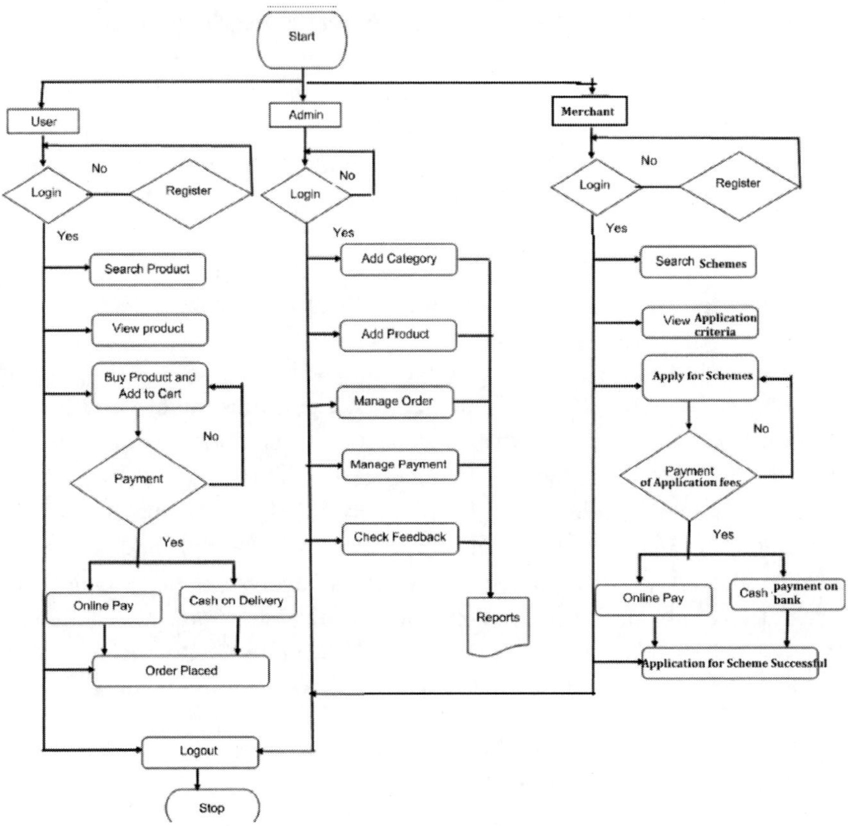

Figure 1. Flowchart of the proposed system.

Explanation of Proposed Framework

System Requirement Analysis

Information Gathering

As an objective of the application, the easiness of its usage with the delivery of a communicating interface alongside an extensive investigation has provided for gathering knowledge on the requirements and performances of several users. The employability of the application has made it suitable and simple to use for every end user.

The classification of any user can be done into two kinds on the basis of their familiarity with the products which suit their requirements. Classification is done on the users who are aware of the product which would satisfy their requirements and on those who have to understand the product which would satisfy their requirements. The search for a product is easier for users who are aware of the product with a single button click after entering a product name as a search item. On the other hand, after a search of a term, a filtering process is to be carried out on various product parameters for users who are totally unaware of their requirements.

Various functionalities are provided to users about the product that include a view of the whole specification of the product, including a view of the images at different angles and various zoom levels. Provisions for reading customer reviews and viewing ratings regarding the product are also given. Options for writing any review on any product and sharing the product with anyone through social networking platforms or mailing services are also being provided.

For an increment in the process of simplicity, users are able to add the product to their shopping cart with a simple drag and drop procedure and edit and remove them on the basis of their requirements. An interactive environment is created once a pop-up appears upon the addition of products to the shopping cart. Notifications are provided to users upon entry to the drop area and when an object can be chosen for addition to the cart.

System Analysis

Subsequent to the careful analysis and judgment on thc ncccssitics and functions of the web application, several significant diagrams have been drawn at thc conclusion of an analysis phase. They are referred to as data flow diagrams, and they form the foundation for the discovery of information flow.

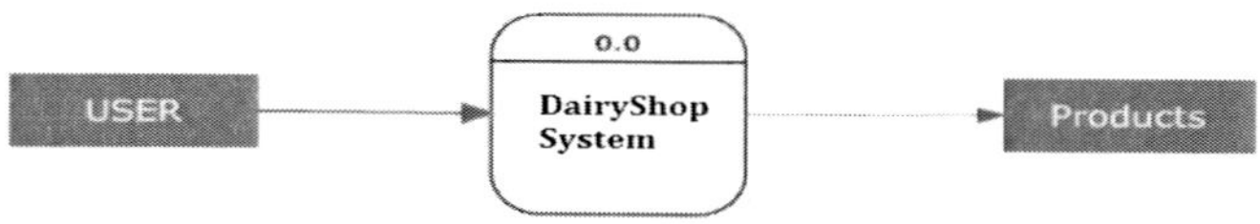

Figure 2. Context Level DFD of the user.

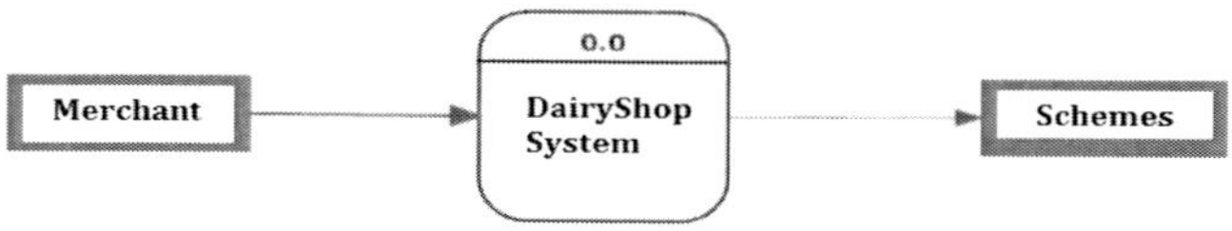

Figure 3. Context Level DFD of the merchant.

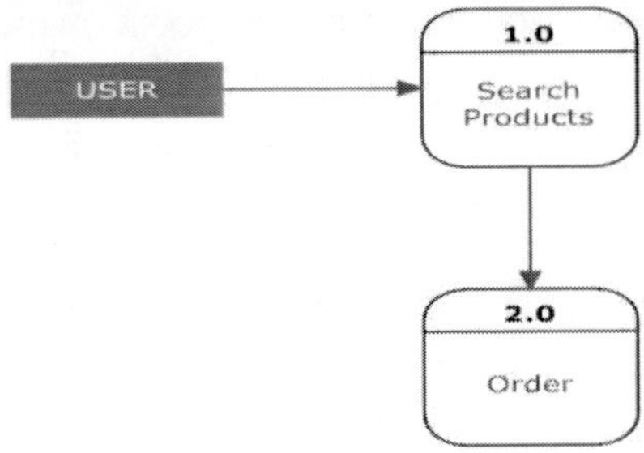

Figure 4. Level DFD 1 of the user.

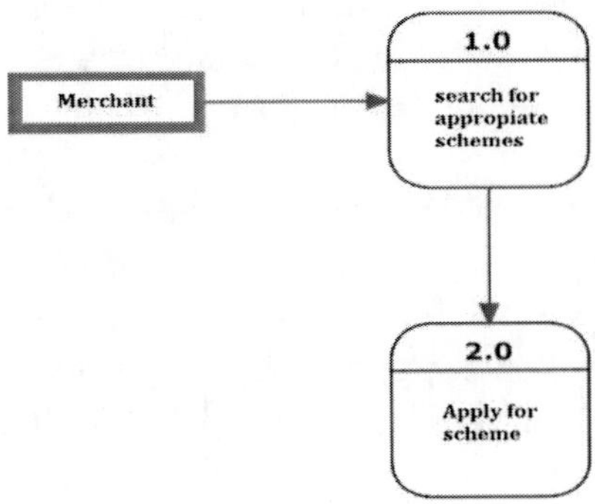

Figure 5. Level DFD 1 of the merchant.

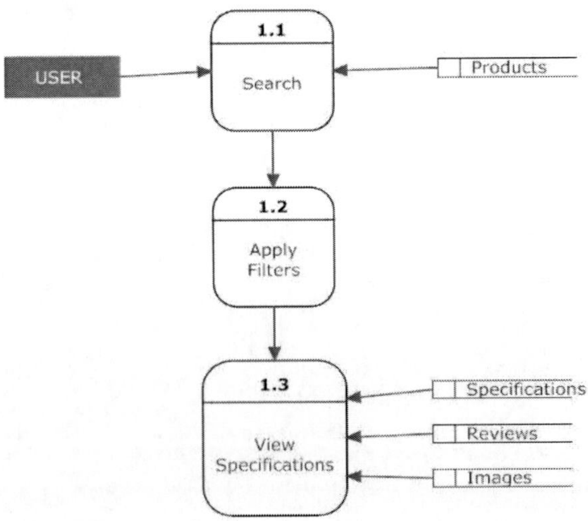

Figure 6. Level DFD 2 of the user.

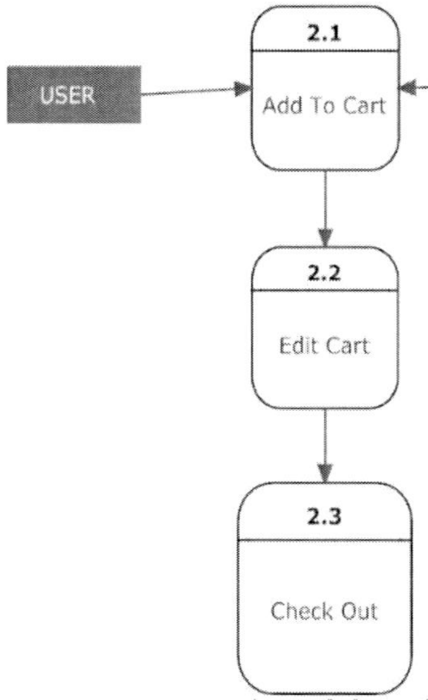

Figure 7. Level DFD 2 of the user.

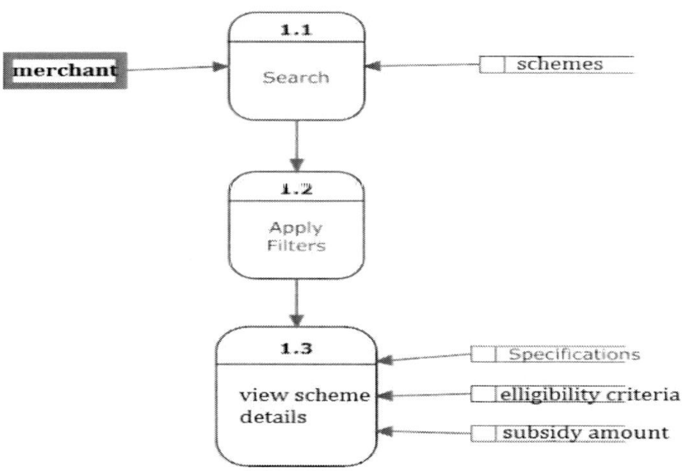

Figure 8. Level DFD 2 of the merchant.

After analyzing the data flow at different levels of the system, use case diagrams of the user and merchant have been drawn to illustrate the different functionalities that are performed by various users of the system.

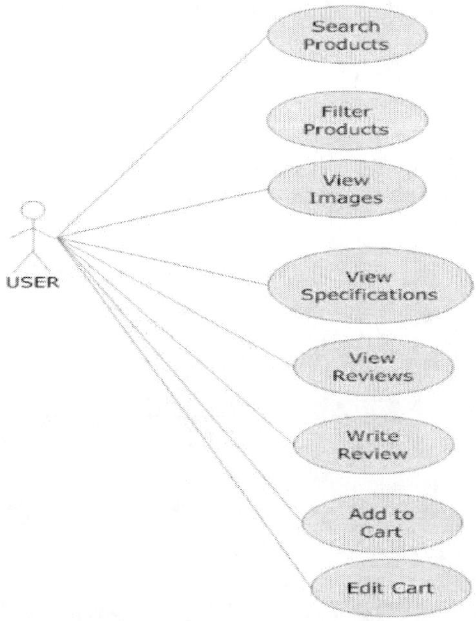

Figure 9. Use Case diagram of the user.

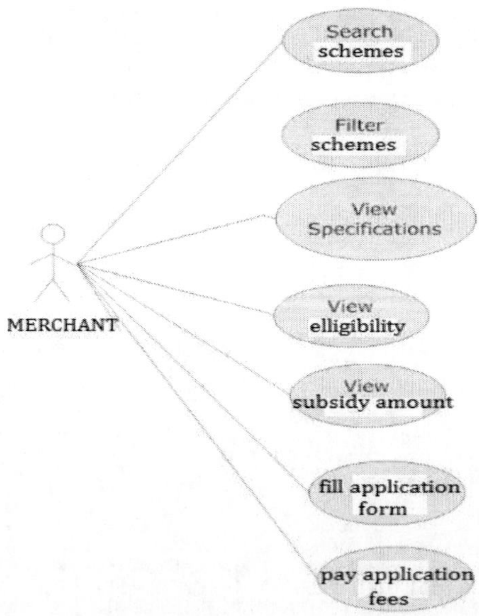

Figure 10. Use Case diagram of the merchant.

Design Objectives

- Designing the web application comprises of designing the forms for cataloguing and searching any products; showcasing the full specification of the product on successful search; and designing a simple and user friendly shopping cart.
- The design of a communicating application enables users to filter products on the basis of various product constraints.
- The design of the application contains characteristics similar to drag and drop.
- The design of an application that prevents data transfer between client and server.

Description of Architectural Design

Within the context diagram, information supplied and obtained from the 'Dairy Shop System' at both user and merchant level is recognized. Incoming and outgoing arrows signify information received and information produced, respectively, by the system. Closed boxes represent a collection of information sources and sinks.

Within the system, it will be observed that interaction between user and application occurs with a graphical user interface. Searching, filtering product specifications, and writing reviews serve as various input forms by the user. Search results and the display of information in repeater and grid views are provided as outputs from the system.

Procedural/Modular Approach

The following is a list of all the modules designed for the DairyShop System.

Shop Products Module

This module commences whenever a user visits a home page or whenever searching is carried out with the entry of a relevant search item. This portion of the application comprises the display of every product that is available or products that match search terms entered by the user. Further, filtering of the products is done on the basis of numerous constraints like manufacturer, product type, operating system supported or a price range. A user can look through the list of available products where each product is demonstrated by means of an image besides its various features. Addition or deletion of a product from the cart is done with a simple drag and drop.

Product Description Module

Commencement of this module is made upon the user's visit to the product description page. The user can view different sized images related to the product. Enlarged images can also be seen within a popup window. A view of the full specifications related to the product can be viewed alongside the production companies. Additional information regarding the discount and exchanges is available for review. Viewing and writing product reviews are also possible.

Shopping Cart Module

Commencement of this module occurs whenever a user observes his shopping cart. Every product that has been added to the shopping cart is listed with the price and quantity, and finally the amount is summed up and displayed. Options are provided for addition, deletion, or any modifications to the existing shopping cart. Updating of the total amount occurs every time the shopping cart is altered.

View Schemes Module

This module commences upon a merchant visiting the Schemes tab or whenever a merchant examines a scheme with an entry of a search term. This portion of the application comprises the display of entire schemes that are accessible or the scheme that matches the search term that is entered by the merchant. The merchant further filters these schemes on the basis of various constraints such as agency, amount, area of application, etc. The user can browse through the schemes, and each scheme will be discussed with a brief description and eligibility criteria. A user can select a scheme for an application and fill out the form with a click on the button. The user would be able to view the application summary after successful submission of the form.

Scheme Description Module

This module commences with the merchant visiting the scheme description page. A merchant is able to see various documents related to the scheme. The user can have a look at the whole requirements of the scheme, such as its features, areas supported, system requirements etc. A user is able to view and write reviews related to the product.

Application Summary Module

This module commences when the merchant sights the application summary. The total list of the schemes which are applied by the merchant is visible in his application summary along with their amount and quantity.

Technical Discussions

Filtration of products is done on the basis of numerous constraints. New items are added to the list of products with the passage of time. Therefore, values for the list of manufacturers and product categories are obtained dynamically with the retrieval from the database. Two different text fields are provided such that users are able to make an entry into the price ranges for better flexibility.

For maintaining equilibrium and affluence in the usage of products, addition and removal of items to the cart are carried out with the simple drag and drop facility. There is an update in the summary label whenever items are added or removed from the cart. However, there is a possibility of the loss of information whenever the internet connection is not present. Therefore, upon dropping a product within the cart area, web services are called which further update session variables for the shopping cart and recalculation of the cart summary is done in this case and displayed to the user. This process will enhance the dependability of the application.

Software Testing

Machine Learning Life Cycle on Software testing

The different phases of machine learning are discussed below with respect to optimizing software testing.

Identifying Data Sources

The sources in software testing are "Test Case Pool," "Testing Tool Details," "Performance Report of Testing," "Test Plan Documents," "Defects Reports" and etc. The different data sources in the testing phase will be used for filtering the required data for preparing the raw data set.

Preparing Data

The collected data from the identified sources in the software testing domain will be used in this phase to prepare a suitable format of data that may help in understanding the data and its different features.

Data Wrangling

In this phase, the issues with data will be solved to make it ready for further processing. The issues may be data type, missing value, unnecessary features, etc. So data preprocessing should be done to improve the quality of the data.

Analyzing Data

This phase has the responsibility of identifying the relationships among the features identified in the collected data. Statistical mathematical approaches are applied to analyze the patterns of values in given features. This helps in designing the algorithm to implement the required action in the testing domain.

Training the Model

After having the suitable algorithm and model, the training data from the data source should be given for gaining knowledge with the help of the algorithm. The performance of the ML model is based on a suitable algorithm, data features, and amount of data. Hence, preferably 70% of the dataset from the original data set should be taken for the training phase. The remaining 30% of the data set is to be used for the 'Testing phase.'

Testing the ML Model

It is about using the test data selected from the original data set to verify whether the trained ML model is able to produce the expected result or not.

Out of which, 30% of the original dataset will be taken for testing the ML model, and that the entire test dataset should be given to the system for collecting output to analyze. If there are any differences in collected output that model behavior is noted for modifying the ML model. In this test, black-box testing from software testing will be suggested to test the ML model for analyzing its different actions when given inputs from the test data set. When the model is ready, then it could be applied to the software testing domain to classify the test cases and "testing tools."

The prediction model such as 'Linear Regression' will be suggested for finding the target variable 'Energy required' for conducting testing from the input variables 'Code Size' or 'Test Suite Size' and 'Number of Functionalities Tested.' From the cumulative answer, the average value will be computed to define the required energy for conducting software testing.

The classification model, like Prediction, suggests categorizing test cases as "Highly Prominent Test Cases," "Less Prominent Test Cases," "High Performance Testing Tools," and "Less Performance Testing Tools." This kind of classification of results may help the testing team choose the perfect test cases and testing tools for performing the required testing procedures.

In the same way, the 'clustering' model may be suggested to group the test cases based on the similar patterns hidden in them. Like that, similar patterns of functionality can be grouped in order to select a suitable test policy for implementing a particular testing type (Jupudi & Ramalingam, 2021).

Sustainable Software Testing Principles

Software testing requires effort to find defects or bugs in the given application. The psychology of a software tester is to detect bugs in implemented software systems by executing different kinds or levels of software testing. Hence, he has to put good effort into testing to uncover bugs in the software application. The machines and men will be deployed to conduct many levels of testing, bringing the bug reports. Testing may be conducted as more iterations on different kinds of features or modules of the proposed application. Reducing efforts or testing procedures may cause some real-life issues. So, it may not compromise with any external factors' recommendations. To reduce energy consumption and resources like paper, machine working hours and human involvement have to be normalized. That can be done only by preparing or producing a perfect testing plan. So, an experienced man will be identified to prepare the testing plan according to the kind of application going to be tested.

The Sustainable Software Testing Process may help us to streamline the testing activity to be performed in the developed product to find all possible hidden bugs or defects, and at the same moment, energy consumption will be controlled (Brajnik et al., 2020). In this section, Green Engineering principles are proposed in order to perform software testing, which may work well and consume less energy by reducing the tool usage cycles by identifying and supplying suitable and perfect input data to test the system's features and functionalities.

Tool selection is a very important and crucial activity because the selected tool has to perform well to test the particular application without causing any side effects during the execution of the testing procedure. As a result, testers must be well-versed in all of the various types of tools available on the market or in an open source pool in order to carry out a specific testing activity (Jupudi & Ramalingam, 2021).

Conclusion

At present, the perception of sustainability has extensively been acknowledged as an obsessive topic for numerous areas. Numerous approaches, devices, procedures, replicas, and methodologies have been established and prolonged so that the performance of sustainability is measurable. Throughout the paper, the authors intend to build a prototype of a web-based platform with the implementation of machine learning techniques. Various aspects in which machine learning can help in building a web-based platform and the ways it is being benefited from are also discussed. The modules are elaborated with the dataflow diagram and use case diagram. The authors intend to implement the prototype with a high-level programming language.

References

Bhatnagar, V., Sharma, S., Bhatnagar, A., Kumar, L. (2021) Role of Machine Learning in Sustainable Engineering: A Review, IOP Conference Series. *Materials Science and Engineering*; Bristol Vol. 1099, Iss. 1, doi: https://doi.org/10.1088/1757-899X/1099/1/012036.

Brajnik, G., Bartolini, M. and Rees, N. (2020) A sustainable software testing process for the Square Kilometre Array project. *INCOSE International Symposium*. Vol. 30. No. 1.

Bruno, G., Leduc, R. and Savard, L. (2012) From a conventional to a sustainable engineering design process: different shades of sustainability. *Journal of Engineering Design* 23. 1: 49-74.

Donald, M., Spiegelhalter, D. J. and Taylor, C. C. (1994) Machine learning. *Neural and Statistical Classification*. 13. 1994 (1994): 1-298.

Jupudi, L. & Ramalingam, K. (2021). Building Sustainable Software Testing using Machine Learning for Green Engineering, *Proceedings of ICICTA 2021*, Reva University, Bangalore.

Portney, K. E. (2015) *Sustainability*. MIT Press.

Sarkar, A. (2018) Top 5 Machine Learning Frameworks For Web Development, *AI Zone - Opinion*, [online] Available at: https://dzone.com/articles/top-machine-learning-frameworks-for-web-developmen [Accessed: Sep. 12, 2018].

Sharma, R. (2020) The Best AI & Machine Learning Frameworks to Learn for Web Development, *Machine Learning*, www.freecodecamp.org, [online] Available at: https://www.freecodecamp.org/news/best-ai-machine-learning-frameworks-for-web-development/ [Accessed: May 14, 2020].

Shaw, N. (unknown) Ecommerce Machine Learning: AI's Role in the Future of Online Shopping, Ecommerce Technology, *bigcommerce,* [online] Available at: https://www.bigcommerce.com/blog/ecommerce-machine-learning/#business-benefits-of-ecommerce-machine-learning.

Chapter 5

Recent Developments in Data Science and Its Challenges

S. Kannadhasan[1,*], R. Nagarajan[2,†], G. Ramya[3] and C. Jisha Chandra[4,‡]

[1]Cheran College of Engineering, Tamilnadu, India

[2,3,4]Gnanamani College of Technology, Tamilnadu, India

Abstract

Big Data is a phrase that refers to a collection of methods that elicit a kind of integration that is used to discover enormous unknown values. These are very complicated values with a massive scope. The phrase "Big Data" was coined in 1997 by NASA scientists in a study report. Big Data is a phrase that refers to data that has a high velocity, diversity, and volume. Big Data is a term used to describe a huge quantity of information. It is possible for this data to be both organized and unstructured. It's also tough to practise using conventional tools and methods. There is a large quantity of big data in the IT industry that is common to many departments, i.e., there is a vast amount of data in the store of industries and no tool exists to manage it before big data comes into play. Big Data Analytics is concerned with the storage and analysis of various, difficult, complicated and massive datasets.

Keywords: data science, big data, 4.0, AI and challenges

---

* Corresponding Author's Email: kannadhasan.ece@gmail.com.

† Corresponding Author's Email: krnaga71@yahoo.com.

‡ Corresponding Author's Email: jishasj.chandra@gmail.com.

In: Computational Intelligence for Sustainable Development
Editor: Brojo Kishore Mishra
ISBN: 979-8-88697-198-9

Introduction

One of the most important properties of large data is its value. It is critical for IT infrastructure systems to store a large quantity of data in their databases. Velocity is a word that refers to the rapid pace at which data is generated. The pace at which data is produced has an impact on its potential. The data flow is enormous and never-ending. The data is provided in both organized and unstructured formats, and it comes from a variety of sources. Video files, e-mail, audio files, world processing files, and other types of data are now being examined. Volume is a phrase used in the context of "Big Data," which refers to very massive data sets. The size of the data determines whether it is considered big data or not. So, among the many factors that should be addressed while dealing with "Big Data," "Volume" is the most essential. The word "variability" refers to the inconsistency of data. When dealing with huge amounts of data, data isn't always accurate. Validity is concerned with data accuracy and correctness. Vulnerability is a term that refers to a problem with data security. After all, a large data data breach is a major breach. Volatility is a Big Data metric that deals with the statistical measurement of dispersion for a given collection of returns. Visualization is a contemporary feature of big data that is concerned with data visualization. In the context of big data, variability refers to the irregularity in the rate at which data is saved in our system. To analyze large data, a variety of tools and methods are used. This paper is structured as follows to explain various tools and techniques: Machine learning may involve a variety of software that can learn from data. It gives computers the capacity to learn without having to be explicitly taught, with the primary emphasis on generating predictions based on known characteristics. Machine learning is currently being used to help distinguish between spam and non-spam communications, as well as to learn user preferences and then make suggestions based on that knowledge [1-5].

It's a software algorithm that's a collection of brain architectures from animals that are similar in nature. The association is seen as a network in this case, and it consists of the following elements: There are three types of layers: input layers, hidden layers, and output layers. Each unit is given a certain weight. Different types of data are sent into the input node, and various methods are employed to modify the weights using a trial-and-error method until it satisfies a set of stopping conditions. Semantria offers us with a one-of-a-kind solution that is based on the collection of different data from various customers. The process of analysing that data is then meticulously applied to produce the most valuable and desirable insights. It is very beneficial in identifying trends and identifying different patterns that are beneficial to achieving success. Semantria is a tool that combines various text analytics in a powerful way. The following advantages are available to consumers as a result of this: gather more trustworthy, useful data Fast configurable, complete system, entity extraction, classification, clustering, visualisation, and 10+ languages are just a few of Semantria's capabilities [6-10].

Sentiment Analysis is a module that analyses open text. It is a kind of classification engine that is used to discover different subjective patterns. It's also used to assess text-based expressions of sentiment. To begin, work on the analysis is done at the subject, sentence, and document levels. Its primary purpose is to determine if portions of text seem realistic. Trackur is a tool that is used to gather data. It looks at the particular terms that the users are monitoring with its automatic sentiment analysis, and then makes judgments. With regard to the associated document, the emotion may be favourable, negative, or neutral. The Trackur algorithm may be used to monitor social media sites and highlight breaking news, as well as gather data via trends and automated sentiment analysis.

SAS also has a sentiment analysis feature that captures sentiments in real time. It accomplishes this goal by using a variety of statistical modelling methods. These processing methods are based on natural language rules. There are several built-in automatic reports that illustrate the different patterns and their responses in detail. Users may correctly improve their models and make various changes connected to classifications with the assistance of continuous assessments. Opinion Crawl is a web-based application for sentiment analysis in current events. It allows different visitors to rate the emotion on the Internet based on a certain subject, such as a business. The user may enter a subject that he wants to evaluate and get an ad-hoc sentiment evaluation on that issue. For each particular subject, the user will get a pie chart displaying the current sentiment. There's also a collection of the most recent news headlines, as well as some pictures connected to the subject. All of these concepts enable users to determine what kind of problems are generated the sentiment, i.e., whether the emotion is positive or negative.

Databases must be used to store the recorded data, which may be organised or unstructured. Some databases are required to handle Big Data. Organizations like as Apache, Oracle, and others have created a number of frameworks that are used as analytics tools to collect and analyse data from these repositories. Apache Hadoop is a system for processing Big Data, which is the unification of large amounts of organised and unstructured data. Apache Hadoop is an open source platform and processing system designed specifically for batch processing. Hadoop was inspired by Google's Map Reduce at initially [11-15]. The whole programme is split into a number of tiny pieces in the Map Reduce software architecture. Fragments are another name for these tiny pieces. These pieces may be run on any of the cluster's systems.

Data Science Analysis

The Hadoop Distributed File System (HDFS) is the software framework's primary component. Hadoop's file system is called HDFS. HDFS is designed to store huge amounts of data. It makes advantage of low-cost, widely distributed hardware. It's a fault-tolerant storage system for big files with sizes ranging from TB to PB. In HDFS, there are two kinds of nodes. There are two types of nodes: name node and data node. It is known as Node and serves as the master node. It includes all of the data node's associated information. It contains information about free space, node addresses, all of the data they hold, active node, and passive node. It also saves information from task and job trackers.

Slave Node: Slave node is another name for data node. The data is stored on a Hadoop data node. TaskTracker's role is to keep track of the tasks that are currently running on the data node, as well as the jobs that are coming in from the name node.

MapReduce: MapReduce is a framework that allows programmers to write programmes that process large amounts of unstructured data in parallel across a distributed network. Many components make up MapReduce, including JobTracker, TaskTracker, JobHistorySever, and so on. Hadoop's native instruction execution engine is another name for this component. It was created in order to process massive amounts of data and store them on commodity hardware. It utilises clusters to store records for processing huge amounts of data. The Map Reduce programming paradigm is built on two functions: the Map and Reduce functions. The Map function is active in the master node. It also accepts the data. The received input is subsequently divided into sub modules, which are then sent to slave nodes. YARN (Yet Another Resource Negotiator) is a Hadoop core service that provides two main services: global resource management (ResourceManager) and application-specific resource management (ApplicationMaster). It is the Hadoop stack's cluster coordinating piece. YARN makes it possible to do so. The MapReduce engine is responsible for Hadoop's functionality.

MapReduce is a framework that runs on less expensive hardware. It has no concept of storing anything in memory. The measurability potential of MapReduce is unfathomable. Thousands of nodes have been created using this method. Different Hadoop scheme modifications will reduce the impact of this to varying degrees; nevertheless, it will always be a factor in rapidly implementing an idea on a Hadoop cluster. Data comparisons between astronomy and Earth science, technique transfer, and metadata and ontologies have all been identified as critical. "Non-homogeneous observables, and diverse geographical, temporal coverage at various resolutions" must be addressed in the convergence or bridging of disciplines. This quote is well-known to us in light of how NoSQL databases, as well as conventional relational databases, are now extensively utilised, and another example is how text mining, social media, and a variety of other areas have grown more significant in a variety of settings.

Then, with computer assistance, "it is the intricacy, not the data amount, that proves to be the greater challenge." Traceability and repeatability are two further advantages of this Data Science convergence. The need of disciplinary convergence and bridging is highlighted as follows: "Methodology transfer can virtually never be unidirectional. Diverse areas benefit from studying techniques used by other disciplines. The most essential thing is to abstract data, which is defined by relevant metadata, which is then linked by a strong ontology." In the case of Data Science, further explanation is required: "We have discussed here a few methods from astroinformatics that are being used in geoinformatics. There are many aspects of earth science that space science might learn from. Other fields, such as bioinformatics, also provide many possibilities for technique transfer and cooperation. Data science is our only hope in the face of ever-increasing data quantities and, more crucially, ever-increasing complexity. All fields will benefit from data science collaboration."

Many academics believe that commercial database management systems are unsuitable for handling very huge amounts of data. They propose a new Big Data database management system that should be both cost-effective and scalable. The adoption of parallelization methods and algorithms is essential for improving Big Data processing scalability and performance. For academics and business, Big Data represents a new challenge. For Big Data management and analysis, researchers are developing new ideas, techniques, and technologies. Machine learning, data mining, and statistical methods for Big Data processing are critical to turning data into usable information. The current data management solutions are incapable of storing the growing amount of Big Data. To address the problem of storing Big Data, a hierarchical storage system is required. Existing data processing algorithms excel at homogenous and static data processing. Today, however, data is constantly generated from a variety of sources. This data is both diverse and dynamic. To handle enormous data, new scalable data processing methods are needed. When it comes to executing a query in Big Data, speed is crucial.

In this situation, indexing is the best option for handling complicated queries. To successfully manage Big Data, parallelization and divide and conquer are excellent algorithmic methods. Using online Big Data apps, businesses are lowering their costs. This approach is lucrative for businesses, but it introduces new security risks. Big Data security is a major issue for both academics and businesses. Big Data security is primarily concerned with how to conduct data mining without revealing sensitive user information. Because Big Data is dynamic in nature, researchers are faced with new difficulties in developing algorithms to deal with such circumstances. Data diversity, volume, analytical workload complexity, and agility are the key difficulties of Big Data. Many businesses are finding it difficult to cope with ever-increasing data quantities. To address this issue, businesses must decrease the quantity of data they keep and take use of emerging storage methods that may enhance performance and storage usage. Every day, we generate 2.5 quintillion bytes of data – so much so that the past two years have produced 90 percent of the data on the planet. Sensors used to collect temperature data, postings to social networking sites, digital photos and videos, buy transaction records, and mobile phone GPS signals, to mention a few sources of data. Big Data is a term that refers to the massive quantity of data that is constantly generated. Big Data decodes previously unseen data to provide new insights that can be incorporated into business processes. The existing methods, however, are becoming outdated as the quantity of data grows exponentially. Dealing with large amounts of data requires extensive coding abilities, subject expertise, and statistical knowledge. Big Data applications are practically widespread, despite its Herculean nature. They range from marketing to scientific research to consumer interests, and so on. Today, we can see Big Data in action nearly everywhere. From Facebook's 40 billion pictures to CERN's Large Hydron Collider (LHC), which produces 15 billion dollars each year, to Walmart, which processes over 1 billion consumer transactions every hour. The World Bank hosted the inaugural WBG Big Data Innovation Challenge over a year ago, which included many innovative Big Data concepts such as using Big Data to forecast poverty, climate smart agriculture, and user-focused identification of road infrastructure condition and safety, among others. The

3Vs – volume, velocity, and variety – which are the driving dimensions of Big Data quantification may be easily stated.

This mostly refers to the massive amounts of data that are produced on a regular basis. Because of the high storage costs, storing such data was first difficult. However, because of lower storage costs, this issue has been held at bay for the time being. However, this is just a temporary solution, and more advanced technology is required. Massive quantities of data are produced by smartphones, e-commerce, and social networking websites, to name a few examples. Structured data, unstructured data, and semi-structured data may all be readily distinguished in this data.

Challenges in Data Science

Big Data is slowly but steadily gaining traction. Big Data analytics may now be used in every industry, including business, health, and general living standards. To put it another way, Big Data is a field that may be applied to any area where a big amount of data can be utilised to one's benefit. Organizations all around the globe are slowly but steadily realising the value of big data analytics. Big Data analytics is a one-stop solution for forecasting consumer buying behaviour patterns, persuading them to make purchases, and identifying fraud and abuse, which was previously an unfathomable job for most businesses. Regardless of the complexity or amount of the data, business professionals should be able to examine and understand it according to their business needs. Data scientists must be able to visualise and communicate data in an understandable way in order to meet this criterion. Data visualisation has been used by companies such as Google, Facebook, Twitter, EBay, Wal-Mart, and others to simplify the management of complicated data. In these types of businesses, data visualisation has had a very beneficial impact. Enterprises may now begin to tap into the enormous potential of Big Data by using data analytics and data visualisation, ensuring better return on investments and company stability.

Since the mid-1980s, the global ability to create, store, and transmit data, mainly in digital forms, has grown at an unparalleled rate. At the same time, access to computing technology such as the personal computer, smart phone, and other portable devices has increased. With these improved data storage and processing capabilities, as well as real-time information transmission through the internet, the average daily data consumption of each person has increased dramatically. According to Asigra, a Cloud Backup business founded in 1986, 90 percent of the data on the planet today was generated in the past two years, and the global number of Internet Protocol (IP) addresses is expected to double by 2015. Furthermore, it is estimated that three billion people would be online in two years, generating up to eight zettabytes of data.

Big Data refers to the reality that today's data is often too big, diverse, and constantly changing to be stored, analysed, and turned into value using traditional methods. Big Data is driven by technological trends: business processes are increasingly being carried out electronically, and consumers are producing more and more data on their own - for example, in social networks, cloud-based processing of petabytes of data can be accomplished in a fraction of the time and cost by implementing big data processing methods. Big Data has quietly crept into our everyday lives, and its impact will continue to increase as inexpensive data storage and smart devices become more widely available, both locally and in emerging nations. This impact has spread to healthcare as well. Due of increasing healthcare expenditures in countries like the United States, big data is in high demand. "After more than 20 years of continuous growth, healthcare expenditures now account for 17.6 percent of GDP —nearly $600 billion more than the anticipated threshold for a country of the United States' size and wealth," according to the McKinsey study. In other words, healthcare expenses are much more than they should be, and they have been steadily increasing for many years. Clearly, we need some data-driven, smart thinking in this area.

Big Data in healthcare typically refers to large amounts of electronic health record, administrative claims, and clinical trial data, as well as data collected from smart phone apps, wearable devices, social media, and personal genomics services; Predictive Analytics refers to innovative methods of analysis developed to overcome challenges associated with big data, such as a Predictive analytics based on big data has been effectively used in many areas of drug management, including identifying difficult patients or those at high risk of noncompliance or harmful effects. In healthcare, big data refers to electronic health data sets that are so vast and complicated that they are difficult (or impossible) to handle using conventional software and/or hardware, as well as with typical data management tools and techniques. (8) Big Data is defined as very big data sets that may be computationally analysed to discover patterns, trends, and connections, as well as visualisation, querying, information privacy, and predictive analytics on a vast and diverse set of data. Exabytes of data are currently being generated in health-care systems (1 exabyte = 1 billion gigabytes). In healthcare, big data is daunting not just because of its sheer amount, but also because of the variety of data kinds and the speed with which it must be handled. Increased telemedicine usage will put the storage capacity for patient data to the test, and doctors' creative use of Google Glass will add to the social and behavioural elements of Big Data. Electronic Medical Records (EMRs) are renowned for facilitating the flow of health care by storing a variety of data such as patient demographics, clinical, and genetic data. Data analytics may be used to organise this data and make it retrievable.

Data is produced from a variety of sources in the digital world, and the rapid adoption of digital technology has resulted in the development of big data. With the gathering of big datasets, it enables evolutionary advances in a variety of areas. It refers to a collection of big, complicated datasets that are difficult to handle using conventional database administration tools or data processing software. These are accessible in petabytes and beyond in organised, semi-structured, and unstructured formats. Formally, it ranges from 3 to 4 volts. Volume, velocity, and variety are the three Vs. The enormous quantity of data that is produced every day is referred to as volume, while velocity relates to the pace of increase and how quickly the data is collected for analysis. Variety describes the many kinds of data, such as structured, unstructured, and semistructured data. Veracity, which combines availability and responsibility, is the fourth V. The primary goal of big data analysis is to use conventional and computational intelligence methods to analyse data with high volume, velocity, diversity, and authenticity.

However, there is no precise definition for big data, and it is thought to be problem-specific. This will allow us to make better decisions, uncover new insights, and optimise our processes while remaining creative and cost-effective. By 2015, it is anticipated that big data would have grown to 25 billion users. Big Data, from the standpoint of information and communication technology, is a powerful push for the next generation of information technology businesses, which are based on the third platform, which includes big data, cloud computing, the internet of things, and social business. Data warehouses have often been used to handle huge datasets. The extraction of accurate information from the available large data is a major concern in this situation. The majority of the data mining techniques that have been described so far have not been able to effectively manage big datasets. The lack of coordination across database systems, as well as analytical techniques like data mining and statistical analysis, is a major issue in big data analysis. These difficulties often emerge when we need to conduct information discovery and representation for practical purposes. How to statistically define the key features of large data is a basic issue. In explaining the data revolution, epistemological implications are required. Furthermore, research into big data complexity theory will aid in understanding the fundamental features and creation of complex patterns in big data, simplifying its representation, improving knowledge abstraction, and guiding the design of big data computing models and algorithms. Various academics have conducted extensive study on big data and related patterns. Smart manufacturing systems need creative ideas to improve the quality and long-term viability of manufacturing processes while lowering costs. Artificial intelligence (AI)-driven technologies, backed by I4.0 Key Enabling Technologies (such as the Internet of Things, sophisticated embedded systems, cloud computing, big data, cognitive systems, virtual and augmented reality), are poised to usher in new industrial paradigms.

In this respect, it's worth recalling that John McCarthy, the founder of artificial intelligence, described artificial intelligence as "the science and engineering of creating intelligent machines, particularly clever computer programmes" in the 1990s. In general, when a computer mimics functions that people connect with other human brains, such as learning and problem solving, the term "AI" is employed. The fields of artificial intelligence are divided into 16 groups on a wide scale.

AI has become a major study topic in many areas in the twenty-first century, including engineering, science, education, health, business, accounting, finance, marketing, economics, stock market, and law, to name a few. Since the intelligence of computers with machine learning skills has had significant effects on industry, governments, and society, the scope of AI has expanded dramatically. Techniques that are both conventional and computationally clever.

They also have an impact on global sustainability trends. Artificial intelligence (AI) has the potential to help address key issues in the field of sustainable manufacturing (e.g., optimization of energy resources, logistics, supply chain management, waste management, etc.). In this regard, there is a tendency in smart production to integrate AI into green manufacturing processes in order to comply with tighter environmental regulations. Indeed, as Hendrik Fink, head of PricewaterhouseCoopers' Sustainability Services, said in March 2019, "if we correctly integrate artificial intelligence, we may accomplish a revolution in terms of sustainability." The fourth industrial revolution will be propelled by artificial intelligence."

As a result, AI subfields including machine learning, natural language processing, image processing, and data mining have become hot topics among today's digital behemoths. Because of the constant development of the technology accessible today, the topic of AI piques the scientific community's attention. The advancement of machine learning as a field of AI is currently extremely rapid. Its use has extended to a variety of areas, including smart manufacturing, medical research, pharmacology, agriculture, archaeology, gaming, and business, among others. In light of the above concerns, this study conducted a comprehensive literature assessment of AI and machine learning research from 1999 to 2019. As a result, in order to have more variance and reflection, it is thought essential to develop a categorization system that refers to publications that jointly address the two subjects. In order to get a better understanding, additional factors were investigated, such as the thematic areas and industries in which the technologies had the greatest impact. The major contribution of this paper is that it offers a summary of previous studies. For many years, a number of remarkable documentations of well-established research techniques and philosophy have been debated. Unfortunately, there is limited cross-study comparability and integration.

Conclusion

This article established a common understanding of AI and machine learning research and its variants. This article does not try to give a comprehensive overview of AI and machine learning research. Rather, it aims to serve as a beginning point for integrating information from many studies in this field and to propose research directions for the future. It delves into research in a variety of new fields, including environmental contamination, medical, maintenance, and manufacturing. More research is required to expand AI's current knowledge boundary by incorporating ideas and philosophies from certain conventional fields into current AI frameworks. The goal of this paper is not to be the catalyst for a rapid expansion of an already concentrated industry; rather, it is anticipated that this study will serve as an essential intellectual instrument for both refocusing effort and generating new intellectual possibilities. This article offers useful suggestions and insights for AI and machine learning research. The ultimate goal was to foresee how the discipline will change in the future.

References

[1] Kakhani, M. K., S. Kakhani and S. R.Biradar, Research issues in big data analytics, *International Journal of Application or Innovation in Engineering & Management*, 2(8) (2015), pp.228-232.

[2] Gandomi, A. and M. Haider, Beyond the hype: Big Data concepts, methods, and analytics, *International Journal of Information Management*, 35(2) (2015), pp.137-144.

[3] Lynch, C., Big Data: How do your data grow?, *Nature,* 455 (2008), pp.28-29.

[4] Jin, X., B. W.Wah, X. Cheng and Y. Wang, Significance and challenges of big data research, *Big Data Research*, 2(2) (2015), pp.59-64.

[5] Kitchin, R., Big Data, new epistemologies and paradigm shifts, *Big Data Society*, 1(1) (2014), pp.1-12.

[6] IHTT. Transforming Health Care through Big Data Strategies for leveraging big data in the health care industry. 2013.

[7] Bellazzi R. Big Data and biomedical informatics: a challenging opportunity. *Yearb Med Inform.* 2014;9:8–13. doi: 10.15265/IY-2014-0024. [PMC free article] [PubMed][Cross Ref].

[8] Frost & Sullivan: Drowning in Big Data? Reducing Information Technology Complexities and Costs for Healthcare Organizations.http://www.emc.com /collateral/analyst-reports/frost-sullivan-reducing-information-technology-complexities-ar.pdf.

[9] Newman H. B., Ellisman M. H., Orcutt J. A. Data-intensive e-science frontier research. *Communications of the ACM* 2003:46(11):68–77.

[10] Müller H., Hanbury A., Al Shorbaji N. Health information search to deal with the exploding amount of health information produced. *Methods Inf Med* 2012. Dec 4;51(6):516–8. [PubMed].

[11] Japec, L., Kreuter, F., Berg, M., Biemer, P., Decker, P., Lampe, C., Lane, J., O'Neil, C., Usher, A. *AAPOR Report on Big Data; Technical Report; American Association for Public Opinion Research (AAPOR)*: Oakbrook Terrace, IL, USA, 2015; 50p. Available online: http://www.aapor.org/Education-Resources/Reports/ Big-Data.aspx (accessed on 18 June 2018).

[12] Abbany, Z. *A Public Transport Model Built on Open Data, News Article.* Available online: http://www.dw.com/en/a-public-transport-model-built-on-open-data/a-41546053 (accessed on 27 November 2017).

[13] Darabi, A. *The UK's Next Census Will Be Its Last—Here's Why, News Report.* Available online: https://apolitical.co/solution_article/uks-next-census-will-last-heres (accessed on 5 December 2017).

[14] Murtagh, F., Orlov, M., Mirkin, B. Qualitative judgement of research impact: Domain taxonomy as a fundamental framework for judgement of the quality of research. *J. Classif.* 2018, 35, 5–28.

[15] Hand, D. Statistical challenges of administrative and transaction data. *J. R. Stat. Soc. Ser. A* 2018, 181, 1–24.

Chapter 6

Intuitionistic Fuzzy Stream Cipher for Protecting Second Layer Lightning Network Transactions in Blockchain

Rajdeep Chakraborty[1,*], Sohini Roy[1] and Arun Sarkar[2]

[1]Department of Computer Science and Engineering, Netaji Subhash Engineering College, Kolkata, India

[2]Department of Mathematics, Heramba Chandra College, Kolkata, India

Abstract

This chapter presents a method to protect the lightning network in Blockchain in stream cipher system where intuitionistic fuzzy logic is used to generate pseudorandom digits. Intuitionistic fuzzy sets allow the assessment of elements by membership and non-membership function which belongs to the real unit interval [0,1] and whose sum is also in the same interval. Stream Cipher encrypts and decrypts data, producing a cipher text by combining the binary digits with pseudorandom digits, one at a time. The lightning network solves the scaling problem of blockchain cryptocurrency by introducing an additional second layer to the already existing blockchain where there exist multiple transaction channels between users. Security is needed to mitigate the risk of cyber-attacks and deceit.

The novelty of this chapter is the fuzzification of stream cipher, here we used two LFSR for generating pseudorandom bits, then these bits are fed to intuitionistic fuzzy system and fuzzification is done. Then using fuzzy rules and defuzzification we obtain a single bit, this bit is exored

* Corresponding Author's Email: rajdeep_chak@yahoo.co.in.

In: Computational Intelligence for Sustainable Development
Editor: Brojo Kishore Mishra
ISBN: 979-8-88697-198-9

with the plaintext bit to obtain the ciphertext bit, which is a stream cipher. This process is continued in every clock cycle to encrypt the complete plaintext, the decryption is just the repetition of the same process where input is the cyphertext. Generally, in LFSR-based stream cipher the non-linearity is obtained either by using more than one LFSR or by using non-linear functions like 2^x, e^x etc. but here we used intuitionistic fuzzy logic to introduce non-linearity, which is the novelty of this work.

The application domain is the blockchain platform, which is now a part of much sustainable devolvement, here we are given the security in the lightning network. It is now seen that for huge nodes in bitcoin, one transaction takes almost five minutes, so, lightning network is second layer network introduced before cryptocurrencies to make the transactions faster. This lightning network enables transactions in various crypto-app in a very short time and a commit protocol is executed over actual cryptocurrency at regular intervals of time.

The security in the blockchain is well established by modern ciphers but providing security in lightning networks is still under research. So, we proposed a model with this stream cipher to provide security, as the longest run stream cipher is the one-time pad and is considered a more secure cipher.

Finally, we evaluated our stream cipher using mono bit frequency test, entropy test, floating frequency test, histogram, N-Gram, Autocorrelation and periodicity to measure the randomness against the widely used stream cipher, RC4.

Keywords: intuitionistic fuzzy system, intuitionistic fuzzy sets, stream cipher, second layer lighting network, blockchain, cryptocurrency

1. Introduction

Blockchain is one of the most important sustainable development technologies that stores data in distributed manner govern by consensus mechanisms like PoW, PoS, PoET etc. Bitcoin is one such blockchain governed by Proof of Work (PoW) and is an example of cryptocurrency. Now, these cryptocurrencies' transaction time has increased a lot and to cope with this situation a second layer of the transaction, known as the lightning layer is now being added to the blockchain network. This layer makes the transactions in a faster manner and the commit protocol is executed in regular intervals to make the transactions in an actual blockchain.

Stream Cipher is one of the most secure manners to protect data if it becomes a one-time pad which decides by the randomness of the output bits in each clock cycle covering the whole plaintext. Now, fuzzy system or fuzzy logic is a truth table based on many-valued systems of variables having real numbers between 0 and 1. This is a system of partial truth and between complete truth to complete false. Intuitionistic fuzzy system is where the input is classical bits say set of {0,1}. The fuzzification is the process to convert the classical bits to an intuitionistic fuzzy system (Xu, Z. and Yager, R.R) [1] this system takes the classical bit input and the intuitionistic fuzzy rules and gives the combination to the inference engine. The novelty of this chapter is the fuzzification of stream cipher, here we used two LFSR for generating pseudorandom bits, then these bits are fed to intuitionistic fuzzy system and fuzzification is done. Then using fuzzy rules and defuzzification we obtain a single bit, this bit is exored with the plaintext bit to obtain the ciphertext bit, which is a stream cipher. This process is continued in every clock cycle to encrypt the complete plaintext, the decryption is just the repetition of the same process where input is the cyphertext. Generally, in an LFSR-based stream cipher the non-linearity is obtained either by using more than one LFSR or by using non-linear functions like 2^x, e^x etc. but here we used intuitionistic fuzzy logic to introduce non-linearity, which is the novelty of this work. The application domain is the blockchain platform, which is now a part of much sustainable devolvement, here we are given the security in the lightning network. It is now seen that for huge nodes in bitcoin, one transaction takes almost five minutes, so, the lightning network is second layer network introduced before cryptocurrencies to make the transactions faster. This lightning network enables transactions in various crypto-app in a very short time and a commit protocol is executed over actual cryptocurrency at regular intervals of time. The security in the blockchain is well established by modern ciphers but providing security in lightning network is still under research. So, we proposed a model with this stream cipher to provide security, as the longest run stream cipher is the one-time pad and is considered a more secure cipher.

Finally, we evaluated our stream cipher using frequency test, entropy test, floating frequency test, histogram, N-Gram, Autocorrelation and periodicity to measure the randomness against widely used stream cipher, RC4.

Frequency test is test of flow where all the 256 ASCII values are calculated and represented graphically, if the frequency is almost equal in ciphertext then it is considered relatively stronger encryption.

Entropy test is the randomness test and it is calculated as negative logarithm of each ASCII value overall ASCII values, and these values are summed up, the maximum entropy is the negative probability of one, and the calculated sum up is closer to maximum entropy, then it is considered better cipher.

Floating frequency is the same as the frequency test, but it is calculated for each 64-bytes, whereas the frequency test is calculated over the ciphertext. Floating frequency reveals some weak points in ciphers.

Histogram is another type of frequency test but it is mainly done for text characters, histogram is generally used as an English dictionary attack in ciphers.

N-gram is the combination test, here various combinations and their frequency is calculated, it is also used as a word-based dictionary attack in ciphertext, there are various N-Grams exist like digram, trigram or four-gram.

Autocorrelation is the correlation test in ciphertext itself, which means if some portion of ciphertext is correlated to another portion in the ciphertext. This attack is mainly performed in ECB-based encryption methods that reveal correlated vulnerabilities in the ciphertext.

Periodicity is simply the repetition of some sequence of bits again and again and it is one of the strongest vulnerabilities in ciphers.

We devised an IF-based stream cipher, which is one of the novels works in the literature and is used to give security in the second layer of cryptocurrency transactions. Section 2 Discusses the Concepts, Section 3 gives a short literature survey, Section 4 gives the modelling and discussion, Section 5 gives a comparative result and Section 6 draws the conclusion and finally, references are listed.

2. The Concepts

In this section, we discuss every concept needed to understand the blockchain, cryptocurrency, transactions and the proposed design.

Section 2. A gives the concept of Intuitionistic Fuzzy system then we discuss blockchain in Section 2. B and followed by second layer lighting network of the blockchain transaction in Section 2. C.

Public key algorithm, the cryptographic hash function and the stream cipher concepts are discussed in Section 2.D, Section 2. E and Section 2. F respectively.

2.1. Intuitionistic Fuzzy System

Fuzzy logic, developed by Lotfi Zadeh, is a mathematical logic that allows intermediate logic values to be defined between conventional equations. It resembles human reasoning. Intuitionistic fuzzy system uses intuitionistic fuzzy logic which extends the concept of classical set theory and fuzzy logic by assessing the two functions of non-membership and membership function through elements. The function belongs to the interval [0,1] and whose sum also lies in the same interval.

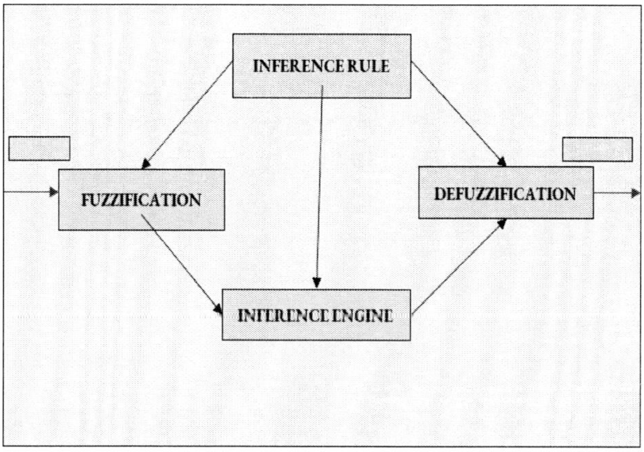

Figure 1. Intuitionistic fuzzy system

Figure 1 shows intuitionistic fuzzy system where the input is taken as a set of [0,1]. Fuzzification is the process of converting the input bits to intuitionistic fuzzy system by taking the input bits and fuzzy rule and giving the combination to the inference engine that converts these classical bits to fuzzy systems with the help of the inference rule. Since the fuzzy system should be understood in the real-world, defuzzification is carried out.

Stream Cipher is used for encryption and decryption of data and obtaining a cipher text by combining each bit or byte of a binary digit with a pseudorandom digit, which is generated by a key generator. Each input bit is combined with a pseudorandom sequence, one at a time. Stream ciphers are very important as it takes very less memory, and it doesn't depend on the size of data, i.e., it can work for both small and large data, security is mentionable and also it has low computing power.

2.2. The Blockchain

In this advancing world, revolutionary ideas are being introduced very often now. One such revolutionary idea that changed the overall lifestyle in the real world is Blockchain. Blockchain [2], the first concept, in 1991, was introduced by a group of researchers which went unused till 2009 when Satoshi Nakamoto [3] again introduced this concept in the first cryptocurrency, bitcoin. In simple words, a Blockchain is a distributed ledger that stores information in the form of blocks. All the blocks are interconnected and they are decentralized. Once some data is stored in the block, it is visible to all the users in the network. Each block has data, hash of its own and a previous block hash. Cryptocurrency, [4] on the other hand, is a digital currency that doesn't require permission from banks or any centralized authority for any kind of transaction. Figure 2 gives the conceptual diagram of a blockchain.

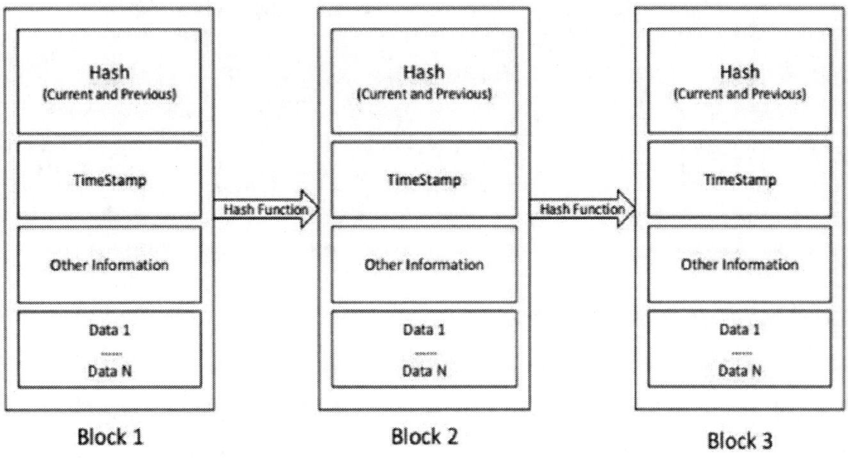

Figure 2. The blockchain.

The blockchain also has many sustainable developments as follows:

- Supply Chain Management,
- Media,
- Capital Markets,
- Trade Finance,
- Money Laundering Protection,
- Cybersecurity,

- Insurance,
- Energy,
- Record Management,
- Peer-to-Peer Transactions,
- International Payments,
- Financial Management and Accounting,
- Healthcare,
- Big Data and IoT.
- Identity Management,
- Electronic Voting,
- Electronic Voting,
- Real Estate,
- Regulatory Compliance and Audit,

As in Figure 2, the immutability is obtained through crypto-hash, SHA-256 is mainly used in many blockchain platforms, the confidentiality is obtained in blockchain by the use of block ciphers like AES, the digital signature is obtained in blockchain by the use of RSA-based Digital Signature Algorithm (DSA). These are the common algorithms but some blockchain platform slightly differ and uses other established algorithms. As so many cryptographic parameters are used in the blockchain platform so, the digital currency developed using blockchain is also known as Crypto-Currency.

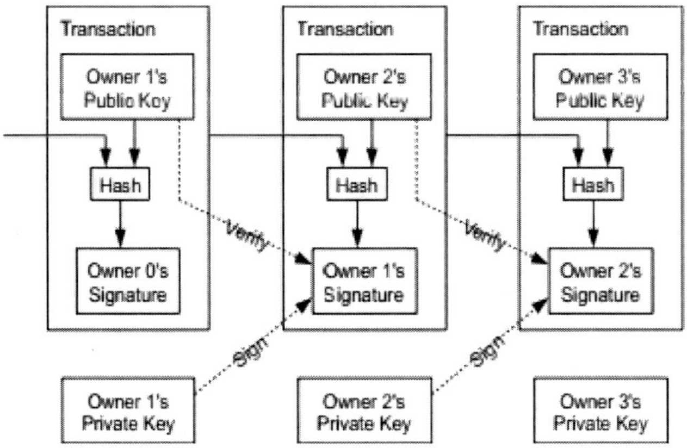

Figure 3. Bitcoin transaction.

Figure 3 shows the abstract diagram of a bitcoin transaction, here each block is in bitcoin blockchain or transaction in bitcoin. This block is having a transaction value that is encrypted by owner's public key and hash ensures the immutability and this transaction is signed by previous owner's private key.

Bitcoin (₿) is a form of digital currency, that is decentralized in nature, without the intervention of a single administrator or a central institution, that is sent from one user to another in a peer-to-peer network without any intermediate medium. All the performed transactions are verified by network nodes through the process of cryptography and stored in blockchain, a public distributed ledger. Although the concept of cryptocurrency had its existence back in 2008 by an unknown person or unknown group of people named Satoshi Nakamoto, its use began in 2009 when it was ultimately released as open-source software.

Bitcoins are created as a reward for the mining process. Other services, currencies, and products can be exchanged for Bitcoin. But the concept of Bitcoin has been criticized for its various improper and illegal use like illegal transactions, carbon footprinting, price volatility, thefts of exchanges, etc. Many a time, economists and investors have criticized Bitcoin by naming it a Speculative Bubble. Despite issuing several investor alerts about Bitcoin by agencies, it has been used as an investment.

Bitcoin is being officially used legally in two countries, El Salvador and the Central African Republic, while it is being used officially in a few areas by the local and central governments. The word, that was defined first through a white paper on 31st October 2008, is actually a concatenation of two words, bit and coin. There exist a nonuniform method for Bitcoin capitalization. Use of Bitcoin as to capitalized that is a technology, network of nodes and the lowercase is used for account unit. The use of lowercase Bitcoin can be seen in the Oxford English Dictionary, The Chronicle of Higher Education, and The Wall Street Journal.

2.3. The Lighting Network

Lightning network was first proposed by J Poon and T Dryja in 2015 [5] and is still in the development phase. It solves the scaling problem, i.e., the slow transaction time and throughput of Blockchain Cryptocurrency by introducing a second layer to the already existing main Blockchain where there exist multiple transactions channels between every user, thus speeding up the transaction time. It uses smart contracts and multiple signatures to achieve this purpose. Transactions happening during the opening and closing of channels are only informed. Thus, all the transactions are not informed in the main blockchain which speeds up the process and also improves the overall throughput.

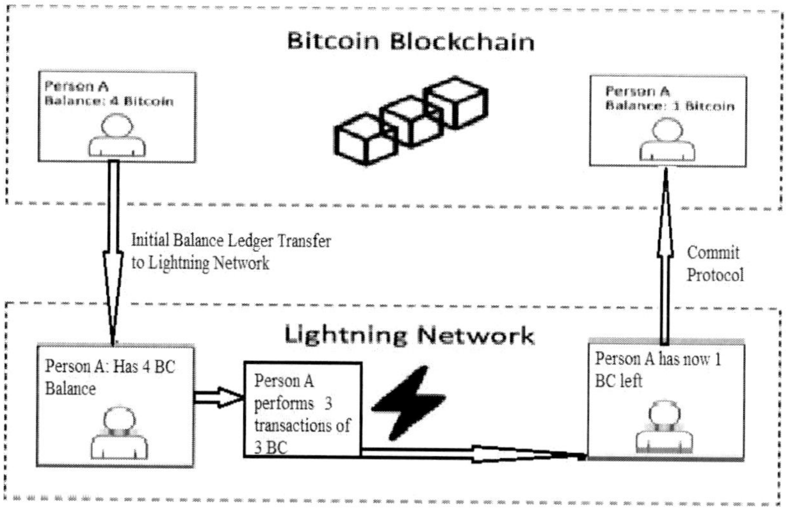

Figure 4. The lightning network.

As in Figure 4, the upper layer is the actual bitcoin network where Person A has 4 bitcoins in the ledger which are then transferred to lightning network ledger. Now, person A performs three transactions in bitcoin (BC) in lightning network expending three more bitcoins. So, now the balance in the lightning network for person A is 1 BC. Finally, a commit protocol is executed to update his transactions in the actual bitcoin network. So, if we consider a five minutes transaction time for bitcoin network, so time taken for three transactions will be about 15 minutes without lightning network. But with this modelling one reduces the time to ten minutes, every five minutes for the initial balance ledger transfer and commit protocol. The transaction time drastically reduces when the number of transactions increases.

The Lightning Network (LN) is a two-layer payment protocol that acts as a top layer to Bitcoin or Litecoin, which is a Blockchain-based cryptocurrency. This layer is designed to enable faster transactions among participating nodes and is a fine solution to Bitcoin's scalability problem. It includes a system of peer-to-peer networks for making cryptocurrency micropayments through payment channel networks which is bidirectional without entrusting the custody of funds. The use of Lightning Network happens in two layers. In layer 1, a payment channel is opened by committing a transaction to the base Blockchain which is relevant. In layer 2, any number of lightning network transactions are being performed, and simultaneously updating the preliminary distribution of the funds with no broadcasting to the Blockchain. In some cases, the payment channel is shut down by broadcasting the final settlement transaction to distribute the channel's funds. The channel is closed for payment settling. The process is started by broadcasting the most up-to-date transaction to the network by a nose. The next process is divided into two parts, one being a cooperative closure where both parties agree to the distribution and funds are settled immediately. The other one is an uncooperative closure which happens in cases where one node is inactive in the network or a node has broadcasted incorrect distribution. In these types of closures, the funds are not settled very easily but a dispute time period is being provided where nodes may contest the broadcast distribution. If the later node provides more up-to-date distribution, then funds are directly transferred to them. This process, also known as the Breach Remedy transaction, ensures that there is no out-to-date transaction broadcast on the network.

2.4. The Public Key Algorithm

The cryptographic system that uses pairs of keys is public-key cryptography. Public key is one pair which is known to all and the other pair is the secret and private key [18]. One way function generates the key pairs and it's also a type of mathematical problem. Private key must be kept private, is the effectiveness, where public key is distributed to all keeping the security [19].

The public key algorithm is widely used for providing security/ confidentiality and digital signature in cryptocurrency transactions, the encryption process and decryption process, are defined in the following paragraphs, Figure 5 gives a block diagram for public key cryptography.

Encryption Algorithm

- The encryption algorithm procedure is discussed in the following steps:
- The receiver's public key (a, b, c) is attained
- Select a random integer n

- Calculate the ciphertext pair C_1 and C_2 using the formulas: $C_1 = b^n \bmod a$
- $C_2 = (mm * c)\, n \bmod a$, where mm is the message to be encrypted.
- The ciphertext pair (C_1, C_2) is sent to the receiver end.

The receiver has the public keys of the sender, let p, g and y. The receiver then selects a random integer, k, and it calculates two ciphertext pairs, C_1 that is gk mod p

Decryption algorithm

- The decryption algorithm procedure is introduced in the following steps:
- Calculate the following parameter: $C_1(a - 1 - b)$ by using private key b
- Recover the message mm by multiplying the previous result by C_2 using the formula:
- $mm = (C_1(a - 1 - b)) * C_2 \bmod a$

On the receiving side, C_1 is calculated first using the formulae $C_1(a - 1 - b)$ using private key b, let the value is nn, then the message, mm, is calculated as, $nn * C_2 \bmod a$, where C_2 is the other part of ciphertext received.

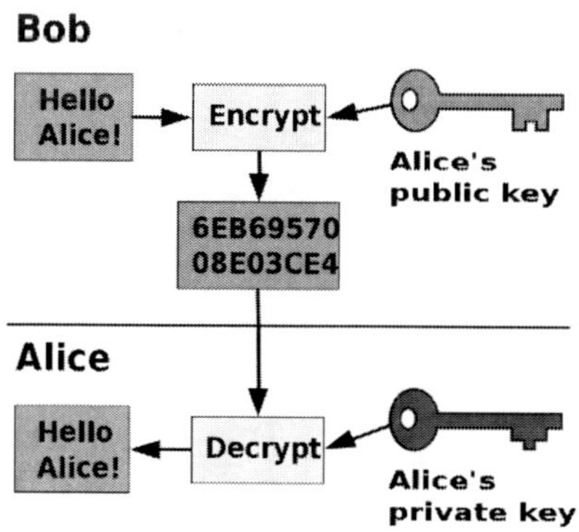

Figure 5. Public key cryptography block diagram.

2.5. The Cryptographic Hash Function

In information security, Hash function plays an important role. A hash function compresses an input numerical value through a numerical value. A fixed value is generated from an arbitrary input value by hash function.

The immutability in a blockchain is one of the most important characteristics provided and it is achieved by cryptographic hash function. In blockchain hash value of the previous node is calculated and stored in the next node, also the total hash value of this node. This is repeated when each block is mined and actually it follows the Markle hash tree. Thus any block is altered, deleted or added then the last hash value will not match and this is the protection of immutability in blockchain.

A mathematical function is a Cryptographic Hash Function (CHF) where a fixed-length value, hash, is generated from an arbitrary length value, a message. This is a one-way function where it is infeasible to reverse or invert the computation. The ideal way to find out whether a message is eligible to produce a given hash is to apply Brute-Force search of all the inputs to check for hash matching or use a rainbow table of matched hashes. Modern-day cryptography uses Hash Functions as a basic function.

For a cryptographic hash function, the same message must always yield the same result. Some of the characteristic properties are:

- For any given message, the hash value should be quick to compute.
- It is not feasible to generate a message that yields a given hash value, that is process reversal that generated the hash value.
- With the same hash value, finding two different messages is not feasible.
- The Avalanche effect should be applicable in cases where a small change in the message should drastically change the hash value so that a new hash value is uncorrelated with the old hash value.

Cryptographic functions have many of it's applications in information security, like in digital signatures, Message Authentication Code (MAC), and other different forms of authentication. Its other applications include ordinary hash functions, indexing data in hash tables, finger printing, detecting duplicate data or unique identifying files, and checksums for detecting accidental data corruption. With reference to information security, they are also sometimes called digital fingerprints, checksums or hash values itself, although these terms have different meanings and properties and are used for more general functions.

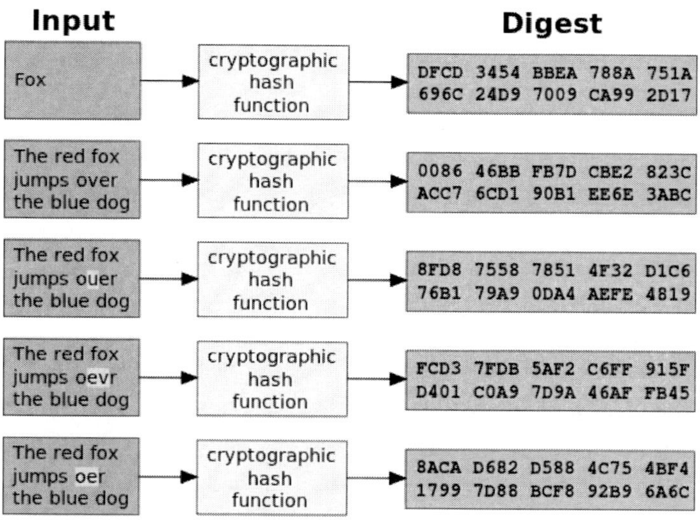

Figure 6. Illustration of Cryptographic Hash Function.

Figure 6, shows an illustration of cryptographic hash function, it states that different phrases give a unique hash value.

In this figure, if we observe the first-word fox which is applied with a cryptographic hash function and produces 160-bit hash value. Now in the next box, the sentence is "The red fox jumps over the blue dog" which is applied to the same cryptographic hash function and produces 160-bit hash value. Now in the next box if we observe the word "over" is replaced by the word "ouer" in the same sentence and applied to the same cryptographic hash function producing a drastic different 160-bit hash value. Again, in the next box if we observe the word "over" is replaced by the word "oevr" in the same sentence and applied to same cryptographic hash function producing again a drastic different 160-bit hash value. Finally, in the last box if we observe the word "over" is replaced by the word "oer" in the same sentence and applied to same cryptographic hash function producing finally a drastic different 160-bit hash value.

2.6. The Stream Cipher

A method is a stream cipher of encrypting text that produces ciphertext, a bit of the key is applied to a cryptographic function and the output bit is XORed with a bit of plaintext. The alternative to a stream cipher is block cipher where a block of plaintext and a block of the key are applied to a cryptographic function that produces a block of ciphertext.

The primary design of this chapter is the stream cipher, Figure 7 depicts the basic diagram of a stream cipher which encrypts bit by bit or byte by byte. The main component is the keystream generator which gives pseudo-random bits and if it is so then it is called a one-time pad.

The main advantages of stream ciphers are:

- It takes low computing power
- It takes low memory
- Stream ciphers work well for large or small chucks of data.
- Stream cipher is equally secure as a block cipher and if it turns out to be a one-time pad then it is most secured

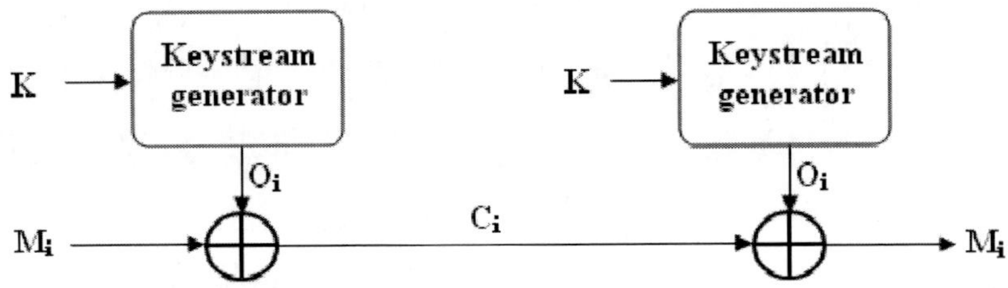

Figure 7. Stream cipher block diagram.

In Figure 7, the keystream generator is the main algorithm of any stream ciphers, the secret key is applied in both sender and receiver side algorithms. This key stream generator then produces a bit or a byte as an output in each clock cycle and then that is exored with the plaintext or message, the resultant is the ciphertext. Since the algorithm and the key are same so this ciphertext on the receiver side if xored with the same output then it will produce the same plaintext again. Generally, in stream cipher Linear Feedback Shift Register (LFSR) and Non-Linear Feedback Shift Register (NFSR) are used.

3. Literature Review

We give three literature reviews in this section. In the first paper, Sattar B. Sadhkan Al Manicky and Sabiha F. Jawad [6] describe a method for security evaluation which is based on fuzzy logic, that takes two parameters into consideration, one is the maximum period of key sequence that is generated from Linear Feedback Shift Register and the other is the entropy of the result from different lengths of shift register, for pseudorandom digit generation mainly in stream cipher system.

In the second paper, Mahdi H. Miraz and David C. Donald discusses the lightning network [7] and how it is useful to clear and settle security and coloured coins. It also discusses the future prospects and new business models utilizing this technology.

In the third paper titled "Solving Intuitionistic Nonlinear Equations", Mohammad Keyanpour and Tahereh Akbarian [8] use Midpoint Newton's Method to propose a numerical method that can be used to solve nonlinear equations which follow intuitionistic fuzzy logic. Here, the Intuitionistic fuzzy [9] quantities are represented in parametric form.

In the fourth paper, Mahdi H Miraz, David C. Donald [10] describes Lapps, which is a Technological, Legal, and Market Potential of Blockchain Lightning Network application. Pioneer Bitcoin footstep is followed, many altcoins, as well as coloured coins, have been being developed and merchandised adopting blockchain as the core enabling technology. However, high and capped (in particular cases) transaction latency is deep-rooted in the architecture of blockchain technology, since interoperability and scalability, in any blockchain-based applications, are by default inherited. Lightning Network (LN) is one of the important supporting technologies that is developed to eliminate this impediment and lankness of blockchain technology by facilitating with instantaneous transferring of cryptos. Since and till now the potentials and advantages of LN is still not known in many spheres, this paper thus investigates all the current states of development with and along all possible non-monetary usage of LN, specifically in settlement-colored of coins such as dividends and securities, also creating of new business models that are based on Lightning Applications (LApps) and microchannel payments and also micro-trades. The legal challenges which should act as an impediment to all the adoption of LN are discussed in detail.

In the fifth paper, FASTA a fully homomorphic encryption stream cipher is proposed by the authors, Various FHE symmetric ciphers have been proposed in recent days, e.g., some of the stream ciphers are Kreyvium, FLIP and Raasta and LowMC is a block cipher. Minimization of the multiplicative complexity of the algorithm is the typical design criterion. But the efficient evaluation in common FHE libraries compromises the real-world performance. The linear layer with the parameter which choose specially and efficient implementation is allowed over and on BGV and helib.

4. Modelling and Discussion

Section 4.1 outlines the stream cipher design and then we give the modelling in Section 4.2 followed by discussion on Section 4.3.

4.1. The IF Stream Cipher

This section first defines what is LFSR mathematical and then the well-established theorem about the length of the stream generated [11] for its tap selection.

Definitions 2. An LFSR of length L over Fq is a finite state automaton which/that produces a semi-infinite sequence of elements of Fq, s = (st)t ≥ 0, satisfying a linear recurrence relation of degree L over Fq

$$s_{t+L} = \sum_{i=1}^{L} c_i s_{t+L-i}, \quad \forall t \geq 0 \ . \tag{1}$$

The L coefficients c1, ..., cL are elements of Fq. They are called the feedback coefficients of the LFSR.

Theorem 1. A sequence (st)t ≥ 0 is generated by an LFSR of length L over Fq with feedback polynomial P if and only if there exists a polynomial Q ∈ Fq[X] with deg(Q) < L such that the generating function of (st)t ≥ 0 satisfies

$$\sum_{t\geq 0} s_t X^t = \frac{Q(X)}{P(X)} \tag{2}$$

So, the polynomial Q is must determine by all the coefficients of P and by the initial state of the LFSR:

$$Q(X) = -\sum_{j=0}^{L-1} X^j \left(\sum_{k=0}^{k} s_k c_{j-k} \right) \tag{3}$$

Proof:

$$\left(-\sum_{i=0}^{L} c_i X^i\right)\left(\sum_{t=0}^{\infty} s_t X^t\right) = \sum_{j=0}^{\infty} X^j \left(-\sum_{k=\max(0,j-L)}^{j} s_k c_{j-k}\right)$$
$$= -\sum_{j=0}^{L-1} X^j \left(\sum_{k=0}^{j} s_k c_{j-k}\right) + \sum_{j=L}^{\infty} X^j \left(\sum_{k=j-L}^{j} s_k c_{j-k}\right).$$

So, after reducing, we get

$$Q(X) = -\left(\sum_{i=0}^{L} c_i X^i\right)\left(\sum_{t=0}^{\infty} s_t X^t\right) \tag{4}$$

So, it is a polynomial of degree < L iff. the second right-hand term becomes zero or vanishes.

$$\sum_{k=j-L}^{j} s_k c_{j-k} = 0 \tag{5}$$

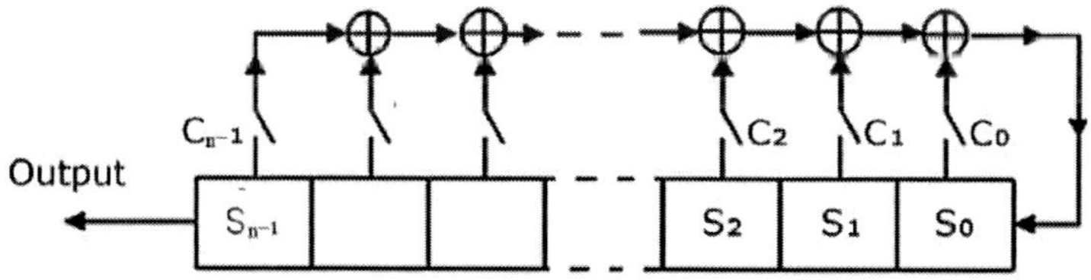

Figure 8. The LFSR.

Here we use the LFSR of eq. (6) and eq. (7)

X61 + X31 + X11 + X7 + 1 (6)

X53 + X23 + X17 + X5 + 1 (7)

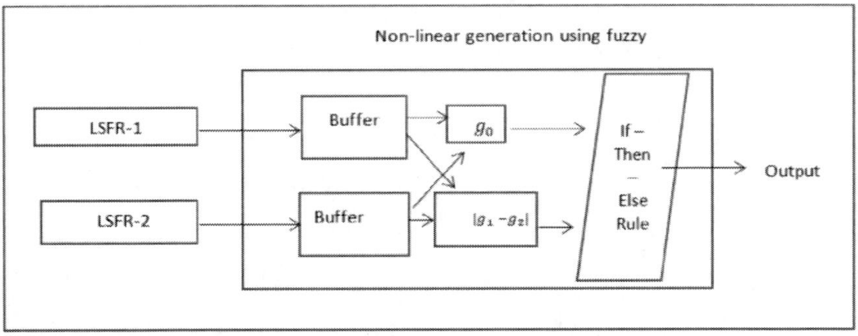

Figure 9. The proposed IF-based stream cipher model.

Figure 8 gives the LFSR used in this chapter and Figure 9 shows the proposed IF-based stream cipher model.

4.2. Security Model for Second Layer Lighting Network

Figure 10 gives the proposed security model for Second Layer Lighting Network.

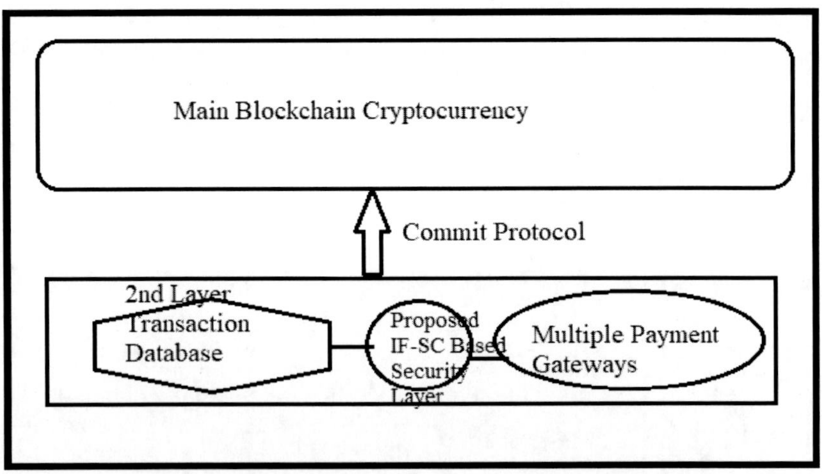

Figure 10. Security model for Second Layer Lighting Network.

In this security model, we placed our proposed intuitionistic fuzzy-based stream cipher (IF-SC) [12] in between the multiple payment gateways and 2nd layer transaction database. This security model acts as end-to-end encryption between each payment gateway and 2nd layer transaction database. The commit protocol is executed after a regular interval of time, say 15 minutes, to reflect the transactions in the main blockchain cryptocurrency.

4.3. Discussion

Zeroes and one's proportion are focused in this test for the entire sequence. To determine whether the number of ones and zeros in a sequence, is the purpose of this test, which should be approximately be the same as these would a truly random sequence and expected. The test also assesses the closeness of the fraction of ones to ½, which means the number count of ones and number count of zeroes in a sequence could be about the same. The subsequent and all tests which should depend on the passing of all these tests.

Table 1. Monobit frequency test of output of the proposed cipher

Frequency test (with significance level alpha value=0.05), Max. allowable value 3.84, F=failed (>3.84), P=passed (<=3.84)										
Samples	***Sample 1***	***Sample 2***	***Sample 3***	***Sample 4***	***Sample 5***	***Sample 6***	***Sample 7***	***Sample 8***	***Sample 9***	***Sample 10***
Pass or Fail	Pass	Fail	Pass	Fail	Pass	Pass	Pass	Fail	Pass	Pass

We performed the frequency monobit test which is the most basic method of testing randomness. This test checks whether the pseudorandom number generated values are uniformly distributed or not. This test focuses on the ratio of zeros and ones present in the entire sequence S against the expectation should the sequences were truly random. It is expected should numbers of zeros and ones to be virtually same. Let, n0 and n1 be the number of zeros or ones in S. The test statistics can be defined as follows:

$$X_1 = \frac{(n_0 - n_1)^2}{2} \qquad (8)$$

here represents a Chi-Square test with one degree of freedom. Equation (9) shows that the higher value generates a higher discrepancy between the observed and expected frequencies of zeros and ones. Table 1 depicts the frequency test results of ten samples. These ten samples are the bit sequence generated in each interval of 128 – bits with a clock cycles of total clock cycles.

Moreover, the stream cipher is considered a fastest algorithm for encryption/decryption, so, this model will perfectly work in Second layer lighting network.

5. Comparative Results

We evaluated our Intuitionistic Fuzzy Stream Cipher (IFSC) using frequency test in entropy test in Section 5.1, floating frequency test in Section 5.2, histogram test in Section 5.3, N-Gram test in Section 5.4, Autocorrelation test in Section 5.5 and periodicity test in Section 5.F to measure the randomness against widely used stream cipher, RC4.

All the results are generated using Cryptool 1.4.41, Cryptool being the cryptography algorithm analyzer used word-wide.

One of the open-source projects, CrypTool, is a worldwide free e-learning software should illustrate cryptographic and cryptanalytic concepts in the cryptology domain.

Almost more than 400 algorithms are implemented by this software which can be adjusted by a user with their own set of parameters. The organization has created different types of graphical interface software which include online documentation, analytic tools and algorithms for the ease of users who are new in this domain.

They contain both classical cipher as well as modern-day asymmetric and symmetric cryptology like digital signatures, RSA, hybrid encryption, ECC, Diffie–Hellman key exchange and homomorphic encryption. From the domain of BB84 Key Exchange Protocol, Quantum Cryptography, WOTS, Post-Quantum Cryptography, SPHINCS, XMSS and XMSS_MT etc. methods are being implemented. For classical ciphers, solvers (analyzers) are also used. Other methods like AES, Huffman code, Keccak, MSS are visualized. Some didactical games like Divider Game, Number Shark, Zudo-Ku and tutorials about primes, lattice-based cryptography and elementary number theory are also included.

5.1. Entropy

In cryptography randomness or diversity of data-generating function plays a major role. In 1948, Claude Shannon invented a mathematical formula for measuring unpredictability of message content. For a system having n states and probability Pi, denotes the occurrences of happening ith steps, the traditional discrete Shannon entropy is defined in. It is

$$H(P) = -\sum Pi \, Log_2 \, Pi \quad (9)$$

where 1<=i<=n, units of H(P) is in bits because of the Log_2 term in the equation. Equation (9) shows that the entropy is directly proportional to the probability of occurrence means high entropy gives high randomness.

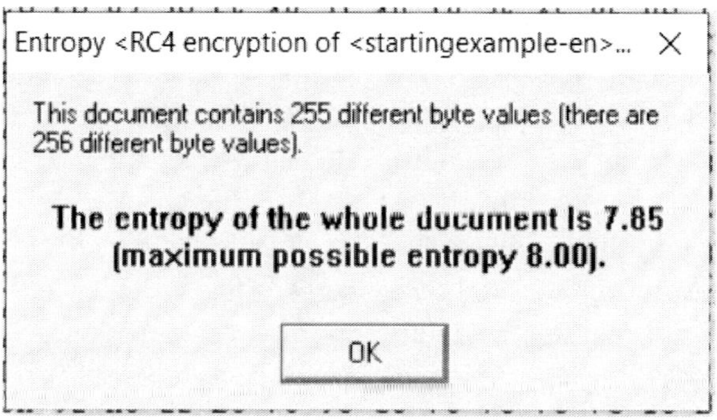

Figure 11. Entropy of RC4.

Figure 12. Entropy of proposed IFSC.

As in Figure 11, the entropy of the ciphertext of RC4 is found to be 7.85 out of 8.00 at maximum equating 98% and As in Figure 12, the entropy of the ciphertext of proposed IFSC is found to be 3.56 out of 4.70 at maximum equating 75%. So, entropy is well comparable of IFSC and RC4. The less the percentage the more is the entropy. The entropy also depend on the choosing of the Pi as in equation (9), so, entropy is the negative summation (1 – Pi) of probabilities of occurrences of the 256 ASCII values in the ciphertext.

5.2. Floating Frequency

The floating frequency of a document represents a number of different character's present in a particular segment of a document. Therefore, it is desired that this number should be as high as possible so that there is no matching character within the content of plaintext and ciphertext.

To measure the floating frequency, first 64 character's long window size is selected and then counts number of different character's present in this "window." Then the window is shifted to one character to the right. This entire procedure will continue till the last byte. The analysis identifies heavy and light-density information places. For documents length n > 64 bytes, n-63 such index numbers occur in its characteristics. This sliding window analysis is truly important because the capturing portion of the cipher gives an attacker hint about the next patterns. Here one input file (pdf type) has been chosen for such analysis.

As in Figure 13, we can observe that floating frequency of RC4 is quite variable where as the floating frequency of the proposed IFSC is almost constant in certain range of values, as in Figure 14. So, the proposed IFSC is well comparable with RC4 in terms of floating frequency.

5.3. Histogram

Histogram of a document refers how frequently the characters are distributed throughout the said document. In order to interdict the attacker from obtaining any statistical similarity between the original and encrypted message, uniform distribution of characters in the ciphertext is a desirable criterion. It ensures that the algorithm is free from frequency analysis attacks.

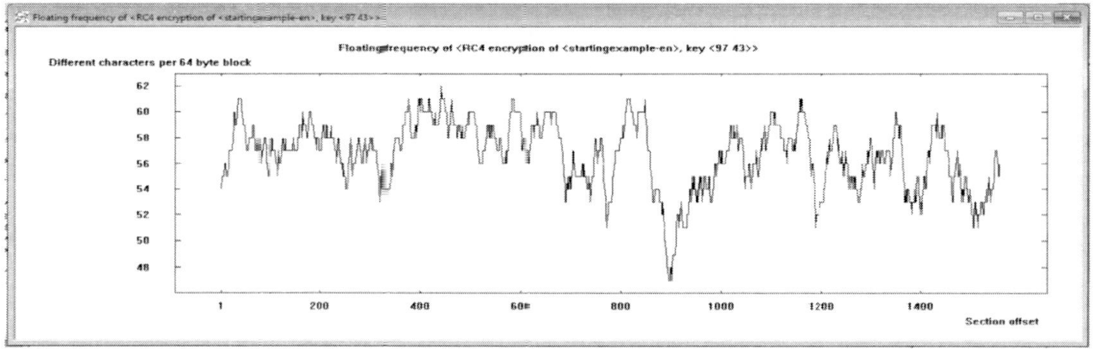

Figure 13. Floating frequency of RC4.

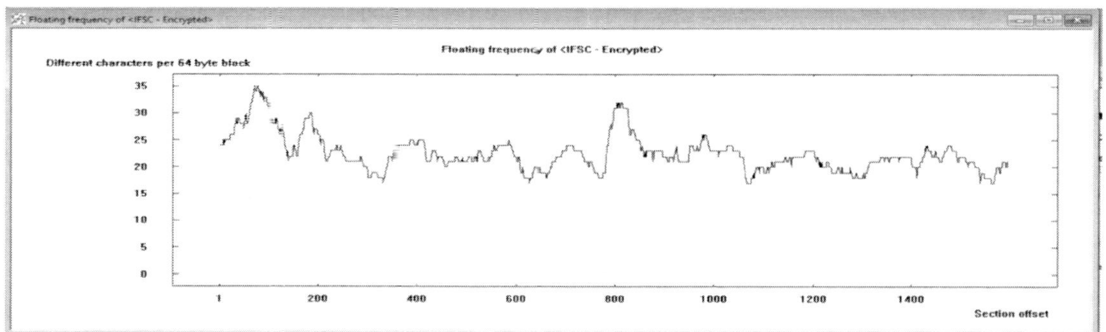

Figure 14. Floating frequency of IFSC.

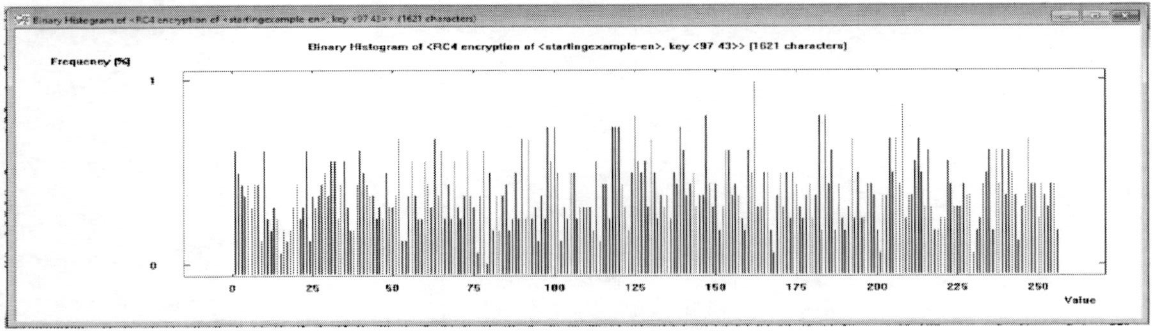

Figure 15. Histogram of RC4.

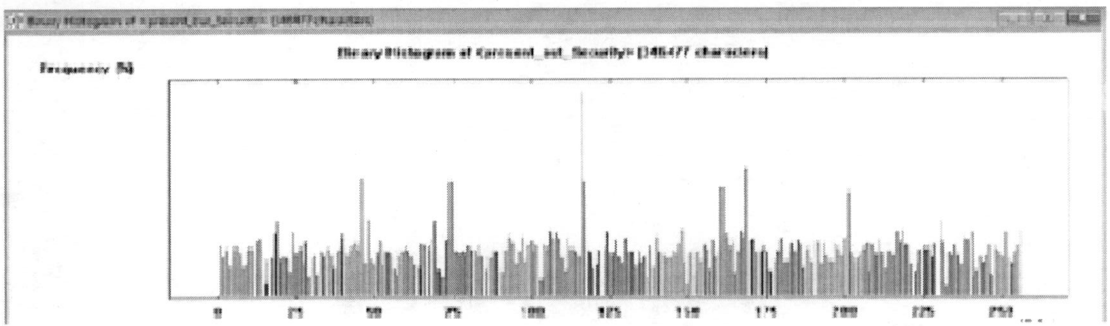

Figure 16. Histogram of IFSC.

As in Figure 15, the histogram of RC4 is not equally distributed over 256 ASCII values where as the histogram of proposed IFSC is equally distributed over 256 ASCII values, as in Figure 16. So, in terms of histogram we can say the proposed IFSC is better than that of RC4.

5.4. N-Gram

An N-grams refers to an n bit substring of a gives sample of text or speech string. This string includes various types like letters, words, syllables and many others based on the application. N-grams are generated by dividing a text string into fixed-length substrings. The length of the substring may be 1,2,3,4 (named as histogram, bigram, trigram, 4-gram) etc. N-grams of texts are extensively used in text mining field and natural language processing. It actually represents a group of co-occurring words within a given window. During computation of N-grams we generally move one word or more than one word (in an advanced scenario) forward. N-grams opcode analysis is used to examine the detection of malware. This analysis helps attackers to crack the substitution types encrypted text when the encryption method is unable to conceal the reoccurring groups of words or letters.

An n-gram or Q-gram which is sequence of n-items in continuity in a speech or text sample were given. Syllables, words, phonemes, letters or base pairs are the items. Speech corpus or text which are collected to form the n-grams typically [16]. The n-grams with words may be called singles.

Two adjacent elements with a string of tokens which are words, letters, syllables, in typical, is known as bigram or digram, so, n = 2, for n-gram which are bigram. In Cryptography, computational linguistic, speech recognition and others, are the applications of bigram in a string found statistically.

N = 3 is for Trigrams, a special case of n-Gram. To perform statistical analysis of text also used in natural language processing, also in cryptography, for cipher use and control use.

N = 4 is for Four-Gram, a special case of n-Gram. To perform statistical analysis of text also used in natural language processing, also in cryptography, for cipher use and control use.

In this section we generated di-gram, tri-gram and four-gram for both the RC4 and proposed IFSC with the reference file, Vigenère cipher text file. After four-gram there no five-gram is available in the ciphertext of RC4 and proposed IFSC.

Table 2. Di–Gram of RC4

No.	Substring	Frequency (in %)	Frequency
1	BS	3.0303	2
2	GW	3.0303	2
3	HI	3.0303	2
4	VZ	3.0303	2
5	YY	3.0303	2
6	AD	1.5152	1
7	AU	1.5152	1
8	AY	1.5152	1
9	BL	1.5152	1
10	BR	1.5152	1
11	CG	1.5152	1
12	CK	1.5152	1
13	CP	1.5152	1
14	CY	1.5152	1
15	DB	1.5152	1
16	DC	1.5152	1
17	DV	1.5152	1
18	HC	1.5152	1
19	HH	1.5152	1
20	HP	1.5152	1
21	IC	1.5152	1
22	ID	1.5152	1
23	IG	1.5152	1
24	IO	1.5152	1
25	JG	1.5152	1
26	JJ	1.5152	1

Table 3. Tri–Gram of RC4

No.	Substring	Frequency (in %)	Frequency
1	AYI	6.6667	1
2	CKB	6.6667	1
3	CPX	6.6667	1
4	ICG	6.6667	1
5	IOH	6.6667	1
6	JJG	6.6667	1
7	KBS	6.6667	1
8	LXH	6.6667	1
9	MDV	6.6667	1
10	OBR	6.6667	1
11	PAU	6.6667	1
12	RMB	6.6667	1
13	ROB	6.6667	1
14	VXX	6.6667	1
15	YRO	6.6667	1

Table 2, shows the di-gram of RC4 and Table 5 shows the di-gram of proposed IFSC, So, the reference di-gram found in RC4 is 26 having almost same frequencies, the reference di-gram found in proposed IFSC is 15 having various frequencies. Thus, proposed IFSC is comparable with RC4 in terms of di-gram.

Table 3, shows the trigram of RC4 and Table 6 shows the trigram of proposed IFSC, So, the reference trigram found in RC4 is 15 having almost same frequencies, the reference trigram found in proposed IFSC is also 15 having decreasing frequencies. Thus, proposed IFSC is well comparable with RC4 in terms of trigram.

Table 4, shows the four-gram of RC4 and Table 7 shows the four-gram of proposed IFSC, So, the reference four-gram found in RC4 is 3 having almost same 33% frequencies, the reference four-gram found in proposed IFSC is also 4 having decreasing frequencies. Thus, proposed IFSC is well comparable with RC4 in terms of four-gram.

Table 4. Four–Gram of RC4

No.	**Substring**	**Frequency (in %)**	**Frequency**
1	CKBS	33.3333	1
2	ROBR	33.3333	1
3	YROB	33.3333	1

Table 5. Di–Gram of IFSC

No.	**Substring**	**Frequency (in %)**	**Frequency**
1	TZ	15.1079	21
2	TS	6.4748	9
3	KL	5.0360	7
4	KK	4.3165	6
5	KZ	4.3165	6
6	LK	4.3165	6
7	TK	4.3165	6
8	YT	4.3165	6
9	ZT	4.3165	6
10	DL	3.5971	5
11	EO	3.5971	5
12	XO	2.8777	4
13	CU	2.1583	3
14	KT	2.1583	3
15	LT	2.1583	3

Table 6. Tri–Gram of IFSC

No.	Substring	Frequency (in %)	Frequency
1	KKZ	15.7895	6
2	KZT	15.7895	6
3	YTS	15.7895	6
4	LKK	13.1579	5
5	TZK	5.2632	2
6	CKK	2.6316	1
7	COC	2.6316	1
8	CUX	2.6316	1
9	DLK	2.6316	1
10	KLP	2.6316	1
11	KTK	2.6316	1
12	KYK	2.6316	1
13	LTZ	2.6316	1
14	PKL	2.6316	1
15	RKL	2.6316	1

Table 7. Four–Gram of IFSC

No.	Substring	Frequency (in %)	Frequency
1	KKZT	46.1538	6
2	LKKZ	38.4615	5
3	CKKZ	7.6923	1
4	RKLP	7.6923	1

5.5. Autocorrelation

Autocorrelation, which is also known as serial correlation, is the correlation of a signal with its delayed copy which is a function of delay in the discrete time-space. It is basically a similarity between observations as a function of the time lag. Autocorrelation is analysed by a mathematical tool which is used to find repeating patterns like the presence of a periodic signal should is masked with noise, in a signal to identify some missing fundamental frequency implied by harmonic frequencies. To analyse function or value series like time-domain signals in signal processing, this can be used. The definition of autocorrelation differs in different fields and all of them are not equivalent. Some domains use autocorrelation with auto variance. Some processes of autocorrelation include autoregressive processes, unit root processes, trend-stationary processes and moving average processes.

Autocorrelation analysis calculates the association of the observations between the different points in time and thus looks for a patterns or trend over the time series. Autocorrelation tests statistics are used to verify the degree of mutual dependency between a sequence (Si) and its shifted sequence (S(i+t)). In equation (10) a constant integer t has been considered which ranges from 1< = t< = n/2 perform a test statistic to check the relationship between (Si) and (S(i+t)) is as follows:

$$C(t) = \sum_{i=0}^{n-t-1} [\![S_i \oplus S(i+t)]\!] \qquad (10)$$

Here, (n-t) > = 10 and function C(t) follows (0,1) distribution. During unknown data analysis it helps us to find out if the data is random or not.

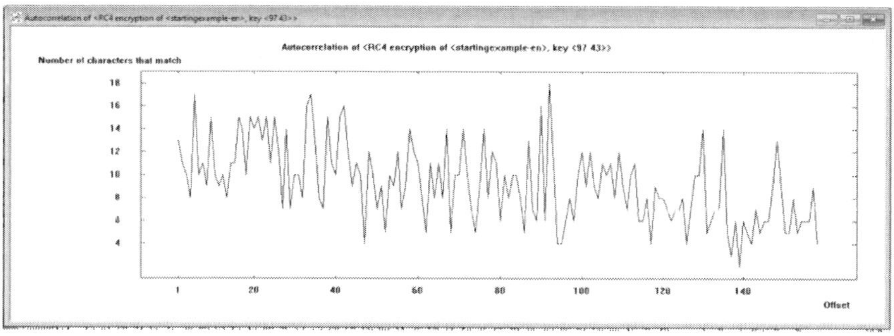

Figure 17. Autocorrelation of RC4.

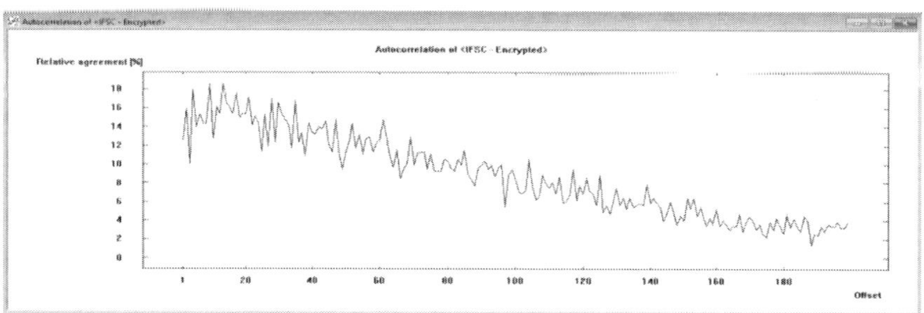

Figure 18. Autocorrelation of IFSC.

Figure 17 gives the autocorrelation diagram of RC4 and Figure 18 gives the autocorrelation of proposed IFSC. The proposed IFSC is having decreasing autocorrelation than that of RC4, Thus, proposed IFSC is well comparable with RC4 in terms of autocorrelation.

5.6. Periodicity

A periodic sequence also cycles, in Mathematics, is such sequence where the same sequence of terms that are repeated again and again:

a1, a2, ..., ap, a1, a2, ..., ap, a1, a2, ..., ap, ...

The number p of repeated terms is called the period (period) [17].

Periodicity is the replication of character's patterns having a certain length L (L> = 1) from a particular position in the document - offset. Two remarkable features of the patterns are offset and period. Here offset refers to the starting position of the patterns and continues through the end of the document. Period refers to the length of the set of repeated characters. A document with L cycles must be at least 2 ^ (L + 1) – 2 bytes long. Identifying repeating (periodic) patterns could reveal important information about the behaviour and trend of the data appearing in the original text. If there is no period then it is difficult to guess the behaviour of the plaintext. The periodicity details were given in.

Here some examples like

kkjjjjj – the cycle of j, offset of 3, period of 1
kkkkmnymny – the cycle of many, offset of 5, period of 3
mmmmmabcd – not a valid period

Figure 19. Periodicity of RC4.

Figure 20. Periodicity of Proposed IFSC.

Figure 19 and Figure 20 give the periodicity result of RC4 and proposed IFSC respectively and both have no periodicity in their ciphertext file(s).

Conclusion

This chapter successfully gives a novel IF-based stream cipher and the pseudo-randomness is already established in the discussion section. This chapter also gives a security model in the 2nd layer lightning network of the blockchain cryptocurrency transaction. Since the proposed model can give a one-time pad (OTP) output and stream cipher is the fastest algorithm of privacy, so, it can be successfully applied at the 2nd layer lighting network transactions of the blockchain cryptocurrency. The future work will be the NIST statistical cryptanalysis of the proposed stream cipher and modelling it at other security solutions.

The proposed cipher is implemented in C programming and compared with RC4 and found to be well comparable in terms of mono bit frequency test, entropy test, floating frequency test, histogram, N-Gram, Autocorrelation and periodicity.

Thus, this proposed IFSC can be applied successfully in the blockchain platform, which is now a part of many sustainable devolvement, here we given the security at the lightning network. It is now seen that for huge nodes in bitcoin, one transaction takes almost five minutes, so, the lightning network is second layer network introduced before cryptocurrencies to make the transactions faster. This lightning network enables transactions at various crypto-app at very short time and a commit protocol is executed over actual cryptocurrency in regular interval of time. The security in the blockchain is well established by modern ciphers but providing security in the lightning network is still under research. So, we proposed a model with this stream cipher to provide security, as the longest run stream cipher is the one-time pad and considered more secure cipher.

References

[1] Xu, Z. and R. R. Yager (2006), "Some geometric aggregation operators based on intuitionistic fuzzy sets", *International Journal of General Systems*, Vol. 35, No. 4, pp. 417-433. A6

[2] Rouhani, S. and R. Deters, "Performance analysis of Ethereum transactions in private blockchain," *2017 8th IEEE International Conference on Software Engineering and Service Science* (ICSESS), 2017, pp. 70-74, doi: 10.1109/ICSESS.2017.8342866.

[3] Nakamoto, Satoshi. Bitcoin: A Peer-to-Peer Electronic Cash System, 2009, pp 1-9, https://bitcoin.org/bitcoin.pdf.

[4] DeVries, Peter D., "An Analysis of Cryptocurrency, Bitcoin, and the Future", *International Journal of Business Management and Commerce*, Vol. 1 No. 2; September 2016, pp – 1-9.

[5] Poon, Joseph and Thaddeus Dryja, "*The Bitcoin Lightning Network: Scalable Off-Chain Instant Payments*", January 14, 2016, pp 1 – 59, https://lightning.network/lightning-network-paper.pdf

[6] Sattar B. Sadhkan Al Manicky and Sabiha F. Jawad, "Fuzzy Logic-Based Security Evaluation of Stream Cipher", Chapter 13 in *Research Methods: Concepts, Methodologies, Tools, and Applications* (4 Volumes), IGI Global, January, 2015, DOI: 10.4018/978-1-4666-7456-1, ISBN 13: 9781466674561, ISBN 10: 1466674563, EISBN 13: 9781466674578, pp 139-157.

[7] Miraz, Mahdi H. and David C. Donald, *LApps: Technological, Legal and Market Potentials of Blockchain Lightning Network Applications*, ICISDM 2019, https://doi.org/10.48550/arXiv.1906.11946, Related DOI: https://doi.org/10.1145/3325917.3325942.

[8] Keyanpour, Mohammad and Tahereh Akbarian, "A new approach to the numerical solution of fractional order optimal control problems", https://www.researchgate.net/publication/267656599_A_new_approach_to_the_numerical_solution_of_fractional_order_optimal_control_problems.

[9] Atanassov, K. T. (1986), "Intuitionistic Fuzzy Sets", Fuzzy Sets and Systems, Vol. 20, pp. 87-96.

[10] Roychowdhury, Shounak and Sujeet Shenoi, "*Fuzzy Rule Encoding Techniques*", Research supported by MPO Contracts MDA904-94-C-6117 and MDA904-96-1-0114 and OCAST Grant AR2-002, 0-7803-3796 4/97/$10.0001997JEE.

[11] Anikin, Igor V. and Khaled Alnajjar. "*Fuzzy Stream Cipher System*". 2015 International Siberian Conference on Control and Communications (SIBCON), 978-1-4799-7103-9/15/$31.00 ©2015 IEEE.

[12] Machado, S., P. D'silva, S. D'mello, S. Solaskar, and P. Chaudhari (2018), "*Securing ATM pins and passwords using Fingerprint based Fuzzy Vault System*", 978-1-5386-5257-2/18/$31.00 ©2018 IEEE.

[13] Cid, Carlos, John Petter Indrøy and Håvard Raddum, "*FASTA - a stream cipher for fast FHE evaluation*", Cryptology ePrint Archive: Report 2021/1205, https://eprint.iacr.org/2021/1205.

[14] *New Version of CrypTool* — CT 1.4.41 released, https://www.cryptool.org/en/posts/2018-03-30/cryptool-1.4.41-released, 2018.

[15] Pearson, K. (1895), "Contributions to the Mathematical Theory of Evolution. II. Skew Variation in Homogeneous Material". *Philosophical Transactions of the Royal Society A: Mathematical, Physical and Engineering Sciences.* 186: 343–414. Bibcode:1895RSPTA.186.343P. doi:10.1098/rsta.1895.0010.

[16] Broder, Andrei Z., Steven C. Glassman, Mark S. Manasse, Geoffrey Zweig (1997), "Syntactic clustering of the web". *Computer Networks and ISDN Systems.* 29 (8): 1157–1166. doi:10.1016/s0169-7552(97)00031-7.

[17] "*Ultimately periodic sequence - Encyclopedia of Mathematics.*" encyclopediaofmath.org. 7 February 2011. Retrieved 13 August 2021.

[18] Shirey, R. (August 2007), *Internet Security Glossary,* Version 2. Network Working Group. doi:10.17487/RFC4949. RFC 4949.

[19] Stallings, William (3 May 1990), *Cryptography and Network Security: Principles and Practice.* Prentice Hall. p. 165. ISBN 9780138690175.

Chapter 7

Synthesis of Bio-Fertilizers with Nutrients (NPK) for Agricultural Use from Organic Matters

Rakesh Kumar Yadav*
and Shivram Krishna Chandrakar
Dr. C. V. Raman University, Kargi Road Kota, Bilaspur, India

Abstract

In the subject of agriculture, new inventions have been made and are continually being made to improve practises. Many chemical composition fertilisers and pesticides, as well as other medications and manure, are commonly utilised for agricultural purposes. Chemical fertilisers are incorporated into our clean ecology in a variety of indirect ways, causing a slew of negative consequences. Organic materials exist in our environment and can fill this void without having a significant impact on ecosystem elements, as opposed to pure chemical compounds. These organic materials might be beneficial while also being environmentally friendly. Bio fertiliser is a synthetic organic substance that contains live microorganisms and is the finest option. When applied to seed, plant surfaces, or soil, bio-fertilizers encourage and increase plant development by improving the availability of nutrients to the plant. There are several techniques to make a bio-fertilizer from readily available natural by-products and organic materials. Composting, landfilling, and a variety of other methods for producing bio-fertilizer and green manure have all been popular. Bio-fertilizers are a low-cost, recyclable/renewable alternative to synthetic fertilisers. Liquid bio

* Corresponding Author's Email: rakeshyadav96@gmail.com.

In: Computational Intelligence for Sustainable Development
Editor: Brojo Kishore Mishra
ISBN: 979-8-88697-198-9

fertilisers have the potential to be a superior alternative to standard chemical fertilisers, and carrier-based bio fertilisers can also help restore soil health. One of the most common methods for making liquid bio-fertilizers is to combine helpful and dangerous plants that are readily available in a certain area. Using Indian Bael (for potassium), banana (for phosphorus), and carrot grass (for nitrogen) with a decomposer solution is one way to generate NPK supplement solution. Because crops such as rice and others require a considerable amount of this NPK bio-organic solution, it is required in enormous quantities. Some crops, for example, have a slow rate of absorption of this bionutrient solution because they are unable to absorb them rapidly. This research provides a catalyst solution that significantly increases the nutrient absorption process. The suggested catalyst solution contains helpful nutrients such as amino acids and other growth hormones, allowing for the use of a little amount of NPK solution. Different plant seeds are also used to create the catalyst solution. Growth rate acceleration has been found using this proposed catalyst solution for several crops. This procedure is environmentally safe, inexpensive, and readily scaled up to boost manufacturing volume. A Many studies have been carried out in this suggested study in order to conclude that bio-fertilizers may give nutrients and certain amino acids, as well as plant hormones as an exogenous source, which can boost the absorption rate and improve the growth rate.

Keywords: growth rate, composition fertilisers and pesticides, synthetic organic substance

Introduction

Agriculture and related industries employ about half of India's workforce. Traditional agricultural methods, which have been recommended and proposed by sages since the Vedic era, are still in use today. New inventions are widespread in today's time, which is a modern and scientific age. Chemical research is no exception to this tendency, and it is developing with the passage of time. In this context, new agricultural inventions have been made and are continually being made in order to improve methods. Many chemical composition fertilisers and pesticides, as well as other medications and manure, are commonly utilised for agricultural purposes. Plants are less dependent on external chemical medications (fertilisers, etc.) for their physical activity and growth, according to plant science (Botany), and rely on the environment for the remaining 98 percent. Hydrogen, carbon, and oxygen are mostly obtained from air and water, whereas macro and micro nutrients are obtained from soil and fertilisers. Chemical fertilisers are being introduced to our natural environment in a variety of ways, impacting the food chain on a small scale and manifesting itself in the form of a variety of ailments and diseases in animals and people. Pesticides have been linked to climate change and global warming in studies. In our environment organic matters are present those can fulfil this deficiency, without affecting much to the elements of ecosystem as compare to pure chemical substances. These organic matters can be useful with having eco-friendly advantages. Crop or plants are made-up of organic matters like other living organisms. The inner and outer structures of a plant formed by many different processes, those processes need nutrients from environment (J.P. Bloomfield et al., 2006).

Plant Nutrient

Plants absorb and use inorganic minerals for nutrition which are present in the soil or subjected/injected to the plants. Inorganic minerals is formed by many complex interactions such as weathering of rock minerals, decaying organic matter, animals, and microbes in soil. Mineral nutrients are absorbed by plant roots as ions in soil water. Many factors could influence nutrient uptake for plants. Ions can be readily available to roots or could be "tied up" by other elements or the soil itself. Soil quality also affects the absorbing quality of these minerals, for instance too high in pH (alkaline) or too low (acid) makes minerals unavailable to plants.

Nutrients Needed for Plant Growth

Nutrients are needed for proper growth and production in a plant life cycle. The ratio of nutrients needed depends on the plants type and soils. The use of soil tests can help to determine the status of plant available nutrients to develop fertilizer recommendations to achieve optimum crop production. The production cost need to be cut below selling price to achieve potential profits for a farmer. Efficient application of the correct types and amounts of fertilizers for the supply of the nutrients is an essential part of obtaining profitable yields. Carbon (C), Hydrogen (H), and Oxygen (O) are produced from carbon dioxide (CO_2) and water, respectively, and are recognised to be responsible for plant development (H_2O). Inorganic salts comprising nitrogen

(N), phosphorus (P), potassium (K), sulphur (S), calcium (Ca), magnesium (Mg), boron (B), chlorine (Cl), copper (Cu), iron (Fe), manganese (Mn), molybdenum (Mo), and zinc (Zn) are often derived from soil. Carbon, hydrogen, and oxygen make up 94% to 99.5 percent of new plant material. The remaining 0.5 to 6.0 percent is made up of other nutrients. Micronutrients are nutrients that are required in minute amounts.

Alternative Methods Present to Supplement Micro-Nutrients

Chemical synthetic fertilizers which are now being used extensively to supplement those nutrients externally, since the green revolution have degraded the soil health. The bio-fertilizers contains all those nutrients are now started being used, which are largely responsible for maintaining soil fertility and providing the necessary nutrients to plants.

Fertility or Nutrition

The term "fertility" refers to the inherent capacity of a soil to supply nutrients to plants in adequate amounts and in suitable proportions. The term "nutrition" refers to the interrelated steps by which a living organism assimilates food and uses it for growth and replacement of tissue. Previously, plant growth was thought of in terms of soil fertility or how much fertilizer should be added to increase soil levels of mineral elements. Most fertilizers were formulated to account for deficiencies of mineral elements in the soil. The use of soilless mixes and increased research in nutrient cultures and hydroponics as well as advances in plant tissue analysis have led to a broader understanding of plant nutrition. Plant nutrition is a term that takes into account the interrelationships of mineral elements in the soil or soilless solution as well as their role in plant growth. This interrelationship involves a complex balance of mineral elements essential and beneficial for optimum plant growth.

Essential vs Beneficial Nutrients

The term essential mineral element (or mineral nutrient) was proposed by Arnon and Stout (1939). They concluded three criteria must be met for an element to be considered essential. These criteria are:

- A plant must be unable to complete its life cycle in the absence of the mineral element.
- The function of the element must not be replaceable by another mineral element.
- The element must be directly involved in plant metabolism.

Apart from the above mentioned criteria beneficial elements are those that can compensate for toxic effects of other elements or may replace mineral nutrients in some other less specific function such as the maintenance of osmotic pressure. The omission of beneficial nutrients in commercial production could mean that plants are not being grown to their optimum genetic potential but are merely produced at a subsistence level.

The Mineral Elements Needed for Plant Growth

There are actually 20 mineral elements necessary or beneficial for plant growth.

- Carbon (C), hydrogen (H), and oxygen (O) are supplied by air and water.
- The six macronutrients, nitrogen (N), phosphorus (P), potassium (K), calcium (Ca), magnesium (Mg), and sulfur (S) are required by plants in large amounts.
- The rest of the elements are required in trace amounts (micronutrients). Essential trace elements include boron (B), chlorine (Cl), copper (Cu), iron (Fe), manganese (Mn), sodium (Na), zinc (Zn), molybdenum (Mo), and nickel (Ni). Beneficial mineral elements include silicon (Si) and cobalt (Co).

The beneficial elements have not been deemed essential for all plants but may be essential for some. The distinction between beneficial and essential is often difficult in the case of some trace elements. Cobalt, for instance, is essential for nitrogen fixation in legumes. Silicon, deposited in cell walls, has been found to improve heat and drought tolerance and increase resistance to insects and fungal infections. Silicon, acting as a beneficial element, can help compensate for toxic levels of manganese, iron, phosphorus and aluminium as well as zinc deficiency. A more holistic approach to plant nutrition would not be limited to nutrients essential to survival but would include mineral elements at levels beneficial for optimum growth. With developments in analytical chemistry and the ability to eliminate contaminants in nutrient cultures, the list of essential elements may well increase in the future.

The Elements of Complete Plant Nutrition

The role of essential and beneficial mineral nutrients are further can be categorised in two parts, those are crucial for growth. Eliminate any one of these elements, and plants will display abnormalities of growth, deficiency symptoms, or may not reproduce normally.

Micronutrients

Micronutrients are needed in small amounts, but thease are essential to maintain the immunity of a plant/tree. Iron is necessary for many enzyme functions and as a catalyst for the synthesis of chlorophyll. It is essential for the young growing parts of plants. Deficiencies are pale leaf color of young leaves followed by yellowing of leaves and large veins. Iron is lost by leaching and is held in the lower portions of the soil structure. Under conditions of high pH (alkaline) iron is rendered unavailable to plants. When soils are alkaline, iron may be abundant but unavailable. Applications of an acid nutrient formula containing iron chelates, held in soluble form, should correct the problem. Manganese is involved in enzyme activity for photosynthesis, respiration, and nitrogen metabolism. Deficiency in young leaves may show a network of green veins on a light green background similar to an iron deficiency. In the advanced stages the light green parts become white, and leaves are shed. Brownish, black, or grayish spots may appear next to the veins. In neutral or alkaline soils plants often show deficiency symptoms. In highly acid soils, manganese may be available to the extent that it results in toxicity. Boron is necessary for cell wall formation, membrane integrity, calcium uptake and may aid in the translocation of sugars. Boron affects at least 16 functions in plants. These functions include flowering, pollen germination, fruiting, cell division, water relationships and the movement of hormones. Boron must be available throughout the life of the plant. It is not translocated and is easily leached from soils. Deficiencies kill terminal buds leaving a rosette effect on the plant. Leaves are thick, curled and brittle. Fruits, tubers and roots are discolored, cracked and flecked with brown spots. Zinc is a component of enzymes or a functional cofactor of a large number of enzymes including auxins (plant growth hormones). It is essential to carbohydrate metabolism, protein synthesis and internodal elongation (stem growth). Deficient plants have mottled leaves with irregular chlorotic areas. Zinc deficiency leads to iron deficiency causing similar symptoms. Deficiency occurs on eroded soils and is least available at a pH range of 5.5 - 7.0. Lowering the pH can render zinc more available to the point of toxicity. Copper is concentrated in roots of plants and plays a part in nitrogen metabolism. It is a component of several enzymes and may be part of the enzyme systems that use carbohydrates and proteins. Deficiencies cause die back of the shoot tips, and terminal leaves develop brown spots. Copper is bound tightly in organic matter and may be deficient in highly organic soils. It is not readily lost from soil but may often be unavailable. Too much copper can cause toxicity. Molybdenum is a structural component of the enzyme that reduces nitrates to ammonia. Without it, the synthesis of proteins is blocked and plant growth ceases. Root nodule (nitrogen fixing) bacteria also require it. Seeds may not form completely, and nitrogen deficiency may occur if plants are lacking molybdenum. Deficiency signs are pale green leaves with rolled or cupped margins. Chlorine is involved in osmosis (movement of water or solutes in cells), the ionic balance necessary for plants to take up mineral elements and in photosynthesis. Deficiency symptoms include wilting, stubby roots, chlorosis (yellowing) and bronzing. Odors in some plants may be decreased. Chloride, the ionic form of chlorine used by plants, is usually found in soluble forms and is lost by leaching. Some plants may show signs of toxicity if levels are too high. Nickel has just recently won the status as an essential trace element for plants according to the Agricultural Research Service Plant, Soil and Nutrition Laboratory in Ithaca, NY. It is required for the enzyme urease to break down urea to liberate the nitrogen into a usable form for plants. Nickel is required for iron absorption. Seeds need nickel in order to germinate. Plants grown without additional nickel will gradually reach a deficient level at about the time they mature and begin reproductive growth. If nickel is deficient plants may fail to produce viable seeds. Sodium is involved in osmotic (water movement) and ionic balance in plants. Cobalt is required for nitrogen fixation in legumes and in root nodules of no legumes. The demand for cobalt is much higher for nitrogen fixation than for ammonium nutrition. Deficient levels could result in nitrogen deficiency symptoms. Silicon is found as a component of cell walls. Plants with supplies of soluble silicon produce stronger, tougher cell walls making them a mechanical barrier to piercing and sucking insects. This significantly enhances plant heat and drought tolerance. Foliar sprays of silicon have also shown benefits reducing populations of aphids on field crops. Tests have also found that silicon can be deposited by the plants at the site of infection by fungus to combat the penetration of the cell walls by the attacking fungus. Improved leaf erectness, stem strength and prevention or depression of iron and manganese toxicity have all been noted as effects from silicon. Silicon has not been determined essential for all plants but may be beneficial for many.

Bio-Fertilizers

Bio-fertilizer is one of the biggest innovations to our agricultural industry and responsible to maintain a healthy relationship with our ecosystem. Bio fertilizer is a synthesized organic matters which contains living micro-organisms. Bio-fertilizers when subjected to seed, plant surfaces, or soil promotes and accelerates the growth by increasing the availability of nutrients to the plant (Mazid et al., 2011a). Bio-fertilizers are the products which is either compost result or some other techniques results containing one or more microorganisms species which have the ability to mobilize nutritionally important elements from non-usable to usable form through biological processes such as nitrogen fixation, etc. IN other words, bio- fertilizers are natural fertilizes which are living microbial inoculants of bacteria, algae, fungi alone or in combination and they improve the accessibility of nutrients to the plants.

How Bio-Fertilizer Differs from Chemical Synthetic Fertilizer?

Bio fertilizers do not supply any nutrients to crops and are cultures of living micro-organism (such as fungi and bacteria), unlike chemical and synthetic organic fertilizer. The production and installation cost of bio-fertilizers are relatively low and simple as compare to that of chemical fertilizers. Fungi and bacteria present in the bio-fertilizers are helpful to improve soil chemical and biological characteristics (ElHabbasha et al., 2007; Yosefi et al., 2011). Microbiological fertilizers are important to environment friendly sustainable agricultural practices (Bloemberg et al., 2000). The bio fertilizer provides many processes to plant from soil such as the nitrogen fixing, phosphate solubilizing and plant growth promoting microorganisms (Goel et al., 1999).

The use of bio-fertilizers are increasing the growth, plant height as well as the grains weight. This experiment has clearly been demonstrated for rice production by Alam et al.

Synthesizing the Bio-Fertilizers

There are many ways to synthesize a bio-fertilizer from the natural byproducts and organic matters those are easily available. Composting, Landfilling and various other techniques have already been in vogue to make bio-fertilizer and green manure. This bio-fertilizer, which is a better alternative of chemical only fertilizer, has been declared as cost-effective, cheap and recyclable/renewable source to replace the chemical fertilizers. The bio-fertilizer can be made in many ways like composing, landfilling etc from organic matters like plants or animal byproducts like cow dung. Biodegradation of different organic waste be performed to make a bio-fertilizer. Bio fertilizer can be prepared in solid and liquid both. Liquid bio fertilizer have the capability to provide a better alternative to the traditional chemical fertilizer and carrier based bio fertilizer can also play an important role in restoring soil health (Uma et al.).

Working of Bio-Fertilizers

Bio-fertilizers work in an eco-friendly and highly balanced way to provide and supply many nutrients to the. Bio-fertilizers have the capabilities to trap surrounding or atmospheric plant- essential elements like nitrogen etc. to the soil and convert them into plant usable forms, for example it can convert the insoluble phosphate forms into plant available forms. They help to stimulate root and overall growth of a plant by producing some auxiliary hormones and antimetabolites. Another bigger advantage of these fertilizers includes that it rebalances the fertility status of a soil and thus improves the soil life. Along with another application of bio fertilizers it completes and boosts nutrient cycling in the soil and starts 'biological buffering' to relax the existing stresses in the soil. The microbes added to the soil as bio fertilizers have capabilities to improve the host/soil immunity and protect the crops from pests and diseases; thus, they can help reducing the use of pesticides to a great extent.

Advantages of Bio Fertilizers

Bio fertilizers have some advantages as follows

- Bio fertilizers have longer shelf life.
- Farmers can use bio fertilizers quite easily.
- Bio fertilizers are completely free of contamination.
- They have a greater capacity to fight with the native population.
- They are not expensive but serve better.
- They bring high commercial revenues.

Types of Bio Fertilizers

Bio fertilizers can be categorized according to the serving purposes, some of them are as follows:

Bio-Fertilizers Categories as per the Serving Purposes

Bio-fertilizers can be developed to serve a specific element or for the group of the nutrients. The contents of bio-fertilizers can be decided by the plant or crop requirements.

Nitrogen Fixing Bio Fertilizers

Nitrogen fixing bio fertilizers contain many microbial inoculants or group of microorganisms which are capable of converting atmospheric N_2 into organic compounds and eventually into plant usable forms. Thus, these bio fertilizers fix N_2 from the atmosphere and supply that to plants in usable forms which ultimately increase soil N_2 level and overall soil fertility level. A number of microorganisms are used as Nitrogen fixing bio fertilizers. For instance Azotobacter, Anabaena, Nostoc, Clostridium etc. are used as free-living N_2 fixing bio fertilizers; while Frankia, Rhizobium, and Anabaena azollae are used as symbiotic N_2 fixing bio fertilizers and Azospirillum is used as associative symbiotic N_2 fixing bio fertilizer.

Phosphorus Solubilizing Biofertilizers

These bio fertilizers solubilize the insoluble phosphates from various phosphate sources and make them plant usable. They lower soil pH by secreting organic acids and thus dissolve the bound phosphates. Some bacteria used as P-solubilizing bio fertilizer are-Bacillus megatherium, Pseudomonas striata, Bacillus circulans, Bacillus subtilis; and some fungus used for this purpose are Aspergillusawamori and Penicilliumspp.

Phosphorus Mobilizing Bio fertilizers

Different fungi and mycorrhiza, which stimulate the transfer of P ions and thus metabolic processes, belong to this group. Microorganisms those are used as P mobilizing bio fertilizers are-Glomus spp., Gigaspora spp., Boletus sp., Laccaria spp., Pisolithus sp., Rhizoctoniasolani, Pezizellaericae etc.

Bio Fertilizers for Micro Nutrients

The microbes used here are silicate and zinc solubilize bacterial species. These microbes also degrade silicates and aluminium silicates. Bacillus spp. is an example of this group.

Plant Growth Promoting Rhizobacteria

These microbes act as biofertilizers as they stimulate plant growth by improving nutrient availability. They are also effective as biopesticides and bioprotectants. Pseudomonas sp., and Bacillus sp. belong to this group.

According to the state, bio-fertilizers can be categorized in two forms:

Bio-Fertilizers Categories as per the State of Matters

Bio fertilizers can have solid form and liquid form both. Both are being developed by the organic matters. For liquid bio fertilizers a carrier solution is needed.

Solid Bio-Fertilizers

Solid Bio–fertilizers may be defined as organic waste water solids that can be reused after suitable sewage sludge treatment processes leading to sludge stabilization such as anaerobic digestion and composting.

Liquid Bio-Fertilizers

The alternatives to chemical and organic fertilizers as well as solid substrate-based bio fertilizers. One of the benefits from bio fertilizer is the contribution from population of microorganisms available. These microorganisms may enhance the plant growth and create healthy rhizosphere.

Advantages of the Liquid Bio Fertilizers

The effectiveness of commercially available biofertilizers is not always the same in pot and field study. In the present work, efficiency of 6 different commercially available liquid biofertilizers containing Azotobacter strains or phosphate solubilizers obtained from three reputed brands was assessed in green house (pot study) and also at field level on growth enhancement of Ciceraeritinum (chickpea). The results obtained in pot level studies showed marginal beneficial effects ranging between 21% and 50% in total final yield. At field level, PPS ($P<0.01$) and consortium of all 6 products ($P<0.01$) showed significant improvements by 144% and 140% respectively than control in total final yield. In pot studies, Azotobacter biofertilizers gave good results when seed germination, average shoot length and leaf chlorophyll contents were analyzed in controlled environments whereas in field studies, in general, phosphate solubilizing biofertilizers gave better overall yield. Though overall yield increase was observed upon application of liquid biofertilizers, the leaf chlorophyll and seed protein content were not different as compared to control. Moreover, where the producer is more inclined to obtain higher yield per hectare, for the consumer the nutrition value is more significant. This stands true for vegetarians, where a large number of people around the world depend on legumes such as chickpea as protein source in their diet as other protein sources are unaffordable for them. Hence to be more effective nutritionally; improvement is required in the already existing liquid bio fertilizers available commercially.

Plant Regulators and Hormones

Plant hormones regulate diverse biological events from growth and development to biotic and abiotic responses at low concentrations. To provide adequate concentrations of the active form of those plant hormones to the cytosol, some are converted to conjugates with glucose or amino acids to inactive forms.

Plant hormones are naturally occurring small molecule compounds and play particularly important roles in the normal functioning of plant. According to structural and chemical diversity, plant hormones are grouped into several classes, including auxins, cytokinins (CKs), abscisic acid (ABA), gibberellins (GAs), ethylene, jasmonic acid (JA), salicylic acid (SA), brassinosteroids (BRs), and strigolactones (SLs). The elucidation of the molecular mechanisms of plant hormone action contributes to understanding of plant biology and improving crops because of the particularly important roles of hormones in the regulation of plant growth and development. Progress toward a better understanding of the biosynthesis, metabolism, and mechanisms of plant hormone function requires detailed information on plant hormone structure, plant hormone distribution, and changing levels in diverse plant organs and tissues. The biological activities of plant hormones depend on their cellular concentration in the plant, so the availability of high-performance and high-sensitivity methods of analysis, especially sensitive and accurate quantification methods, is essential for facilitating our understanding of biosynthesis and mechanisms of action of plant hormones.

Plant Regulator

A chemical which in small amounts regulates (promotes, inhibits or otherwise modifies) a physiological process. A broad term which includes the hormones (endogenous regulators) as well as exogenous regulators.

Plant Hormone

An endogenous regulator. To be a hormone, a chemical must be produced within the plant, transported from a site of production to a site of action, and be active in small amounts.

Vitamin

An organic chemical required in small amounts from external sources for the normal growth and maintenance of an organism. Green plants (autotrophic organisms) do not require vitamins. Non-green plants and animals (heterotrophic organisms) usually require one or more vitamins.

Inhibitor

A regulator which inhibits or retards a physiological process.

- *Growth Substance* — a term used variously, usually as a synonym for auxin.
- *Growth Factor* — usually an exogenous regulator, often equivalent to vitamin or hormone. Frequently used with reference to the culture of micro-organisms, tissues, or organs.

Growth Hormones: Auxin

The generic name for the indole hormones. These promote growth by stimulating cell enlargement. This early years of 18th century culminated in the simple but conclusive experiments of a clever scientist Went which showed that the growth of the oat coleoptile requires a chemical which is produced in the tip of the coleoptile. The chemical was called auxin. It is widely distributed in plants and appears to be universally required for growth, hence Went's Law of plant growth: 'Without auxin, no growth.'

Indoleacetic acid is now considered to be the principal auxin, but there are several other auxins. All are closely related chemically, and more than one is often present in a tissue. The best known of these are indoleacetaldehyde, indoleacetonitrile, and ethyl indoleacetate.

Auxin has several physiological properties which account for most of its functions. It promotes cell enlargement by increasing the plasticity of the cell wall, thus facilitating the uptake of water. It is transported basipetally (from tip toward base) in organs. This transport is modified by gravity and light: auxin tends to move in the direction of the force of gravity, and away from a source of light. Further, auxin in the plant is inactivated by light. Also, the different organs of plants differ in their sensitivity to auxin; roots are the most sensitive, stems are the least sensitive. Through the interaction of these properties, growth and growth movements of plants are regulated. (The growth movements include phototropism, geotropism (both positive and negative), plagiotropism, thigmotropism, epinasty, and photonasty.).

Auxin has other functions. Its accumulation at the base of a cutting is necessary for the initiation of roots. Auxin from the apical bud inhibits the development of lateral buds. The auxin gradient across a leaf or fruit abscission zone regulates abscission (a relatively high gradient retards abscission, and a low gradient accelerates abscission). These three functions at first glance would appear unrelated, but they may all be expressions of another postulated property of auxin, that is, to mobilise food to regions where auxin is in high concentration. The existence of this property has not yet been established, but there is considerable supporting evidence. If it does exist, it would account for much of what is known concerning the auxin regulation of root inhibition, bud inhibition and abscission. Another function of auxin is the initiation of cell divisions in the cambium.

Continued research on auxin has made it apparent that auxin physiology is much more complicated than it first seemed. Auxin appears to be present in all living parts of the plant, mature as well as immature. The amounts present are effected by at least three general processes: auxin production, auxin transport, and auxin inactivation. Many of the early investigations did not recognise the existence of these three processes and their results must be re-evaluated. For example, many studies of auxin transport did not take into account the probability of considerable auxin inactivation during the course of transport. Auxin is produced principally in young tissues, but can also be produced by mature tissues. The amino acid tryptophan, a common constituent of proteins, is the precursor of auxin, but the precise chemical steps of its conversion to auxin are not yet settled. The transport of auxin can be through the parenchyma, as it is in the oat coleoptile, but in more mature tissues transport is largely in the phloem. In the coleoptile transport is correlated with the streaming of protoplasm. Auxin inactivation is accomplished by an oxidative enzyme which can function either in the dark or under the influence of light. Mature tissues have relatively high auxin-inactivating capacities. In addition to these general processes other factors, still obscure, also influence the auxin in tissues. The interaction of these processes and factors determines the level of auxin which is available to influence growth and morphogenesis.

Growth Hormones: Gibberellins

The gibberellins produce effects on growth, particularly cell elongation, which are very similar to the effects of auxin, but they function in situations where auxin does not promote elongation. Although physiological and biochemical knowledge of them is still fragmentary, they are growth factors which are probably hormones and hence should be included here. The chemicals derive their name from the fungus Gibberella, from which they can be obtained. Immature seeds are also very rich sources.

One of the most interesting series of experiments with the gibberellins was conducted with a dwarf corn (maize). This particular mutant dwarf had been the subject of an intensive auxin study, and its auxin physiology was found to be completely normal. That is, auxin production, transport and inactivation were identical with those of normal corn, and applications of additional auxin did not affect its growth; the plants never grew more than a few inches tall. However, weekly sprays of gibberellins stimulated the mutant to the normal rate of growth and practically normal appearance. The results of a similar experiment conducted several years earlier, which were at the time puzzling, can now be interpreted as due to gibberellins: an extract from immature bean seeds was applied to a bush variety of beans (Phaseolus); the stems then elongated in the manner characteristic of the tall varieties of beans. In other experiments, gibberellins sprayed on pasture grasses have induced abnormally rapid growth.

Another effect of gibberellins is in relation to both growth and flowering. Hyocyamus is one of the typical 'long-day plants'. It grows as a rosette with its leaves clustered about the very short stem until it has been exposed to a period of cold followed by a period of long days. Then the stem rapidly elongates and produces flowers. It has been found that gibberellins can replace the cold treatment; sprays followed by long days stimulate stem elongation with flowering.

Wound Hormone

Following an injury to a plant, the parenchyma cells underlying the injured area are stimulated to divide and form a protective callus. Under the stimulus, cells divide which would otherwise remain intact to the death of the plant. Early experiments showed that if the injured area is washed immediately, cell division is prevented; this suggested that a hormone might be involved. Such a hormone was isolated by Bonner and English. Starting with 100 pounds of string beans they isolated a small amount of a chemical which they called traumatic acid (chemically, decenedicarboxylic acid) which is the wound hormone of beans. However, this compound does not stimulate cell division in other species. So there remain other chemicals yet to be identified as wound hormones.

Root Growth Hormones

Knowledge of root growth hormones has come largely from experiments with the culture of isolated roots. The repeated attempts to culture isolated tissues of plants were successful in 1933 with tomato roots and a culture medium consisting of sucrose, salts, and yeast extract. Yeast extract is a very complex mixture of chemicals and attention was immediately given to determination of the active components. These were soon found to be thiamin and pyridoxin which in small amounts (a few parts per million) could completely replace the yeast extract. Thus tomato roots, which in the field would live only a few months, have been kept growing in culture in a synthetic medium since soon after 1933. Thiamin and pyridoxin were first called growth factors, since their role in the intact plant was not known. However, Bonner showed that they are produced in leaves and transported downward to roots, thus establishing them as hormones.

Other experiments showed that niacin is a root growth factor, and is presumably also a root growth hormone. In various combinations thiamin, pyridoxin or niacin will support the indefinite growth of isolated roots of many species. For a few species other factors are required such as the amino acids glycine, lysine and arginine.

Although the roots of many plants will grow rapidly (at rates at least equal to the rates of roots on intact plants) and indefinitely in synthetic culture media, important problems still remain unsolved. One is the culture of isolated roots of monocotyledonous plants. In spite of numerous attempts, these have never been established in culture. Another is the development of the cambium, which has not been induced in roots of established cultures. Further, branching of cultured roots is often abnormal. Thus the knowledge of root growth physiology is far from complete and much work lies ahead.

Experiments with root cultures brought to light an important interrelationship of vitamins and hormones. The chemicals thiamin, pyridoxin and niacin are vitamins, necessary in the diet of animals and other heterotrophs for normal growth and maintenance. In the green plant these same chemicals function in the physiological role of hormones. And within the cells of organisms they each function as a part of a vital enzyme. Thus the same chemical may function in any of three physiological roles: vitamin, hormone, enzyme.

Leaf Growth Hormone: Phyllocaline

In a search for hormones other than auxin Went performed an extensive series of grafting experiments. He worked with varieties of garden peas which differed markedly in their growth habits. The results showed, for example, that leaves of different varities differed in their ability to stimulate root growth. Similar differences among roots and buds were observed. Went postulated that these differences in growth were the result of differences in production of special hormones by the varieties. One of these postulated hormones was called phyllocaline. It is produced in cotyledons and mature leaves, and stimulates the growth of young leaves. This hormone was isolated and identified as adenine. Another property of adenine was later discovered; tissue cultures of plant callus ordinarily grow indefinitely as an undifferentiated, or at best, slightly differentiated mass of cells. In the culture medium adenine stimulates the differentiation of leafy buds.

Flowering Hormone: Florigen

Flowering is influenced by many factors including mineral and carbohydrate nutrition, temperature, photoperiod, and a postulated hormone, florigen. This hormone is produced in leaves (under particular conditions) and is transported to buds where it brings about the conversion of a vegetative stem apex to a reproductive stem apex (flower bud).

Reproductive Hormones

In the lower plants a number of hormones influencing reproductive processes have been described, as well as nutritional factors which can be called reproductive vitamins.

Growth Factors

Experiments have demonstrated growth factor requirements for many plant parts. Many, possibly all, of these growth factors are plant hormones, but present knowledge is too fragmentary in most cases to permit positive statements.

Pollen germination and tube growth factors. Pollen of some species will germinate and grow well in artificial media; pollen of others will grow poorly or not at all. Stigmatic exudates are usually very stimulatory and presumably provide hormones required by the pollen. Chemicals which have been found to promote germination or tube growth of various species include: boric acid, manganous sulphate, ascorbic acid, aminobenzoic acid, indoleacetic acid, inositol, lactoflavin, guanine, pyridoxin, thiamin.

Growth factors of tissue and organ cultures. Since the successful establishment of root cultures, other organs and several types of tissues have been successfully cultured including embryos, shoots, and callus. Often successful culture has required the use of complex mixtures such as malt extract, young seed extracts, or coconut milk. The latter is a potent source of important growth factors; its use has enabled the culture of very small embryos, but the active chemicals in coconut milk have not been identified. Growth factors which have been identified include: ascorbic acid, adenine, biotin, indoleacetic acid, niacin, pantothenic acid, thiamin. It is of interest to note that each of these is already known to have functions as a vitamin and/or hormone.

Endogenous and Exogenous Plant Hormones

Endogenous plant hormones are naturally occurring low molecular weight compounds that have various physiological functions in the regulation of plant growth and development. There are several groups or families of hormones each with different physical and chemical properties. The ability of sensitive, fast, accurate and quantitative analysis of plant hormones is very important to elucidate metabolic and signaling processes in plants.

Exogenous Source of Hormones with Essential Nutrients

An experiment was conducted in a greenhouse to investigate the availability of phosphate fertilizer in acidic soils, and to evaluate the effects of phosphorous and synthetic plant hormones on yield and nutrient uptake of maize (cv. Golda FAO-240). In the part of this experiment it has been observed that if exogenous source of plant hormones are their along with essential nutrients, then absorption of essential nutrients have increased, so the growth and development rate of the crop have rapidly increased.

Partheniumhysterophorus is a species of flowering plant in the aster family, Asteraceae. It is native to the American tropics. In India, it is locally known as carrot grass, congress grass or Gajar Ghans. This present in a huge quantity in Achanakmar Tiger reserve, which is near to Bilaspur. Partheniumhysterophorus invades disturbed land, including roadsides. It infests pastures and farmland, causing often disastrous loss of yield, as reflected in common names such as famine weed]. In some areas, heavy outbreaks have been ubiquitous, affecting livestock and crop production, and human health. As an invader it first appeared as a contaminant in imported wheat. The plant produces allelopathy chemicals that suppress crop and pasture plants, and allergens that affect humans and livestock. It also frequently causes pollen allergies. It is being investigated as a means of removing heavy metals and dyes from the environment, control of aquatic weeds, commercial enzyme production, an additive in manure for biogas production, as a bio pesticide, and as green manure and compost. Contact with the plant causes dermatitis and respiratory malfunction in humans, and dermatitis in cattle and domestic animals. The main substance responsible is parthenin, which is dangerously toxic. It also is responsible for bitter milk disease in livestock when their fodder is polluted with Parthenium leaves. Among other allelopathic effects of the species, the presence of Parthenium pollen grains inhibits fruit set in tomato, brinjal, beans, and a number of other crop plants.

Figure 1. A carrot grass plant.

Figure 2. A mature carrot grass plant.

According to the chemical composition of carrot grass, amino acids containing Nitrogen are present in a rich amount.

Figure 3. Soyabean oil cakes.

Figure 4. Mustard oil cakes.

Amla

Phyllanthusemblica, also known as emblic, emblicmyrobalan, myrobalan, Indian gooseberry Malacca tree, or amla from Sanskrit amalaki is a deciduous tree of the family Phyllanthaceae. It has edible fruit, referred to by the same name. Amla leaves are a good source of P, K, Mn.

Figure 5. Amla fruits and leaves.

Termite Mounds

Phosphorus did not differ among mound sections, but it was the nutrient that most differed between the termite mound and the adjacent soil, with higher values in the mound than in the soil. Although there was no significant difference, the P values at the distance of 0.50 m were almost the double of the values at 1.50 m from the mound, which indicates the influence of the mound on its adjacent soil. quantified and systematized phosphorus forms in different sections of termite mounds (exterior wall, internal wall, and center) of different trophic guilds and soils in seven Brazilian ecosystems and concluded that termite activity results in a gross enrichment of phosphorus in labile form in nets, and the composition of P in termite constructions is reflected in their diet. According to Oliveira et al. (2012), the higher amount of P in termite mounds than in the adjacent soil is related to the amount of clay used in their construction, which hinders the loss of P available.

Figure 6. A termite mound.

Banana

The result of mineral content analysis shows the concentration of potassium to be highest (78.10mg/g). The concentration (mg/100g) of calcium, sodium, iron, and manganese were 19.20, 24.30, 0.61 and 76.20 respectively. Phosphorus also present in a good amount in the pulp but less as compare to other abovementioned elements.

Cow Dung

Cow dung is high in organic materials and rich in nutrients. It contains about 3 percent nitrogen, 2 percent phosphorus, and 1 percent potassium (3-2-1 NPK). In addition, cow manure contains high levels of ammonia and potentially dangerous pathogens.

Combined application of cow dung and NPK fertilizer significantly ($P < 0.05$) increased organic matter, available P, exchangeable K and Effective Cation Exchange Capacity compared to the control in both locations. Uptake of N, K, Ca and Mg were higher with treatments in which cow dung was combined with NPK fertilizer compared to the control at both locations. Also, combined application of cow dung and NPK fertilizer significantly ($P < 0.05$) increased sweet potato vine length, number of branches and tuber yield relative to the control. From the results obtained, application of 1.9 t/ha cow dung + 200 kg/ha NPK fertilizer is recommended for sweet potato production in Umudike while application of 0.9 t/ha cow dung + 300 kg/ha NPK fertilizer is recommended in Uturu.

Bael

Aeglemarmelos L., commonly known as bael or golden apple, (Japanese bitter orange, stone apple or wood apple) is a species of tree native to the Indian subcontinent and Southeast Asia.

The bael fruit typically has a diameter of between 5 and 12 cm. It is globose or slightly pear-shaped with a thick, hard rind and does not split upon ripening. The woody shell is smooth and green, gray until it is fully ripe when it turns yellow. Inside are 8 to 15 or 20 sections filled with aromatic orange pulp, each section with 6 (8) to 10 (15) flattened-oblong seeds each about 1 cm long, bearing woolly hairs and each enclosed in a sac of adhesive, transparent mucilage that solidifies on drying. The exact number of seeds varies in different publications.

It takes about 11 months to ripen on the tree and can reach the size of a large grapefruit or pomelo, and some are even larger. The shell is so hard it must be cracked with a hammer or machete. The fibrous yellow pulp is very aromatic. It has been described as tasting of marmalade and smelling of roses. Boning (2006) indicates that the flavor is "sweet, aromatic and pleasant, although tangy and slightly astringent in some varieties. It resembles a marmalade made, in part, with citrus and, in part, with tamarind" [10]. Numerous hairy seeds are encapsulated in a slimy mucilage. The values for various minerals found in bael fruit pulp as P, K, Ca, Mg, Fe, Cu and Zn were 51.6, 603, 78, 4.0. 0.55. 0.19, 0.28 mg per 100g respectively.

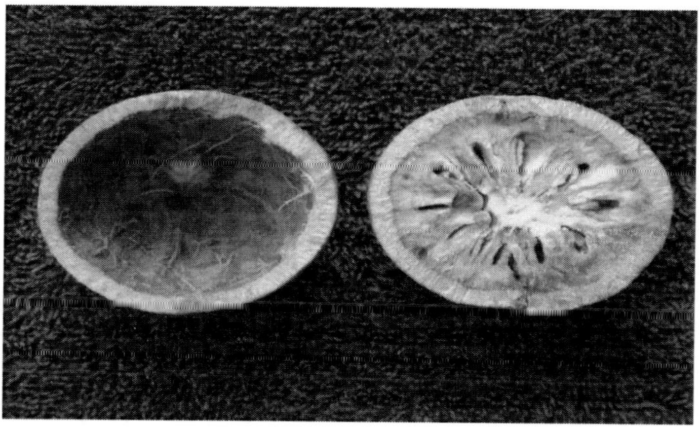

Figure 7. A bael pulp and opened fruit.

Jaggery

A product of sugarcane, is such a product which is rich in important minerals (viz Calcium-40-100 mg, Magnesium-70-90 mg, Potassium-1056 mg, Phosphorus-20-90 mg, Sodium-19-30 mg, Iron-10-13 mg, Manganese-0.2-0.5 mg, Zinc-0.2-0.4 mg, Copper-0.1-0.9 mg, and Chloride-5.3 mg per 100 g of jaggery), and vitamins (viz Vitamin A-3.8 mg, Vitamin B1-0.01 mg, Vitamin B2- 0.06 mg, Vitamin B5-0.01 mg, Vitamin B6-0.01 mg, Vitamin C-7.00 mg, Vitamin D2-6.50 mg, Vitamin E-111.30 mg, Vitamin PP-7.00 mg, and protein-280 mg per 100 g of jaggery).

Different Plant Seeds

- Narrow leaf plant seeds: Rice, wheat, corn etc.
- Pulses seeds: Gram, peas etc.
- Oil crop seeds: Soya, sunflower, peanuts etc.
- Fruit seeds: papaya, guava, orange etc.
- Vegetable seeds: pumpkin, ladyfingers etc.
- Flower seeds: marigold, rose etc.

Different plant seeds can serve the purpose of supplying micronutrients if mixed with bio-fertilizer raw matters. The different plant seeds have different ration of nutrients available, so there is a possibility that the above mentioned group can supply all the micronutrients.

Aerobic Composting

Aerobic composting is decomposition of organic matter using microorganisms that require oxygen. The microbes responsible for composting are naturally occurring and live in the moisture surrounding organic matter. Oxygen from the air diffuses in to the moisture and is taken up by the microbes. As aerobic digestion takes place the by-products are heat, water and carbon dioxide (CO2). While CO2 can be classified as a greenhouse gas it's evolution from the composting process is not counted in emissions. Additionally, CO2 is only 1/20th as harmful to the environment as methane (the main by-product of anaerobic degradation).

The heat produced in aerobic composting is sufficient to kill harmful bacteria and pathogens as these organisms are not adapted to these environmental conditions. It also helps support the growth of beneficial bacteria species including psychrophilic, mesophilic, and thermophilic bacteria which thrive at the higher temperature levels.

From start to finish, the HotRot in-vessel aerobic composting process takes only 8-10 days. No leachate is produced as any surplus moisture is extracted as water vapour which can be condensed and used for watering nearby vegetation.

Anaerobic Composting

Anaerobic composting is decomposition that occurs using microorganisms that no not require oxygen to survive. In an anaerobic system the majority of the chemical energy contained within the starting material is released as methane. The process is characterised by very strong odours and only a small amount of heat is generated meaning decomposition takes much longer and doesn't reach sufficient temperatures to safely kill plant pathogens, weed and seeds. To overcome these limitations external (artificial) heat is normally added.

As the material is broken down by anaerobic digestion, it creates a sludge-like material that is even more difficult to break down. This material, digestate, typically requires aerobic composting to complete the stabilization process.

Anaerobic organisms work at slower rates than their aerobic counterparts, and it's impossible to monitor their progress without digging into the hole and poking around. Anaerobic organisms exude smelly gas as a byproduct of their exertions. And because of the colder conditions, weed seeds and plant pathogens aren't destroyed.

Raw manures can have issues related to their direct use with the right composting or anaerobic fermentation and composting waste with some noxious nature are transformed into valuable inputs and their issues with disposal are greatly minimized by the 80 to 90% consumption of the organic materials. This is critical for our soil fertility and environmental issues related to synthetic nitrogen and use of raw manures which can contaminate our water systems through the eutrophication which results from excess soluble products in the water which ends in robs water of oxygen needed for fishery health and best recreational use.

How to Make Bio-Fertilizers

Producing biofertilizers is a job requires great care. Generally making biofertilizer consists of 6 major steps. They are:

- Choosing active organisms is the first step of making biofertilizers. You must decide which type you prefer from nitrogen-fixing bacteria, organic acid bacteria, and a combination of some other organisms.

- Isolation and selection of target microbes are the second step where the target microbes are separated from their habitation. Generally, the microbes are isolated from plants root or by luring them with a decoy.
- Selection of method and carrier material is the next step of biofertilizer preparation where the isolated organisms are grown on a petri plate, shake flask, and then glasshouse to choose the right one. Along with the selection of right method, selection of right carrier material is also equally important.
- Selection of propagation method is the fourth step of biofertilizer making process. It refers to finding out the optimum condition for the organisms.
- The next step is making and testing prototype.
- The final step is large scale testing, where the fertilizer is widely tested under different environmental conditions to analyze its performance under various conditions.

How to Apply Bio-Fertilizers

- For main field application, 1 kg of inoculants is sufficient for a field of 1 acre. That amount of inoculants is mixed with 20 kg of powdered farm yard manure and then broadcasted on the land. Biofertilizers are applied to the main field just before transplanting a crop.
- For transplanted crops, the roots of the seedlings are dipped in a mixture containing inoculants. Around 400g of inoculants is added to 40 liters of water to prepare the mixture in an amount sufficient for the seedlings to be planted on a 1-acre land. Seedlings required for a 1-acre land are taken and their roots are dipped into the mixture for 5-10 minutes before transplanting them.
- Bio fertilizers are also used for seed treatment after making a slurry with them. 200g of inoculants and 200 ml of rice kanji are mixed together to make the slurry in sufficient amount for a 1-acre land. A uniform coating of inoculants are created over the seeds by treating them with the slurry and then shade drying them for 30 minutes. After shade drying, the seeds need to be sown within 24 hours.

How to Store

- The packets containing biofertilizers must be kept in a cool place which is protected from heat and direct sunlight.
- Room temperature or cold storage condition, any of these two can be used in storing inoculants packed in poly bags or gunny bags.
- The inoculant population in the inoculant containing packets should be checked after each 15 days, as a minimum number of cells must be present per gram of inoculants to keep the fertilizer effective.

Formation of the Waste Decomposer Solution

This is needed to improve the soil health and useful as plant protection agent. A decomposer is in solid form (30 gm), which is a consortium of micro-organism extracted from desi cow dung, is mixed with jaggery and waters (2:200). To make a de-composer solution, that is helpful to make a composting from organic waste, proper mixing (with the help of stirring process) is required for 7-8 days. This solution need to be kept covered for that duration.

(Waste Decomposer made by using NCOF technology Ghaziabad, has been used here, appendix A has mentioned the details of this waste decomposer)

Solutions to most of the problems faced in organic farming like Unproductive soil, Bad Soil health, Pests, Unsatisfactory Fruit quality. Waste decomposer works as Bio fertilizer, Biocontrol and as well as Soil Health Reviver.

It contains beneficial microorganisms from Desi Cow Dung for Soil Health Reviver. It can control all types of soil borne, foliar diseases, insects and pests as plant protection agent.

It can be used in various ways such as quick composting of bio wastes, foliar spray as bio pesticide against most of the plant diseases for all types of agricultural and horticultural crops.

Waste decomposer application can eliminate up to 90% uses of all types of pesticides, fungicides, and insecticide since it controls both root diseases and shoots diseases.

- Step 1: Take one bottle of Waste Decomposer and mix it well in 200 L of water and 2 Kg Jaggery. Keep the solution for a week.
- Step 2: After a week use the solution to spray at your farms/water your land.

Solution can be used again and again from the above formed mixture. For that, 20 Litres of the above solution is added to a drum with 2 kg of Jaggery and 200 Litres of water. And in such a way this can be used again and again.

As Manure

- 18-20 cm thick layers of 1-ton bio-waste such as agricultural waste, kitchen waste, cow dung etc are piled on the ground.
- Wet the waste with solution of the waste decomposer.
- Again another 18-20 cm thick layer of bio-waste is spread and again wet with waste decomposer solution.
- The above processes are repeated till the piling goes 30-45 cm higher.
- Turn the pile at every 7 days interval for uniform composting and add more solution at every turning.

Maintain 60% moisture during the entire period of composting. If required again add the solution. The compost gets ready to use after 30-40 days.

As Bio-Pesticide

The mass multiplied liquid waste decomposer culture is diluted in the ratio of 1:3 with water and applied as a foliar spray to control pest and diseases. It can control all types of soil borne, foliar diseases, insects and pests. Farmers are investing huge finances in purchasing fertilizer, pesticides, herbicides and other manures to revive the soil quality, but all this can be eradicated by one pocket saving solution.

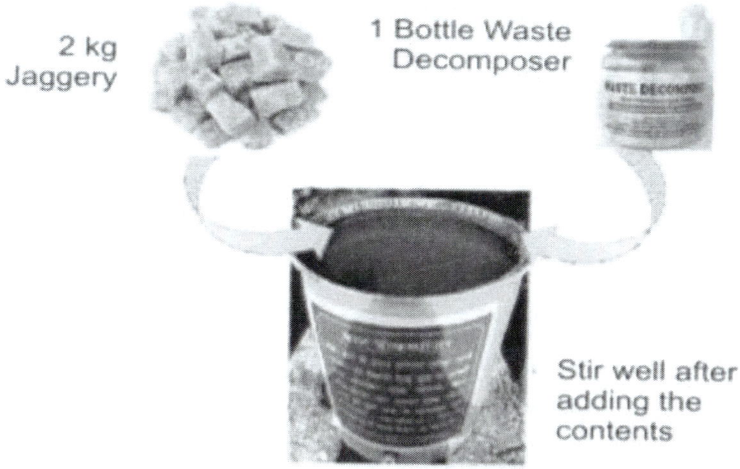

Figure 8. Making of a bio decomposer solution.

Nitrogen Alternate (N) Solution

The carrot grass is easily available in Bilaspur region. The carrot grass due to its harmful ful effects need to be abolished, thus is a good candidate for raw organic material to make bio-fertilizers.

2Kg carrot grass paste can be mixed with 10 Liter of the waste decomposer solution and keep it for 10 days. Once the carrot grass paste is liquefied/melted, filter the solution. This solution fulfils the Nitrogen related nutrients. (The mentioned amount is sufficient for 1 acre).

Phosphorus Alternate (P) Solution

Similar to Nitrogen solution, raw materials need to be easily available to farmers to make P alternate solution. 4 Kg soil beneath the Banyan tree or the Peepal tree, 2kg termite home soil/mounds (or Amla leafs), 2kg remaining of oil seeds after extraction of oil (oil cakes), 2kg cow dung, 250gm Jaggery, 250gm curd and 2-3 kg banana can be mixed with 30 liter of the waste de-composer solution and keep it for 10 days. Once mixture is liquefied/melted, filter the solution. This solution fulfils the Phosphorus related nutrients. (The mentioned amount is sufficient for 1 acre).

Potassium Alternate (K) Solution

Half kg Indian Bael powder, half kg Indian Bael leaf, 5 kg Indian Bael pulp, half kg jiggery can be mixed with 20 liter of the waste de-composer solution. After 10 days, it can be filtered and sprayed. (The mentioned amount is sufficient for 1 acre). This solution fulfils the Potassium related nutrients.

The bio fertilizer used here can contains plant growth hormones or the chemicals replicating the plant hormone functionalities, however plant tissues cannot store these plant growth hormones due to their unstable nature, but in contrast to the same fact Miezah et al. successfully developed a method to identify and found the traces of plant growth hormones present in compost and co-compost organic materials. Looking in to the fact (Miezah et al.) came up with in the part of this proposed ork germinated seeds have been mixed with waste decomposer and NPK alternate solution.

Table 1. Growth rate impact due to catalyst solution with NPK solution

Crop	Area	Improvement in the growth	Plant height and growth	Increment in the production volume
Chickpeas	5-6 Acre	4-5 days early flowering	1-2 cm more height	1-2 quintal per acre
Wheat	5-6 Acre	-	2-3 cm more height	2-3 quintal per acre
Rice	20-25 Acre	More grains per plant	1-2 cm more height	5-6 quintal per acre
Soya bean	2-3 Acre	-	-	2-3 quintal per acre

Conclusion

This proposed catalyst solution is a new approach to enhance the absorbing capability of a plant while using the liquid bio-fertilizers. A small scale experiment has been done for different crops and the expected outcomes have likely observed. A large scale approach is being experimented on different crops. This catalyst solution is useful when bio-fertilizers are being used for a large scale crops, because proposed catalyst solution makes those bio-fertilizers solution very effective even if it is supplied in small quantity.

References

Aliyu, O. M., Adeigbe, O. O. & Awopetu, J. A. "Foliar application of the exogenous plant hormones at pre-blooming stage improves flowering and fruiting in cashew (Anacardiumoccidentale L.)"; *J. Crop Sci. Biotechno,* 14: 143, 2011.

Amarjeet Kaur and Manoranjan Kalia. "Physico Chemical Analysis of Bael (AegleMarmelos) Fruit Pulp, Seed and Pericarp"; *Chemical Science Review and Letters, Chem Sci Rev Lett,* 6(22), 1213-1218, 2017.

Anhwange, Benjamin & Ugye, Joseph & Nyiatagher, T. D. "Chemical Composition of Musa sepientum (Banana) Peels"; *Electronic Journal of Environment, Agriculture and Food Chemistry.* 2009 Vol. 8.

Asawalam, D. O. and Onwudike, S. U. "Complementary Use of Cow Dung and Mineral Fertilizer: Effect On Soil Properties, Growth, Nutrient Uptake and Yield of Sweet Potato (Ipomeabatatas)" PAT June vol. 7 (1) 36-48, 2011.

Bloomfield, J. P., R. J. Williams, D. C. Gooddy, J. N.Cape, P. Guha. "Impacts of climate change on the fate and behaviour of pesticides in surface and groundwater- a UK perspective" *Sci. Total Environ.,* 2006 369 (1–3), pp. 163-177.

Bloomfield, J. P., R. J. Williams, D. C. Gooddy, J. N.Cape, P. Guha; "Impacts of climate change on the fate and behaviour of pesticides in surface and groundwater- a UK perspective"; *Sci. Total Environ.,* 369 (1–3), pp. 163-177; 2006.

El-Habbasha S. F., Hozayn M., Khalafallah M. A. "Integration effect between phosphorus levels and biofertilizers on quality and quantity yield of faba bean (Viciafaba L.) in newly cultivated sandy soils." *Research Journal of Agriculture and Biological Science* 3(6) 966-971.2017.

Frederick T. Addicott; "Plant Hormones"; Wellington, Tuatara: *Journal of the Biological Society Tuatara:* Volume 6, Issue 3; 1957.

Getachew Agegnehu 1 and GirmaTaye, "Effect of plant hormones on the growth and nutrient uptake of maize in acidic soils of the humid tropics"; *Ethiop. J. Sci.,* 27(1).17–24; 2004.

Goel A. K., Laura R. D. S., Pathak G., Anuradha G. and Goel A., "Use of bio-fertilizers: potential, constraints and future strategies review"; *International Journal of Tropical Agriculture,* 17 1–18; 2008.

Gudadhe, Nitin., "Effect of biofertilizers on growth and yield of mustard (Brassica juncea L.)"; *Journal of soils and crops. 1*5. 160-162; 2005.

Kachari, Manisha & Korla, B. N. "Effect of biofertilizers on growth and yield of cauliflower cv. PSB K-1," *Indian Journal of Horticulture.* 2009 66. 496-501.

Kachari, Manisha & Korla, B. N.; "Effect of biofertilizers on growth and yield of cauliflower cv. PSB K-1"; *Indian Journal of Horticulture.* 66. 496-501; 2009.

Kumawat, Narendra. "Role of Biofertilizers in Agriculture." Popular kheti, 2017 5, 63-66.

Mazid, M., Khan, T. A., Mohammad, F. "Potential of NO and H_2O_2 as signaling molecules in tolerance to abiotic stress in plants." *Journal of Industrial Research & Technology,* 2011 1(1), 56-68.

Miezah, K., J. Ofosu-Anim, G. K. O. Budu, L. Enu-Kwesi and O. Cofie; "Isolation and Identification of Some Plant Growth Promoting Substances in Compost and Co-Compost"; *International Journal of Virology,* 4: 30-40; 2018.

Miransari, Mohammad & Smith, Donald; "Plant hormones and seed germination"; *Environmental and Experimental Botany*. 99; 2013.

Mohammed Faisal Ansari Devayani R.Tipre Shailesh R. Dave; "Efficiency evaluation of commercial liquid biofertilizers for growth of Ciceraeritinum (chickpea) in pot and field study"; *Biocatalysis and Agricultural Biotechnology* Volume 4, Issue 1, Pages 17-24; 2015.

Ngampimol, Hasarin & Kunathigan, Viyada; "The Study of Shelf Life for Liquid Bio fertilizer from Vegetable Waste"; *Au J Technol.* 11; 2008.

Patel, Seema; "Harmful and beneficial aspects of Partheniumhysterophorus: an update"; *Biotech* vol. 1, page 1-9; 2011;

Rokhzadi A., Asgharzadeh A., Darvish F., Nourmohammadi G. and Majidi E.; "Influence of plant growth-promoting rhizobacteria on dry matter accumulation and yield of chickpea (Cicerarietinum L.) under field condition"; *Am-Euras. J. Agric. Environ. Sci.* 3(2): 253-257; 2005.

Ross McKenzie "Crop Nutrition and Fertilizer Requirements Soil Fertility/Crop Nutrition"; Alberta Agriculture, Food and Rural Development, Lethbridge.1998.

Uma Maheswari, N., M. Kalaiyarasi; "Comparative Study of Liquid Biofertilizer and Carrier Based Biofertilizer on Green Leafy Vegetables"; *Int. J. Pharm. Sci. Rev. Res.,* 33(1), Article No. 42, Pages: 229-232; 2015.

Yasothai, R., "Chemical composition of sesame oil cake – review"; International Journal of Science, *Environment and Technology,* Vol. 3, No 3, 827-835; 2014.

Chapter 8

IoT-Based Security System Using ESP-32 and Lasers and It's Various Applications

Payal Bansal[1,*], Rajeev Kumar[2,†], Brojo Kishore Mishra[3,‡] and Devendra Somwanshi[1,§]

[1]Poornima College of Engineering, Jaipur, India
[2]Chandigarh University, Chandigarh, India,
[3]GIET University, Gunupur, India

Abstract

The agriculture business depends on wind with restricted water assets, so water systems are utilized in the farming area. In the water system framework, contingent upon the kind of soil, water is accommodated planting. In farming, two things are vital, first and foremost to get data about soil richness and besides to gauge soil dampness. Today, different techniques are accessible which are utilized to decrease precipitation reliance. Furthermore, particularly this cycle is driven by electrical power and turn on/off setting. In this interaction, the water level pointer situated in the water stockpiling region and the dirt dampness sensors is supplanted by the plant roots and close to the module and entryway unit it catches the sensor data and sends the information to a regulator that controls the progression of water through valves.

[*] Corresponding Author's Email: payal.bansal@poornima.org.
[†] Corresponding Author's Email: Rajeev2009mca@gmail.com.
[‡] Corresponding Author's Email: brojomishra@gmail.com.
[§] Corresponding Author's Email: imdev.som@gmail.com.

In: Computational Intelligence for Sustainable Development
Editor: Brojo Kishore Mishra
ISBN: 979-8-88697-198-9

1.1. Food Innovation

To develop maintainable interest and decrease the stock of food needs, it is essential to quickly work on the creation of food innovation. Farming is the main hotspot for this. This is a significant component in human social orders in the development and interest for adaptability in food creation. Horticulture assumes a significant part in the economy and advancement, as in India. Because of the absence of water and the deficiency of groundwater causing a diminishing in groundwater volume, the rancher utilizes the water system. The water system can be characterized as the study of watering in the dirt or in the dirt which implies that relying upon the kind of soil, the plant ought to be provided with water.

1.2. Area of Utility

- The primary focus of this project is to help the farmers and reduce their work.
- This module can be implemented in perennial plant irrigation land and gardening land.

1.3. Embedded Sensors

In the water system field, soil dampness sensor, temperature sensors are embedded into the base of the plant and the microcontroller handles tangible data and communicates information. One calculation was created to quantify the temperature scope of the temperature sensor and the dirt dampness sensor which was customized into a little volume regulator for water volume.

A robotized water system framework model in view of microcontroller and sunlight-based power was utilized uniquely for power source. Different sensors are set in the paddy field. The sensors sense the water consistently and furnish the rancher with data by means of cell. The rancher controls the motor utilizing the mobile phone without going to the paddy field. Assuming the water level arrives at a degree of risk, the motor will close down naturally without the rancher's assent.

1.4. Basics of IoT

1.4.1. Internet of Things

Web of Things (IoT) is an organization of articles, vehicles, structures and other inserted objects and electronic gadgets, programming, sensors, and organization associations that permit these items to gather and trade information. IoT permits objects to be heard and somewhat controlled across existing organization foundation, sets out open doors for virtual direct mix into PC based frameworks, and prompts further developed execution, exactness and monetary advantages, where IoT is added with sensors and actuators, innovation turns into a good example the most well-known of digital actual frameworks, which likewise incorporate advances like savvy matrices, shrewd homes, brilliant vehicle and savvy urban areas. Everything looks exceptional with its inserted PC program yet can associate inside existing Internet foundation.

1.4.2. IoT Cloud Interface

Thus, Internet of Things or IoT is an engineering that contains particular equipment sheets, Software frameworks, web APIs, conventions which together establishes a consistent climate which permits brilliant installed gadgets to be associated with web to such an extent that tactile information can be gotten to and control framework can be set off over web and gadgets can be associated with the web utilizing different means like Wi-Fi, Ethernet and then some. Moreover, gadgets may not be needed to interface with the Internet autonomously. Rather a bunch of gadgets (for instance network sensor) might be made and the essential channel or group head can be associated with the Internet. This prompts more equivocal development of correspondence processes going from high to low.

Most intriguing is that these things must be found in an alternate manner. To get various gadgets on the Network, they need to have an alternate IP address. IoT gadgets really have an IPv6 address framework. These gadgets have virtual IP or Subnet addresses covered by v6 type. Interesting IP tends to make IoT gadgets accessible online as a private space. This is the main idea you ought to have as a primary concern to comprehend IoT.

1.4.3. What IoT Is All About?

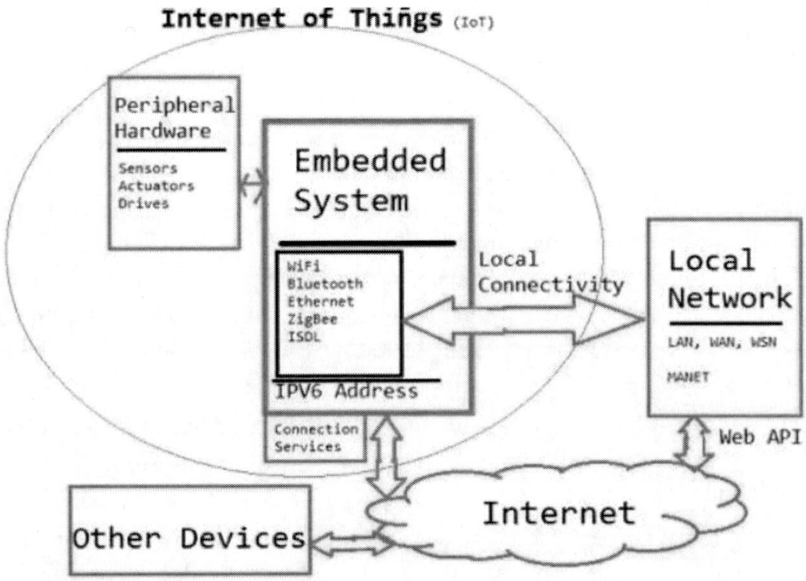

Figure 1. Internet of Things basic architecture.

1.4.4. What Devices Makes It to IoT?

Since IoT is inserted projects and shrewd web associations with a special IP address that can be gotten to and associated with the web. We additionally saw that IoT gadgets might have outer peripherals like Actuator and Sensors.

1.4.5. Are Mobile Phones IoT Devices?

One of the most widely recognized things in day-to-day existence is PDAs. Portable is an installed framework with a processor in setting with a presentation and keypad. They support an assortment of sensors, for example, encompassing light sensors, Accelerometer, Gyroscope and then some. They are associated with the Internet. Mobile phones get IP locations, and they can get to the Internet. As such it rises to each IoT definition.

IoT is thusly a web association for savvy gadgets and an implanted framework with next to no cell phones that can be associated with outer equipment and Mobiles, Tablets, Laptops and PCs are the IoT control/passageway.

1.4.6. IoT Devices

The most common and popular technologies in IoT will give an overview device. IoT gadgets are in two general classes: Approachable and IoT inserted Microcontroller/Microprocessor gadgets. A few inserted gadgets like the ArduinoLillypad are little and can be utilized consistently to make a wearable arrangement.

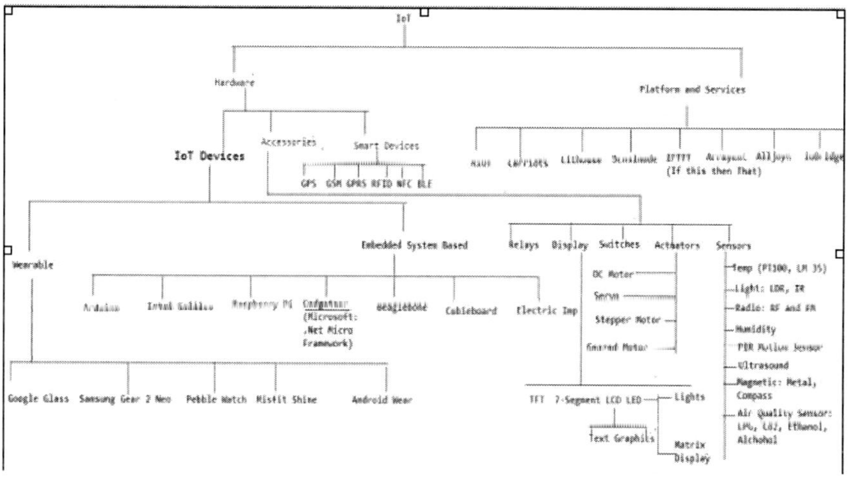

Figure 2. Common IoT devices & techniques.

However, the wearables incorporate top notch equipment and IoT just have programming engineer programming. A portion of the PC equipment you might require is IoT equipment inserted. The applications can be utilized with well-known jungle gyms, IoT Embedded stage might consolidate numerous advancements like Raspberry Pi, Arduino or Galileo, and so forth.

1.4.7. IoT Platforms

IoT improvement can be isolated into two comparative advances: Clothes and inserted. Designers can fabricate applications for custom wearable gadgets, for example, Peeble, Samsung Gear or they can for the most part assemble their own foundation utilizing Embedded Solution and can work on the presentation of that stage.

1.4.8. Wearable Platform

Tizen is possibly becoming the most well-known versatile stage and cell phone. The Tizen SDK accompanies a removable format that makes it simple to foster compact answers for the Tizen stage.

As proposed, the bigger Android Wear gadget is currently being updated and promoted. Shrewd watches are turning out to be increasingly more well-known each day. Android Wear applications can be updated and tried on Eclipse. This is an Android Developer Guide to assist you with setting up an Android Wear advancement site on Eclipse.

Salesforce is one more stage that accompanies an astounding advancement stage, APIs in wearable innovation. Their answer goes from Google Glass.

1.4.9. Embedded Platform

Arduino is most likely the best previously inserted IoT based site. Essential Arduino sheets don't accompany Ethernet safeguard or Wi-Fi safeguard and for Arduino to work as an IoT gadget, it's necessary to choose Arduino with Ethernet safeguard or Wi-Fi safeguard. Then again there is the Arduino UNO board that comes furnished with an Ethernet safeguard.

The Raspberry Pi is probably the best thing that occurs in DIY IoT. For an assortment of Data-driven applications, for example, Home Automation Server to Home Multimedia server, File Server can be updated with Pi. Arduino-like PI has standard IO pins. Be that as it may, consistent work with the nerves annoys Pi.

Another powerful IoT board is Intel Edition which consolidates BLE and Wi-Fi among numerous different elements. Upholds a wide scope of industry-standard PC equipment (north of 30) utilizing a 70-pin virtual connection point.

Intel Galileo is another extraordinary Intel offering that upholds a similar safeguard as Arduino Uno. It can along these lines be supposed to be the principal Intel controlled gadget viable with Arduino. It has in addition to other things a USB dealing with regulator like the Raspberry Pi which makes this an alluring equipment. Galileo additionally has an inherent ethernet safeguard.

1.4.10. Cloud Platform

IoT can genuinely bring a few organizations (like an electronic portion doorway), several gear stages (like an embedded dashboard) and splendid contraptions and data like NFC, GPS in a predictable environment.

It can now fuse an electronic portion into a reward sweets machine, if one uses the close by drink expert association, then, using the region and portion organization ought to be conceivable. One can get clinical logical data like an ECG (procured by another embedded electronic clinical contraption board) in a cloud with the objective that a couple of experts can see it and make an exhaustive viewpoint on the patient's condition.

Taking everything into account, they are in general possible. To some degree cognizance of the web and programming design can take your mind to the fogs. Like the Mechanical Web, Machine to Machine (M2M) or Machine to Material (M2O) and any similar affiliations several typical modules and a couple of modules anticipate that data should be open for sharing.

Cloud APIs help in this. For example, to make a device available on the web, then, give a respectable IP address, track and follow a couple of social capacities. Yaler is an inconceivable outline of what organizations and frogs can offer that may be valuable. This gives the network as an assistance to make your device adequately open and connected with the web missing a great deal of issue and with immaterial security.

1.4.11. Axeda Provides Infrastructure for M2M Architecture

Open IoT is an IoT open-source stage that gives an individual of a sort Sensing organizations as an assistance. Google has at this point fused neighborhood resources with its cloud. Region wiped out from your contraptions is discreetly associated with your notification on facebook and twitter and used for individual request. So cloud APIs have exceptional IoT limits at all plan levels from firmware to state of the art designing.

1.5. Introduction of Hardwares

1.5.1. Arduino

Arduino is an open-source real enlisting stage reliant upon essential data/yield board and a progression environment that executes the Processing language. Arduino can be used to make free instinctive things or can be related to programming on your PC. The sheets can be assembled the most difficult way possible or purchased preassembled; the open-source IDE (Integrated Development Environment) can be downloaded for nothing from. Introduction to Arduino Boards.

Arduino is a designer that joins the Atmel microcontroller family with standard gear into a board with inbuilt bootloader for fitting and play embedded programming. Arduino Software goes with an IDE that helps forming, researching and consuming framework into Arduino. The IDE furthermore goes with a Serial Communication window through which can without a doubt get consecutive data from the board.

1.5.2. Arduino Uno

The Uno is a microcontroller board reliant upon the ATmega328P. It has 14 progressed data/yield pins (of which 6 can be used as PWM yields), 6 basic data sources, a 16 MHz quartz pearl, a USB affiliation, a power jack, an ICSP header and a reset button. All of the 14 progressed pins can be used as information or result, using pin Mode, digital Write, and digital Read limits. They work at 5 volts. Each pin can give or get 20 mA as proposed working condition and has an inside pull-up resistor (withdraw obviously) of 20-50k ohm. A restriction of 40mA is the value that should not be outperformed on any I/O pin to avoid enduring mischief to the microcontroller.

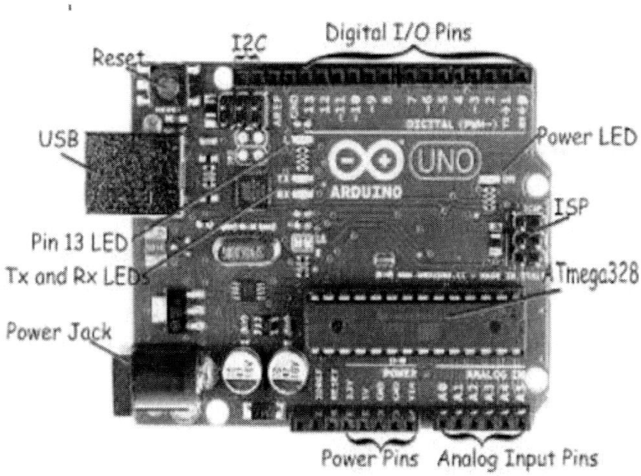

Figure 3. Arduino Uno.

In addition, a couple of pins have explicit limits:

- Successive: 0 (RX) and 1 (TX). Used to get (RX) and send (TX) TTL successive data. Outside Interrupts: 2 and
- These pins can be set up to trigger a ruin on a low worth, a rising or falling edge, or a change of critical worth.
- SPI: 10 (SS), 11 (MOSI), 12 (MISO), 13 (SCK). These pins support SP correspondence using the SPI library. Driven: 13. There is an intrinsic LED driven by automated pin 13.
- TWI: A4 or SDA pin and A5 or SCL pin. Support TWI correspondence using the Wire library.

The Uno has 6 basic wellsprings of data, checked A0 through A5, all of which give 10 bits of objective (for instance 1024 unmistakable characteristics). As is normally done they measure from ground to 5 volts, but is it possible to change the upper completion of their scope using the AREF pin and the analog Reference () work.

There are a few distinct pins on the board, AREF Reference voltage for the basic wellsprings of information. Used with simple Reference (). Reset. Bring this line LOW to reset the microcontroller. Regularly used to add a reset button to shields which block the one on the board.

1.5.3. Arduino Uno Technical Specifications

Microcontroller	ATmega328P
Operating Voltage	5V
Input Voltage (recommended)	7-12V
Input Voltage (limit)	6-20V
Digital I/O Pins	14 (of which 6 provide PWM output)
PWM Digital I/O Pins	6
Analog Input Pins	6
DC Current per I/O Pin	20 mA
DC Current for 3.3V Pin	50 mA
Flash Memory	32 KB (ATmega328P) of which 0.5 KB used by bootloader
SRAM	2 KB (ATmega328P)
EEPROM	1 KB (ATmega328P)
Clock Speed	16 MHz
Length	68.6 mm
Width	53.4 mm
Weight	25 g

Figure 4. Technical specifications.

1.5.4. ESP8266

1.5.4.1. Introduction to ESP8266

ESP8266 is a gotten done and autonomous Wi-Fi network course that can pass on programming applications, or through another application processor uninstall all Wi-Fi sorting out capacities. When the contraption is mounted and as the fundamental utilization of the application processor, the flicker memory can be started directly from an external Move. Implied store memory will help with additional creating system execution and lessening memory necessities. Another situation is when far off Internet access acknowledge the task of Wi-Fi connector, you can add it to any microcontroller-based arrangement, and the affiliation is direct, just by SPI/SDIO point of association or central processor AHB length interface. Dealing with and limit on ESP8266 solid piece, it will in general be facilitated through GPIO ports sensors and various applications unequivocal stuff to achieve the most diminished first thing in the new development and movement of at minimum incorporate framework assets. The ESP8266 essentially arranged chip, including radio wire switch balun, power the board converter, so with irrelevant external equipment, and joins front-end module, including the entire game plan expected to restrict the space required by PCB. The system is outfitted with ESP8266 showed driving components are: energy saving VoIP quickly switch between the rest/wake plans, with low- power action flexible radio inclination, front-end signal taking care of limits, examining and radio structures correspond credits kill cell/Bluetooth/DDR/LVDS/LCD block.

1.5.5. Block Diagram of ESP8266

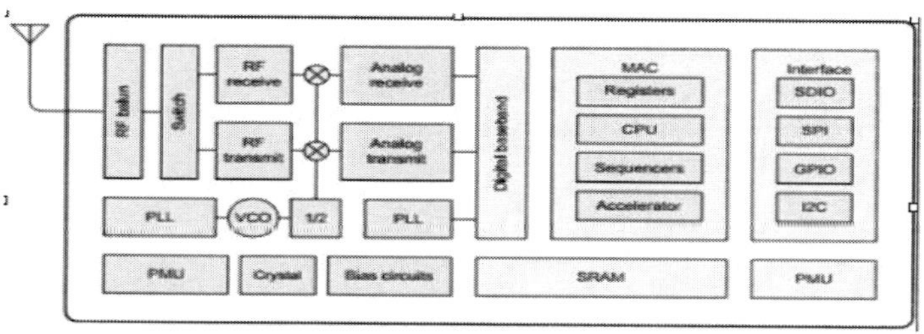

Figure 5. Block diagram of ESP8266.

1.5.6. Characteristics of ESP8266

- 802.11 b/g/n
- Wi-Fi Direct (P2P), delicate AP
- Underlying TCP/IP convention stack
- Underlying TR switch, balun, LNA, power intensifier and matching organization
- Implicit PLL, voltage controller and power board parts
- 802.11b mode + 19.5dBm result power
- Inherent temperature sensor
- Support receiving wire variety
- Off spillage current is under10uA
- Inherent low-power 32-digit CPU: can serve as an application processor

- SDIO 2.0, SPI, UART
- STBC, 1x1 MIMO, 2x1 MIMO
- A-MPDU, A-MSDU total and the 0.4 Within wake
- 2ms, interface and move information bundles
- Reserve power utilization of under 1.0mW(DTIM3)

1.5.7. Schematic Diagram ESP8266-EX

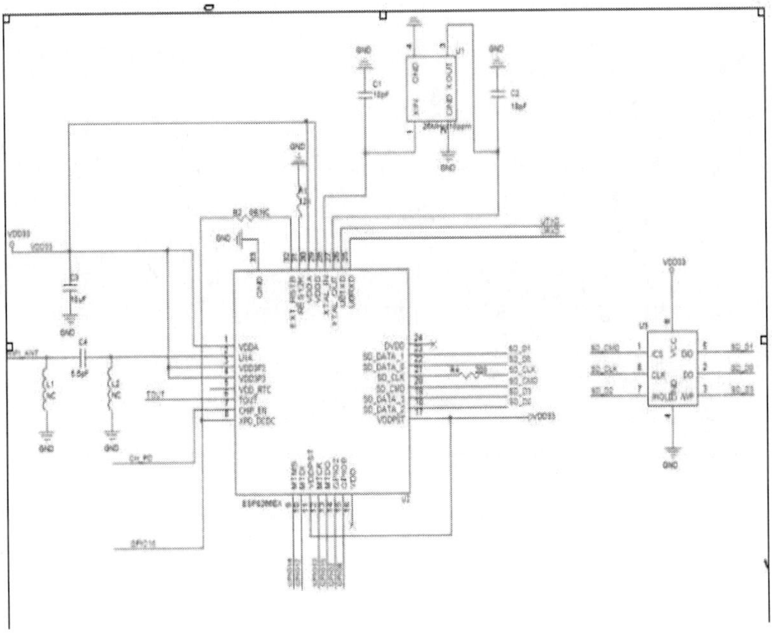

Figure 6. Schematic diagram of ESP8266.

1.5.8. ESP Modules

The ESP8266 is a minimal expense Wi-Fi chip with full TCP/IP stack and microcontroller capacity conveyed by Shanghai-based Chinese maker, Espress if. These were the main series of modules made by outsider producer, AI-Thinker with the ESP8266 and remain the most comprehensively available.

1.5.9. ESP8266 Applications

- Smart Power Plug
- Home Automation
- Industrial wireless control
- Baby Monitor

1.5.10. Explore ESP8266 Wi-Fi Module

The ESP8266 ESP12E Wi-Fi Module is easier to use with the Explore ESP8266 Wi-Fi Module. It fits on a breadboard with all pins taken out. The module goes into programming mode with a solitary reset switch.

Features:

- All pins of ESP12E takeout.
- Separate serial pins break out compatible with FTDI cable layout.
- On-board LM1117-3.3V regulator.
- Works with Arduino IDE for ESP8266.
- Programs can easily dump using USB to TTL converter.
- ESP8266 ESP12Efeatures.

1.5.11. Schematic Diagram of Explore ESP8266 Wi-Fi Module

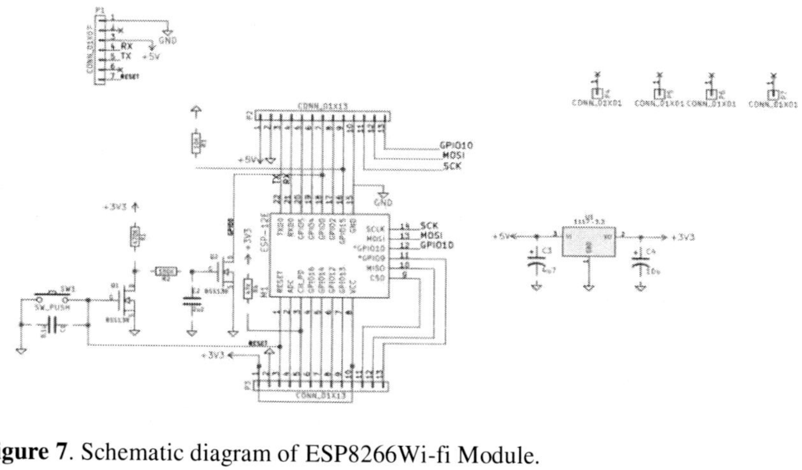

Figure 7. Schematic diagram of ESP8266Wi-fi Module.

1.5.12. AT Commands

ESP8266, in its default setup, boots up into sequential modem mode. In this mode you can speak with it utilizing a bunch of AT orders. AT orders depend on the Hayes Command Set.

Index of all known AT commands is given in the table (Figure 9) below.

Basic	Wi-Fi layer	TCPIP Layer
AT	AT+CWMODE	AT+CIPSTATUS
AT+RST	AT+CWJAP	AT+CIPSTART
AT+GMR	AT+CWLAP	AT+CIPSEND
AT+GSLP	AT+CWQAP	AT+CIPCLOSE
ATE	AT+CWSAP	AT+CIFSR
	AT+CWLIF	AT+CIPMUX
	AT+CWDHCP	AT+CIPSERVER
	AT+CIPSTAMAC	AT+CIPMODE
	AT+CIPAPMAC	AT+CIPSTO
	AT+CIPSTA	AT+CIUPDATE
	AT+CIPAP	+IPD

Figure 8. AT command table.

1.5.13. Soil Moisture Sensor

This moistness sensor can examine how much suddenness' present in the soil including it. It's a low-tech sensor, but incredible for actually looking at a metropolitan nursery, or your pet plant's water level. This is a verifiable prerequisite to have instrument for a related nursery. This sensor uses the two tests to go current through the soil, and a short time later it scrutinizes that insurance from get the suddenness' level. More water makes the soil direct power even more adequate (less obstruction), while dry soil conducts power insufficiently (more resistance). It will be valuable to remind you to water your indoor plants or to screen the soil.

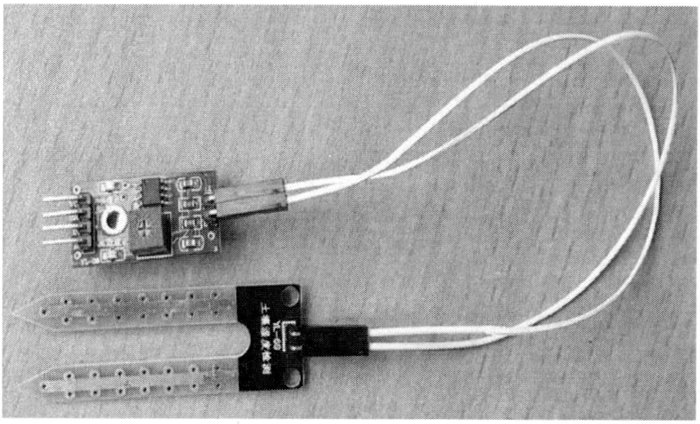

Figure 9. Soil Moisture Sensor YL-69 YL-38.

A Chinese manufactured YL-69 sensors (Figure 9) go with a 'middle man' circuit which licenses to get two outcomes: one is a basic readout of the obstacle between the sensor's tests and the second is a high-level outcome (essentially, HIGH or LOW, 5v or 0v) dependent upon whether the moistness is above or under an edge which can along these lines be changed by a basic POTS. The YL-69 sensor has two pins which ought to be wired to be the two pins on the YL-38 Bridge. On the contrary completion of the YL-38 have four pins which address VCC, GND, D0 and A0. VCC and GND are power pins which should set to 3.3/5V and ground independently. A0 is a straightforward outcome. D0 is a high-level outcome.

1.5.14. Submersible Water Pump

A submarine siphon (or sub siphon, electric sub siphon) (Figure 3.8) is a device which has a hermetically sealed fixed motor close-coupled to the siphon body. The whole social affair is brought in the fluid down to be siphoned. The essential advantage of this sort of siphon is that it thwarts siphon cavitation's, an issue related with a skyscraper qualification among siphon and the fluid surface. Little DC Submersible water siphons push fluid to the surface instead of flying siphons pulling fluids. Subs are more useful than stream siphons.

Figure 10. Submersible water pump.

Specifications:

- Voltage: 2.5-10V
- Maximum lift: 40-110cm /15.75”-43.4”
- Flow rate: 80-120L/H
- Outside diameter: 7.5mm /0.3”
- Inside diameter: 5mm / 0.2”
- Diameter: Approx. 24mm /0.95”
- Length: Approx. 45mm /1.8”
- Height: Approx. 30mm /1.2”
- Material: Engineering plastic
- Driving mode
- DC design
- Magnetic driving
- Continuous
- Working life for 500hour.

Conclusion

The agribusiness field is being checked by Thing Speak API. The ESP8266 is the device at field end which receives the messages from Thing Speak API association and controls it and will fill the job referred to in message. After it will send the messages to Thing Speak API association and subsequently it will be conveyed to the (customer end). The ESP8266 is the best contraption for IoT projects. Since it is nearly nothing, moderate, lightweight, viably programmable, and adequately installable and has enough GPIO pins to use them.

In like manner, the splendid water framework structure is feasible and useful for smoothing out the water resources for agrarian creation. This kind of water framework structure grants improvement in places with water deficiency consequents.

References

Dattatraya, P. Y., J. Agarkhed and S. Patil, "Cloud helped execution upgrade of brilliant applications in Wireless Sensor Networks," *2016 International Conference on Wireless Communications, Signal Processing and Networking* (WiSPNET), Chennai, 2016, pp. 347-351.

Deals, N., O. Remédios and A. Arsenio, "Remote sensor and actuator framework for brilliant water system on the cloud," *2015 IEEE second World Forum on Internet of Things* (WF-IoT), Milan, 2015, pp. 693-698.

Gubbi, Jayavardhana, RajkumarBuyya, SlavenMarusic, MarimuthuPalaniswami, "Web of Things (IoT): A dream, structural components, and future headings," *Future Generation Computer Systems*, vol 29, ELSEVIER 2013, 1645-1660.

Nakutis, Z., Deksnys, V., Jaruevicius, I., Marcinkevicius, E., Ronkainen, A., Soumi, P., Nikander, J., Blaszczyk, T., & Andersen, B. "Far off Agriculture Automation Using Wireless Link and IoT Gateway Infrastructure," *2015 26th International Workshop on Database and Expert Systems Applications (DEXA)*, Valencia, 2015, pp. 99-103.

Pavithra D. M. S. Srinath "GSM based Automatic Irrigation Control System "for Efficient utilization of Resources and harvest Planning by utilizing an Android Mobile" IOSR *Journal of Mechanical and Civil Engineering* (IOSR- JMCE), e-ISSN: 2278-1684, p-ISSN: 2320-334X, Volume 11, Issue 4 Ver. I (Jul-Aug. 2014), PP49-55.

Sensor based Automated Irrigation System with IoT" *International Journal of Computer Science and Information Technologies*, ISSN : 0975-9646, Vol. 6 (6) 2015,5331-5333.

Sharma, Deepak, Amol P Bondekar, AmritcshOza, Awdhesh Kumar Shukla, C Ghanshyam, "A Technical Assessment of IoT for Indian Agriculture Sector," *47th Mid-Term Symposium on Modern Information and Communication Technologies for Digital India*, Chandigarh; Research Gate, April2016.

Sneha Angal, "Raspberry pi and Arduino Based Automated Irrigation System" *Worldwide Journal of Science and Research (IJSR)* ISSN (Online):2319-7064.

Wang, Ning, Naiqian Zhang, Maohua Wang, "Remote sensors in agribusiness and food industry-Recent turn of events and future viewpoint," *Computers and Electronics in Agriculture* 50, pp. 1-14, 2006.

Yogesh G. Gawali, Devendra S. Chaudhari, Hitendra C. Chaudhari "Robotized Irrigation System utilizing Wireless Sensor Network" *Global Journal of Advanced Research in Electronics and Communication Engineering (IJARECE) ISSN :* 2278 - 909X Volume 5, Issue 6, June201

Chapter 9

Column Test Methodology and Its Versatile Applications in Geo-Environmental Engineering

J. Sumalatha*

Ramaiah Institute of Technology, MSR Nagar, Bengaluru, India

Abstract

Migration of pollutants from soil to waterbodies is a major problem which affects the ecosystem. The plants grown on the polluted lands and the direct contact with this land causes serious health problems. The fate of these pollutants in the soil can be evaluated through column tests. The soil column tests can be performed as different variants depending on the duration restriction, resources and its applications. The outcomes of column tests can be used to calculate the rates of contaminant transport in soil, which are required to determine the thickness of the soil barrier system to prevent the migration of contaminants with a view to design appropriate soil type and its thickness. Further, the results of column test can be used to select appropriate fluid to remediate contaminated soil by soil washing, efficiency of selected fluid, duration of washing and to estimate the volume of fluid required for washing the field soil. Contaminant transport parameters (diffusion coefficient and retardation factor) are commonly determined through column tests. Breakthrough curves are obtained through column tests which are useful for the calculation of transport parameters. The transport parameters are predicted by comparing the experimental breakthrough curve with the theoretical curve generated using an advection dispersion model and with assumed transport parameters. The transport parameters, thus obtained

* Corresponding Author's Email: latha.msrit@gmail.com.

In: Computational Intelligence for Sustainable Development
Editor: Brojo Kishore Mishra
ISBN: 979-8-88697-198-9

are taken as the parameters for any given contaminant and for the given medium. The transport parameters can find many different applications. This paper presents their applications for the design of clay liner for landfills, appropriate flushing fluid for the remediation of contaminated soil by washing and suitable amendment to immobilize the contaminants in the soil.

Keywords: column test, column leaching test, dispersion coefficient, distribution coefficient, red soil, HCl, EDTA, lime, cement

Introduction

The problems of decline in the soil fertility due to soil erosion and the soil degradation due to metal contamination are the major threats to the future world. The industrial activities, agricultural activities and waste disposal are the major sources of soil contamination. The heavy metals in the contaminated lands is a serious threat to the human health as they can easily enter the food chain by initially entering into the plants which are grown on polluted lands. Apart from the heavy metals, nano-plastics also migrate in the environment for longer distances due to their smaller size and low density. The bio-availability of these nono-plastics which have stronger affinity to organic matter is another serious threat to the living beings (Yasir et al., 2022; Zhang et al., 2022). The contamination of soil due to veterinary antibiotics was also observed in the farmlands due to the application of fertilizers in the field. They can easily migrate with rainfall and impacts the ecosystem (Gbadegesin et al., 2022). The containment of waste facilities and remediation of polluted lands require the knowledge of contaminant transport through soils. The field conditions can be simulated using laboratory column test to assess the migration processes and rates of the pollutants. Column tests are widely used to find the rates of contaminant transport, which are having wide applications in Geo-environmental engineering, particularly in the case of protecting soil and ground water from pollution. One of the important ways by which soil and ground water contamination occurs is by migration of leachates from waste piles and impoundments apart from direct discharge of effluents from industries and domestic sewage. To restrict the contaminants of land fill migrating into nearby waterbodies, a barrier system is required around the waste disposal facilities. Soil liners usually form an important component of the barrier system. The thickness of the mineral liner required for landfills depends on the rates of migration of the critical pollutants of the leachate.

The processes involved in contaminant transport through a soil liner includes advection, dispersion and are affected by retardation of solute mass. The Advective-Dispersive transport can be used to assess two important parameters i.e., Dispersion coefficient (D) and Distribution coefficient (K). These two parameters are called contaminant transport parameters which are essential for contaminant transport modelling. This modelling helps geo-environmental engineers to design the thickness of soil liners for landfills and slurry walls and also for the design of soil remediation techniques. The column tests can be conducted under two test conditions, constant source concentration and variable source concentration. In the variable source concentration test condition, the variations in influent concentrations can also be used to estimate the transport parameters. The contaminant transport through soil was studied by many researchers in the past (Yong et al., 1992; Shackelford, 1993; Moncada, 2004; Ochola et al., 2004; Safari et al., 2012; Ekeleme et al., 2021; Mirza et al., 2021). The contaminant transport through composite liner system considering the consolidation was studied by Yan et al., 2022 and Xie et al., 2022. The contaminant transport parameters are usually estimated through laboratory experiments such as column tests (Jaramillo et al., 2019; Pietrzak et al., 2019; Hameed et al., 2021; Mondal et al., 2021; Schubl et al., 2021). In this study, the transport parameters of metal ion zinc were determined through locally available red soil using column test. This paper reports the use of transport parameters in the determination of not only the thickness of the liners, but also for the remediation of contaminated soils.

Remediation of sites subjected to heavy metal contamination is essential as they cause threats and hazards to humans (McLaughlin et al., 2000; Khalkhaliani et al., 2006; Ling et al., 2007; Hirokazu et al., 2010; Wuana and Okieimen, 2011). The use of compost to remediate the soil contaminated with organic pollutants is an economical soil remediation method where the microbes are supplemented to the soil in the form of compost (Ren et al., 2018; Hijri and Marc, 2019; Li et al., 2019; Liu et al., 2019; Ye et al., 2019). Various green remediation techniques can also be used to treat the polluted lands. Phulpoto et al., 2020 studied the use of micro-organisms to treat the oil contaminated soil. David et al., 2019 studied the use of sulphur modified rice husk biochar to treat the mercury contaminated land. Nanotechnology in soil remediation was studied by Kam et al., 2009, Qian et al., 2020, Souza et al., 2020; Ebrahimbabaie et al., 2021; Kristanti et al., 2021 and Yadav 2021. Soil washing is one of the important methods of remediation of contaminated soil and various chelating agents and acids can be used to remove the heavy metals from the soil (Korf et al., 2011; Akahane et al., 2013; Kogbara and Reginald, 2014; Ma et al., 2014; Bilgin et al., 2016; Piccolo et al., 2019). In the present study, soil washing with 0.1NHCl, 0.1N EDTA, 0.1N EDTA + 0.1NHCl and Ferric Chloride ($FeCl_3$) solutions, has been studied on a zinc contaminated soil by conducting column leach tests. The transport parameters were also determined for the selected fluid which are beneficial to calculate the quantity of fluid required in the field and to predict/ model the contaminant transport at the site if untreated.

Though the soil washing technique has advantages such as complete removal of metals from the soil when compared with immobilization technique, it is not a feasible method for less permeable soils such as clay. In clayey soils, since the permeability is very less, it is very difficult to remove the pollutants by washing technique. In such soils immobilization technique can be effectively used to treat the contaminated soil by converting the metals to their hydroxide form. In the stable hydroxide form, the metals will remain immobilized and do not cause any threat to the environment. The immobilization of metals in soil was studied by several researchers (Czupyrna et al., 1988; Lin et al., 1998; Trakal et al., 2011; Zeng et al., 2010; Hong et al., 2022; Qian et al., 2022; Sumalatha et al., 2022). The conversion of metals into their hydroxide form mainly happens due to the increase in the pH value of the soil. Amendments (either organic or inorganic) which increase the pH of the soil can be effectively used to immobilize the metals. In the present study, the immobilization technique on landfill soil is studied using the chemicals, lime, sodium hydroxide (NaOH) and cement. The efficiency of amended soil mixtures was studied to know the suitability of the amendments in the field where the rainfall on the amended soil should not cause reversal of the process or migration of the immobilized metals. To know the efficiencies of the selected amendments, Column Leaching tests were conducted on the amended soils. This paper discusses in detail the importance of column leaching tests in remediation of contaminated sites.

Materials and Methods

Soils

The properties of red soil and contaminated soils are presented in Table 1. The red soil as per Indian Standard (IS) classification was identified as low plasticity clay (CL). This soil was used to conduct the series of column tests and estimate the migration rates of zinc under different test conditions and densities. The industrial soil which was contaminated with zinc was studied to know the efficiency of soil washing method of remediation. The zinc content in the zinc contaminated industrial soil was measured as 22090 mg/kg of soil which is relatively high. This zinc contaminated soil as per IS classification was classified as Silty sand (SM). This soil is used to study the soil washing technique as it is having good permeability to flush the metals from the soil. The landfill soil which is having various heavy metals was identified as low plasticity clay (CL) as per the IS classification system. The quantities of heavy metals such as cadmium, chromium, copper, iron, lead, nickel and zinc in this landfill soil were, 1.8, 21.2, 89.2, 213.6, 5.4, 34.8 and 101.3 mg/kg respectively. Since it is a clayey soil, this soil is used to study the immobilization technique with different inorganic chemicals.

Table 1. Characteristics of red soil and zinc contaminated soil

Soil	Specific Gravity	Liquid Limit (%)	Plastic Limit (%)	Shrinkage Limit (%)	Maximum dry density (g/cc)	Optimum water content (%)
Red Soil	2.67	44.6	23.8	15.8	1.78	18.4
Zinc contaminated soil	2.77	38.2	22.6	18.5	1.80	19.3
Landfill soil	2.68	31.2	24.2	14.6	1.69	22.5

Column Tests

The column test apparatus (Figure 1) is used under two types of test conditions, constant source concentration and decreasing source concentration. The concentration of chemical solution in the overhead tank needs to be maintained constant for conducting the constant source concentration test where as in decreasing source concentration test, the level of source solution is maintained by adding distilled water to the chemical solution which decreases the concentration of source solution with time. The type of test condition can be chosen based on the field condition. To conduct the test, the soil of required density is placed in the column and it is allowed to saturate with distilled water. After the saturation, the chemical solution of required concentration was passed through the soil. The effluent from the soil column was collected periodically to measure the concentrations with time and to plot the experimental breakthrough curve. The theoretical breakthrough curve is then prepared with assumed values of D and K and the iterative calculation was done to match with the experimental curve and thus the transport parameters were obtained.

In case of low permeability soils where the effluent volume was very low, quick column test can be conducted and the concentration profile can be prepared by using soil sectioning method. For high plasticity clays where the conduction of column test takes several months, the density of soil in the column can be reduced and the transport parameters can be estimated. To get the transport parameters corresponding to maximum density, it requires a series of column experiments with different densities and the corresponding transport parameters need to be estimated for all the samples. A best fitting curve can thus prepared for D and K corresponding to different densities and then it can be used to obtain the parameters corresponding the required density of soil at the site. This technique along with the soil sectioning will drastically reduce the time required for conducting the column test.

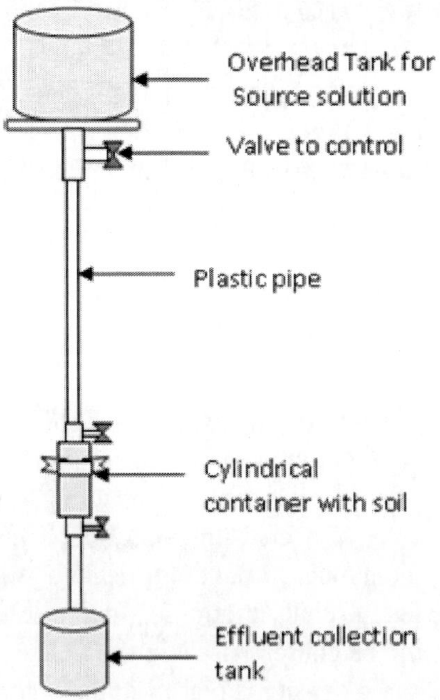

Figure 1. Schematic diagram of column test apparatus.

Determination of Transport Parameters from Column Test Results

The one dimensional ADE for solute transport in a saturated porous medium for reactive solution is as given in Eq. 1 (Sharma and Sangeeta, 1994):

$$R\frac{\partial C}{\partial t} = D\frac{\partial^2 C}{\partial Z^2} - v_s\frac{\partial C}{\partial Z} \qquad \text{(Eq. 1)}$$

where R is the retardation factor which is equal to $1+\rho K/n$

ρ is the dry density of soil in g/cc, n is the porosity, v_s is the seepage velocity in cm /s and K is the distribution coefficient in cc/g.

There are various analytical and numerical solutions available to solve the above equation which requires iterative calculation. For the present study, the analytical solution given by Ogata and Robert, 1961 (Eq. 2) was used to solve the ADE and by solving this, the theoretical breakthrough curves were prepared for the trial values of transport parameters. The theoretical breakthrough curve was matched with the experimental breakthrough curve to get the transport parameters.

$$C(z,t) = \frac{C_0}{2}\left[\operatorname{erfc}\left(\frac{Rz - v_s t}{2\sqrt{DRt}}\right) + \exp\left(\frac{v_s z}{D}\right)\operatorname{erfc}\left(\frac{Rz + v_s t}{2\sqrt{DRt}}\right)\right] \qquad \text{(Eq. 2)}$$

where,

C (z, t) = concentration of effluent at time 't'

Z = thickness of soil sample

C_0 = initial concentration of influent

t = time

Column Leaching Tests

Column test under leaching condition is called as column leaching test which can be used to evaluate the efficiency of a chemical solution to leach out the metals from a contaminated soil. The diluted acids and other chelating agents can be effectively used to wash out the pollutants from the soil. To conduct the test, the contaminated soil sample was placed in the column test apparatus and the chemical solution was passed through the contaminated soil sample. The effluent that was coming out of the sample was collected and the effluent concentrations were measured periodically. The effluent's volume and the mass of contaminant in the effluent (Δm) with time were recorded. From the volume of effluent, the number of pore volumes of flow (T^1) with time were estimated. The ratio of Δm to initial solute mass (M_0) is taken as Leaching Mass Ratio (LMR) and its cumulative value is noted as Cumulative Leaching Mass Ratio (LMR_m). The LMR_m values were plotted against the $T^1$and thus the experimental elution curves (i.e., breakthrough curves for leaching) were obtained (Shackelford and Michael, 1997). The column leaching tests were also conducted to evaluate the immobilization remediation of contaminated soil. The leaching tests on the amended soil samples were performed to assess the efficiency of the chemicals used for immobilizing the metals in soil. For this, the distilled water was passed through the stabilized soil column and the effluent concentrations were observed for a period of three months to find out the quantity of metals leached and to evaluate the efficiency of the immobilizing chemical agent.

Determination of Transport Parameters from Column Leaching Test Results

The governing equation for modelling the column leaching test as suggested by Shackelford and Michael, 1997 is given below (Eq. 3).

$$\text{LMRm} = \frac{\Sigma\Delta m}{M_0} = \frac{T^I}{R} - \frac{1}{2}\{(\frac{T^I}{R} - 1)erfc(\frac{R-T^I}{2\sqrt{\frac{T^I R}{P_L}}}) + (\frac{T^I}{R} + 1)\exp(P_L)\, erfc(\frac{R+T^I}{2\sqrt{\frac{T^I R}{P_L}}})\}$$

(Eq. 3)

where P_L = column Peclet Number = VsL/D;

A MATLAB program for the above equation was prepared to perform the iterative calculation. The trial values of D and R and experimental values of $T^1$and P_L were given as input to the program to plot the theoretical elution curve. The theoretical elution curve was then matched with the experimental one and this process was continued till the theoretical and experimental curves match. When they are matched, the corresponding trial values were taken as transport parameters (D and K).

Results and Discussions

Applications of Transport Parameters in the Design of Liner Thickness

For the red soil tested on six samples for constant and variable test conditions, the theoretical breakthrough curves were generated using MATLAB code for Eq. 2. The theoretical curve with trial values of transport parameters was compared with the experimental curve (Figures 2-7) obtained from column test and thus the transport parameters were determined. For the variable source concentration test, the transport parameters can be predicted from the variations in the influent concentrations also. Hence, more accurate values can be obtained with the variable test condition. The transport rates obtained along with the physical properties of test samples are reported in Tables 2 and 3.

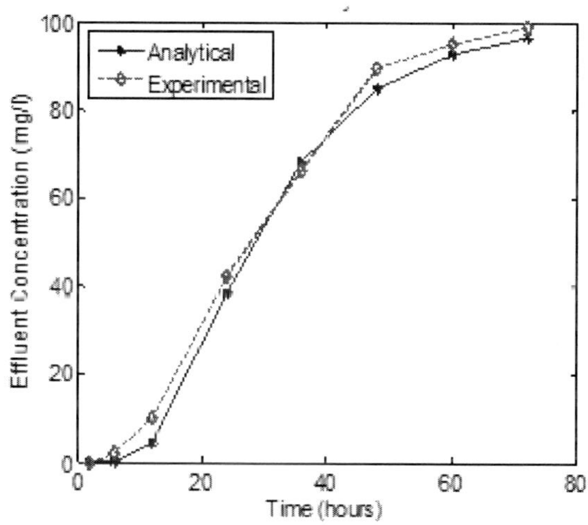

Figure 2. Breakthrough curves of sample RE/ZN/C-1.

The column tests were conducted on soil samples of different densities and the transport parameters were estimated corresponding to each density using the above procedure. The properties of the tested soil samples and the estimated values of transport parameters are given in Table 2 and Table 3. The sample code for example RE/ZN/C-1 represents Red soil/Zinc/Constant source concentration test – sample 1 and the code RE/ZN/V-1 represents Red soil/Zinc/Variable source concentration test – sample 1. The seepage velocities through the soil samples compacted to different densities were in the range of 2.5E-3 to 5.91E-3 cm/s. From Table 3, it can be observed that the Dispersion coefficients of Zinc through Red soil were in the range of 2.10 x 10^{-03}–1.50 x 10^{-02} cm^2/s whereas the Distribution coefficients of Zinc through this soil were in the range of 6.9–11.5 cm^3/g.

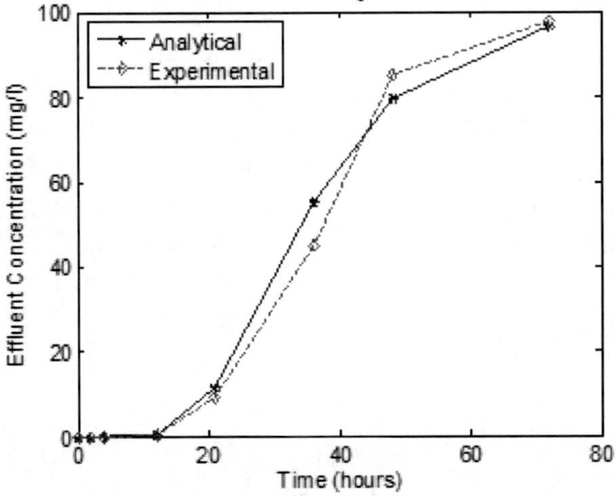

Figure 3. Breakthrough curves of sample RE/ZN/C-2.

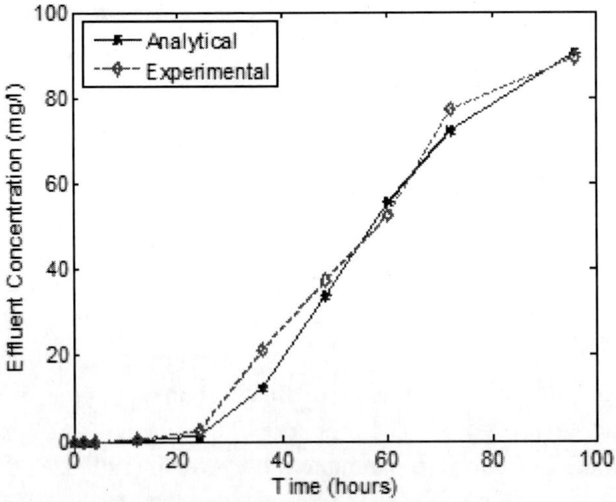

Figure 4. Breakthrough curves of sample RE/ZN/C-3.

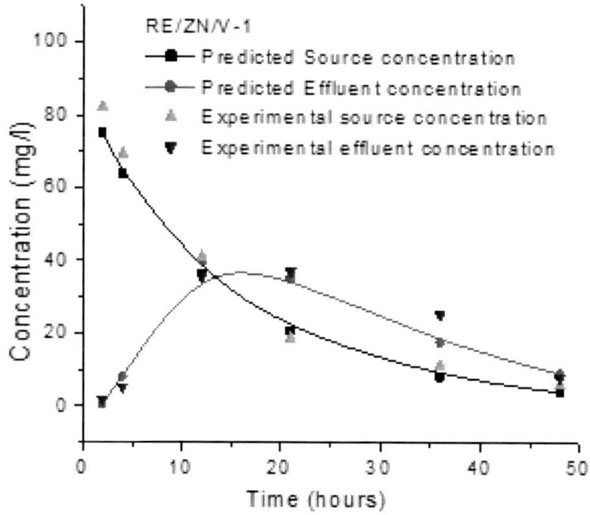

Figure 5. Breakthrough curves of sample RE/ZN/V-1.

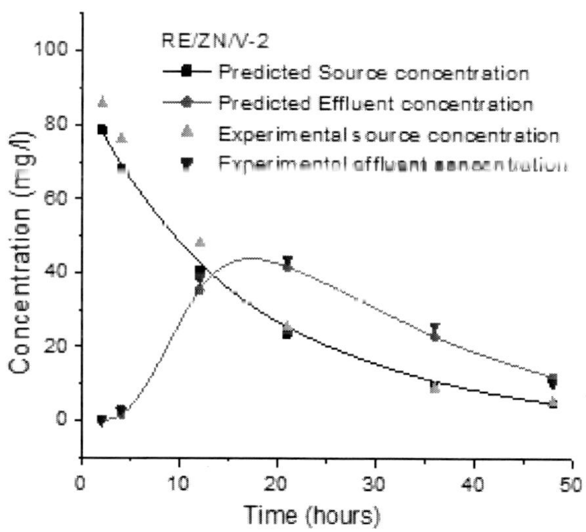

Figure 6. Breakthrough curves of sample RE/ZN/V-2.

The variations in seepage velocity, D and K are shown in Figures 8-10. These values are corresponding to the constant and variable source concentration test conditions for the migration of zinc ion through red soil.

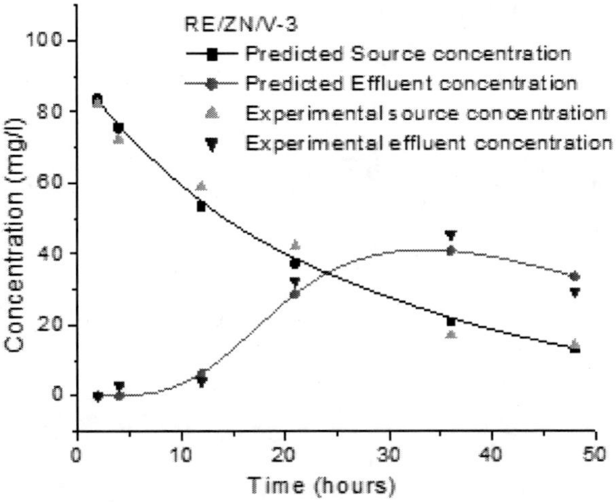

Figure 7. Breakthrough curves of sample RE/ZN/V-3.

Table 2. Physical and hydraulic properties of red soil samples with zinc

Sample Code	Type of test	Dry Density (g/cc)	Porosity (n)	Coefficient of permeability (cm/s)	Darcy velocity (v) (cm/s)	Seepage velocity, (v_s) (cm/s)
RE/ZN/C-1	Constant source concentration test	1.59	0.40	1.60×10^{-4}	1.60×10^{-3}	4.00×10^{-3}
RE/ZN/C-2		1.62	0.39	1.40×10^{-4}	1.40×10^{-3}	3.59×10^{-3}
RE/ZN/C-3		1.70	0.36	9.00×10^{-5}	9.00×10^{-4}	2.50×10^{-3}
RE/ZN/V-1	Decreasing source concentration test	1.50	0.44	2.60×10^{-4}	2.60×10^{-3}	5.91×10^{-3}
RE/ZN/V-2		1.53	0.43	2.35×10^{-4}	2.35×10^{-3}	5.46×10^{-3}
RE/ZN/V-3		1.60	0.40	1.50×10^{-4}	1.50×10^{-3}	3.75×10^{-3}

Table 3. Contaminant transport parameters of zinc through red soil

Sample Code	Dispersion Coefficient (D) (cm^2/s)	Distribution Coefficient (K) (cm/s)	Retardation factor (R)
RE/ZN/C-1	6.20E-03	10.8	43.9
RE/ZN/C-2	3.20E-03	11.3	47.9
RE/ZN/C-3	2.10E-03	11.5	55.3
RE/ZN/V-1	1.50E-02	06.9	24.5
RE/ZN/V-2	9.00E-03	07.4	27.3
RE/ZN/V-3	4.80E-03	08.8	36.2

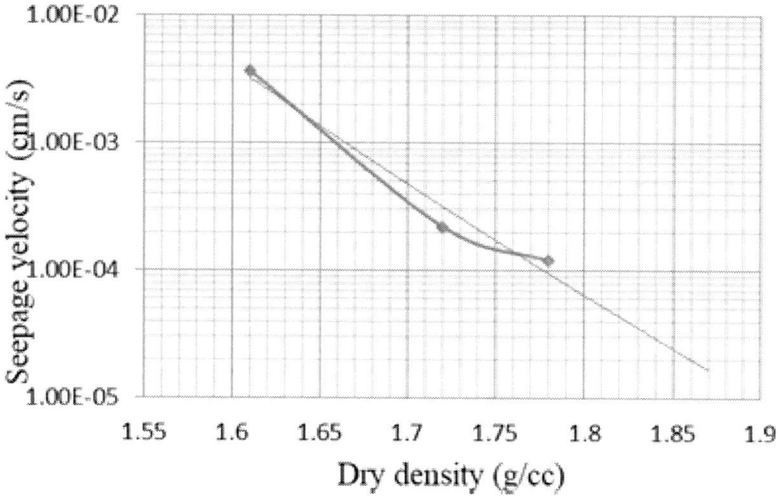

Figure 8. Variation in seepage velocity with density.

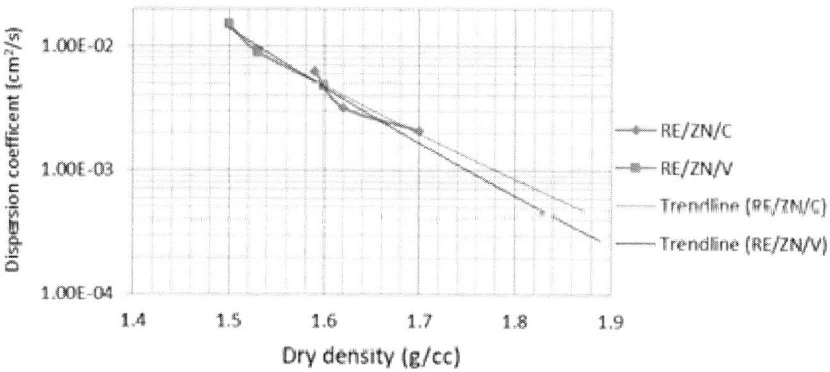

Figure 9. Variation in dispersion coefficient with density.

The best fitting lines are prepared to find the transport rates corresponding to the required density of soil sample. From these plots, it can be observed that the seepage velocity and D values decreases with increasing density whereas the K value increases with the soil density. The increase in K value with density is due to the less seepage velocity which allowed the solute to react with the soil for a relatively more time and caused sorption.

The transport parameters obtained from the column test and transport modelling can be used to estimate the thickness of soil liners required at a site. The thickness required with respect to each metal ion expected at the site needs to be estimated and the maximum value will be taken as the final thickness of the liner. The transport rates corresponding to the maximum dry density were predicted from the trendlines of seepage velocity, D and K (Figures 8-10). These transport parameters were substituted in ADE to estimate the liner thickness corresponding to different design periods. A model chart was prepared to design the thickness of liner for different breakthrough times and relative concentrations (Figure 11). The transport parameters are thus useful to estimate the liner thickness for a landfill facility corresponding to the required design period.

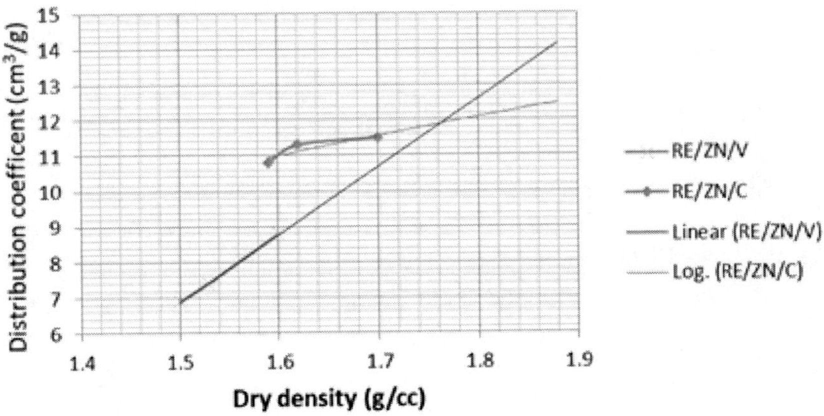

Figure 10. Variation in distribution coefficient with density.

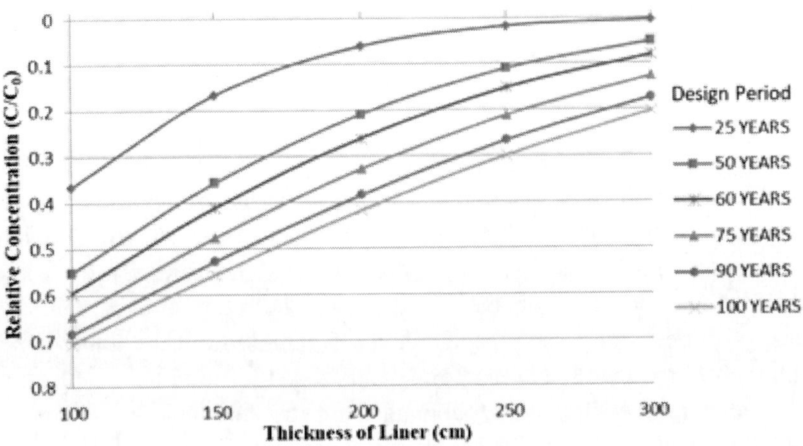

Figure 11. Thickness of liner with red soil for different breakthrough times.

Applications of Transport Parameters in Soil Washing

The column leaching tests were conducted on zinc contaminated soil to find the efficiency and the quantity of leaching solution. The chelates/solvents used for these soils 0.1NHCl, 0.1N EDTA, 0.1N EDTA + 0.1NHCl and Ferric Chloride ($FeCl_3$). The experimental elution curves are as shown in Figure 12.

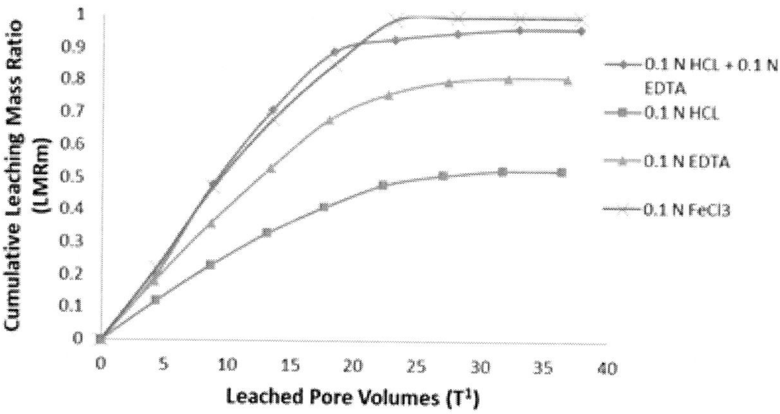

Figure 12. Leaching test results on zinc contaminated soil.

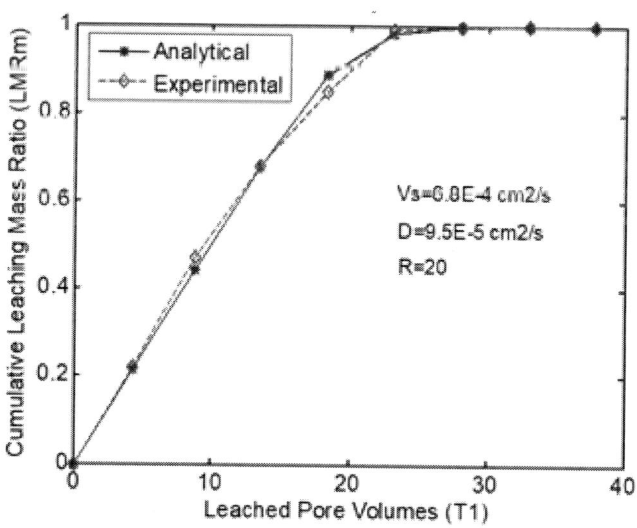

Figure 13. Theoretical Elution curve for zinc contaminated soil.

The trial values of D and R and experimental values of $T^1$and P_L (from the column leaching tests) are used with Eq. 3 to determine the transport parameters (Figure 13). These transport rates can be effectively used to calculate the volume of leaching solution required in the field (using Eq. 3). Thus the column leaching tests are effective in identifying the best chemical solution and to design the soil remediation using soil washing technique.

Applications of Transport Parameters in Immobilization Technique

The landfill soil was studied using three inorganic amendments, Lime, NaOH and Cement. The soil samples were amended with the chemical agent to obtain the required pH of the soil mixture. The mixtures were allowed to react for about one month and then the column leaching tests on the amended samples were conducted to know the efficiency of immobilization with selected chemicals. For this, the stabilized soil was flooded with distilled water and effluents were monitored to know the leachability of different metals from this mixture. Lesser the quantity of metals leached out from the soil, higher will be the efficiency of the amendment to immobilize the metals in soil.

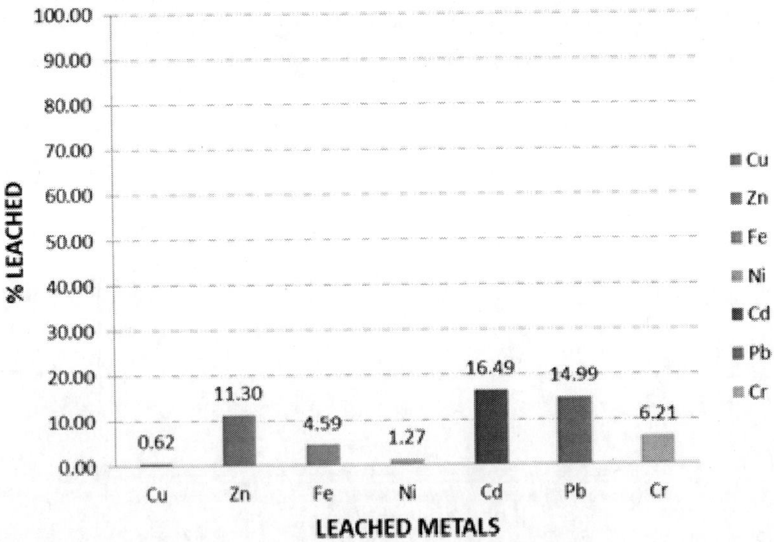

Figure 14. Metals leached from the landfill soil after lime stabilization.

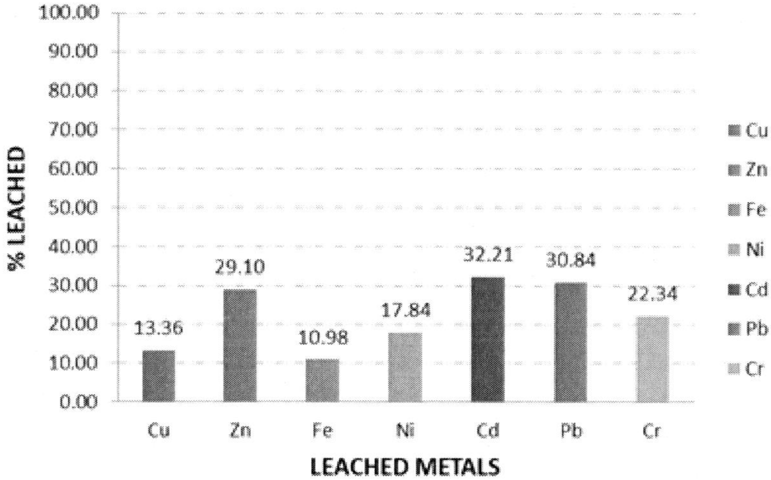

Figure 15. Metals leached from the Bingipura soil after stabilizing with NaOH.

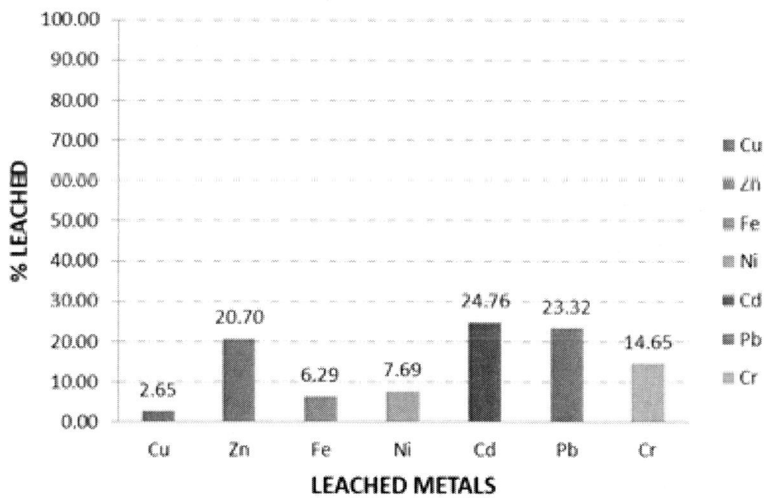

Figure 16. Metals leached from the Bingipura soil after stabilizing with Cement.

From these leaching tests, the optimum results were obtained at pH value of 10. The metals leached out from the amended landfill soil with respect to the additive were shown in Figures 14-16. The percentage leached was in the range of 0.62 to 32% for different metals and amendments and the values of lime were much lower than the remaining amendments. It is thus clear that lime addition can greatly reduce the leachability of metal contaminants. The long term efficiency of an immobilizing agent is thus estimated through column leach test results. The paper thus bring out the use of transport parameters determined with different variants of column tests can be effectively used for design of liner thickness and to remediate the contaminated soil using soil washing or by immobilization of contaminant within the cemented matrix.

Conclusion

Soil column tests can be performed in different variants viz., constant or variable source head. The tests also can be modified to reduce the time taken to obtain breakthrough curves in cases where the duration can extend to very long duration. The results from column breakthrough curves which are used to obtain contaminant transport parameters can be used for many different and versatile applications. The transport parameters have application in the design of waste containment facilities such as design of liner thickness and remediation of contaminated soil and sludge. The transport rates determined are also useful to select the suitable leaching solution for removing the contaminants by soil washing technique and its quantity required in the field. The fluid which gives high rates of transport for the contaminant is selected with the least amount of fluid is preferred for the removal of the maximum amount of contaminant. Another application of column leaching test results is for selecting the suitable chemical agent to immobilize the heavy metals present in the contaminated soil. The column leaching tests on stabilized soil gives the amount of metals leached after immobilizing with the selected additive. The amendment which gives lower rates of transport for the contaminant is selected for immobilization which is in principle opposite of flushing.

References

Akahane, Ikuko, Tomoyuki Makino, Yuji Maejima, Takashi Kamiya, Hiroyuki Takano, Toshiyuki Ibaraki, and Makoto Inahara. "Remediation of cadmium-contaminated paddy soils by washing with ferric chloride (FeCl3): effect of soil washing on the cadmium concentration in soil solution and spinach." *Japan Agricultural Research Quarterly: JARQ* 47, no. 3 (2013): 273-281.

Bilgin, Melayib, and S. Tulun. "Removal of heavy metals (Cu, Cd and Zn) from contaminated soils using EDTA and FeCl3." *Global Nest J* 18 (2016): 98-107.

Czupyrna, G., R. D. Levy, A. I. Maclean, and H. Gold. *In situ immobilization of heavy-metal contaminated soil*. FOSTER-MILLER INC WALTHAM MA, 1988.

Ebrahimbabaie, Parisa, and John Pichtel. "Biotechnology and nanotechnology for remediation of chlorinated volatile organic compounds: Current perspectives." *Environmental Science and Pollution Research* 28, no. 7 (2021): 7710-7741.

Ekeleme, Anthony Chibuzo, Benjamin Nnamdi Ekwueme, and Jonah Chukwuemeka Agunwamba. "Modeling contaminant transport of nitrate in soil column." *Emerg Sci J* 5, no. 4 (2021): 471-485.

Hameed, Rashida, Cheng Lei, Jing Fang, and Daohui Lin. "Co-transport of biochar colloids with organic contaminants in soil column." *Environmental Science and Pollution Research* 28, no. 2 (2021): 1574-1586.

Hirokazu, Ozaki, Ida Shingo, Pu Jian, Zhang Yanfeng, Jiang Chunxiao, Sun Hongwen, and Fukushi Kensuke. "Cadmium extraction from contaminated soil using an iron chloride solution in asian countries." In *Sustainability in Food and Water*, pp. 417-424. Springer, Dordrecht, 2010.

Hong, Young-Kyu, Jin-Wook Kim, Sang-Phil Lee, Jae-E. Yang, and Sung-Chul Kim. "Effect of combined soil amendment on immobilization of bioavailable As and Pb in paddy soil." *Toxics* 10, no. 2 (2022): 90.

Jaramillo, Marcela, Thomas Grischek, Hilmar Boernick, and Jaime I. Velez. "Evaluation of riverbank filtration in the removal of pesticides: an approximation using column experiments and contaminant transport modeling." *Clean Technologies and Environmental Policy* 21, no. 1 (2019): 179-199.

Karn, Barbara, Todd Kuiken, and Martha Otto. "Nanotechnology and in situ remediation: a review of the benefits and potential risks." *Environmental health perspectives* 117, no. 12 (2009): 1813-1831.

Khalkhaliani, Z. N., A. R. Mesdaghinia, A. Mahvi, J. Nouri, and F. Vaezi. "An experimental study of heavy metal extraction, using various concentration of EDTA in a sandy loam soils." *Pakistan Journal of Biological Sciences* 9, no. 5 (2006): 837-842.

Kogbara, Reginald B. "A review of the mechanical and leaching performance of stabilized/solidified contaminated soils." *Environmental reviews* 22, no. 1 (2014): 66-86.

Korf, Eduardo Pavan, Antonio Thomé, Nilo Cesar Consoli, Rafael de Souza Tímbola, and Gláucia Carine dos Santos. "Metal transport parameters in residual soil with an undisturbed and remolded structure percolated by an acid solution." *Journal of Environmental Protection* 2, no. 08 (2011): 1076.

Kristanti, Risky Ayu, Rachael Mei Yen Liong, and Tony Hadibarata. "Soil Remediation Applications of Nanotechnology." *Tropical Aquatic and Soil Pollution* 1, no. 1 (2021): 35-45.

Li, Mingyue, Liheng Ren, Jiachao Zhang, Lin Luo, Pufeng Qin, Yaoyu Zhou, Chao Huang, Jiayi Tang, Hongli Huang, and Anwei Chen. "Population characteristics and influential factors of nitrogen cycling functional genes in heavy metal contaminated soil remediated by biochar and compost." *Science of the Total Environment* 651 (2019): 2166-2174.

Lin, Cheng-Fang, Shun-Shin Lo, Herng-Yuh Lin, and Yichin Lee. "Stabilization of cadmium contaminated soils using synthesized zeolite." *Journal of Hazardous Materials* 60, no. 3 (1998): 217-226.

Ling, Wanting, Qing Shen, Yanzheng Gao, Xiaohong Gu, and Zhipeng Yang. "Use of bentonite to control the release of copper from contaminated soils." *Soil Research* 45, no. 8 (2007): 618-623.

Liu, Ling, Siqi Wang, Xiaoping Guo, and Hongjie Wang. "Comparison of the effects of different maturity composts on soil nutrient, plant growth and heavy metal mobility in the contaminated soil." *Journal of Environmental Management* 250 (2019): 109525.

Ma, Fujun, Qian Zhang, Duanping Xu, Deyi Hou, Fasheng Li, and Qingbao Gu. "Mercury removal from contaminated soil by thermal treatment with FeCl3 at reduced temperature." *Chemosphere* 117 (2014): 388-393.

McLaughlin, Michael J., B. A. Zarcinas, D. P. Stevens, and N. Cook. "Soil testing for heavy metals." *Communications in Soil Science and Plant Analysis* 31, no. 11-14 (2000): 1661-1700.

McLaughlin, Michael J., R. E. Hamon, R. G. McLaren, T. W. Speir, and S. L. Rogers. "A bioavailability-based rationale for controlling metal and metalloid contamination of agricultural land in Australia and New Zealand." *Soil Research* 38, no. 6 (2000): 1037-1086.

Mirza, Itrat Abbas, Muhammad Saeed Akram, Nehad Ali Shah, Shehraz Akhtar, and Mirfat Muneer. "Study of one-dimensional contaminant transport in soils using fractional calculus." *Mathematical Methods in the Applied Sciences* 44, no. 8 (2021): 6839-6856.

Moncada, M. P. H. "Laboratory Study of Collapse Characteristics and Solute Transport Associated with the Infiltration of Caustic Liquor in a Laterite Soil." *Master's Thesis, Departamento of Civil Engineering, Pontifícia Universidade Católica do Rio de Janeiro, Rio de Janeiro, in Portuguese* (2004).

Mondal, Abhisek, Brajesh Kumar Dubey, Meenakshi Arora, and Kathryn Mumford. "Porous media transport of iron nanoparticles for site remediation application: A review of lab scale column study, transport modelling and field-scale application." *Journal of Hazardous Materials* 403 (2021): 123443.

Ochola, Charles, and Horace Moo-Young. "Evaluation of metal attenuation through paper clay utilized for containment of contaminated groundwater." *Journal of Environmental Engineering* 130, no. 8 (2004): 873-880.

Ogata, Akio, and Robert Blackburn Banks. *A solution of the differential equation of longitudinal dispersion in porous media: fluid movement in earth materials*. US Government Printing Office, 1961.

Phulpoto, Irfan Ali, Zhisheng Yu, Bowen Hu, Yanfen Wang, Fabrice Ndayisenga, Jinmei Li, Hongxia Liang, and Muneer Ahmed Qazi. "Production and characterization of surfactin-like biosurfactant produced by novel strain Bacillus nealsonii S2MT and it's potential for oil contaminated soil remediation." *Microbial cell factories* 19, no. 1 (2020): 1-12.

Piccolo, Alessandro, Riccardo Spaccini, Antonio De Martino, Francesco Scognamiglio, and Vincenzo di Meo. "Soil washing with solutions of humic substances from manure compost removes heavy metal contaminants as a function of humic molecular composition." *Chemosphere* 225 (2019): 150-156.

Pietrzak, Damian, Jarosław Kania, Ewa Kmiecik, and Katarzyna Wątor. "Identification of transport parameters of chlorides in different soils on the basis of column studies." (2019).

Qian, Wei, Jing-Yi Liang, Wen-Xuan Zhang, Shi-Ting Huang, and Zeng-Hui Diao. "A porous biochar supported nanoscale zero-valent iron material highly efficient for the simultaneous remediation of cadmium and lead contaminated soil." *Journal of Environmental Sciences* 113 (2022): 231-241.

Qian, Yuting, Caidie Qin, Mengmeng Chen, and Sijie Lin. "Nanotechnology in soil remediation– applications vs. implications." *Ecotoxicology and Environmental Safety* 201 (2020): 110815.

Ren, Xiaoya, Guangming Zeng, Lin Tang, Jingjing Wang, Jia Wan, Jiajia Wang, Yaocheng Deng, Yani Liu, and Bo Peng. "The potential impact on the biodegradation of organic pollutants from composting technology for soil remediation." *Waste management* 72 (2018): 138-149.

Robichaud, Kawina, Catherine Girard, Dimitri Dagher, Katherine Stewart, Michel Labrecque, Mohamed Hijri, and Marc Amyot. "Local fungi, willow and municipal compost effectively remediate petroleum-contaminated soil in the Canadian North." *Chemosphere* 220 (2019): 47-55.

Safari, Edwin, Mahdi Jalili Ghazizade, and Mohammad Ali Abdoli. "A performance-based method for calculating the design thickness of compacted clay liners exposed to high strength leachate under simulated landfill conditions." *Waste management & research* 30, no. 9 (2012): 898-907.

Schübl, Marleen, Aleksandra Kiecak, Katrin Hug, Jutta Lintelmann, Ralf Zimmermann, and Christine Stumpp. "Sorption and biodegradation parameters of selected pharmaceuticals in laboratory column experiments." *Journal of Contaminant Hydrology* 236 (2021): 103738.

Shackelford, Charles D., and Michael J. Glade. "Analytical mass leaching model for contaminated soil and soil stabilized waste." *Groundwater* 35, no. 2 (1997): 233-242.

Souza, Lilian Rodrigues Rosa, Luiza Carolina Pomarolli, and MárciaAndreia Mesquita Silva da Veiga. "From classic methodologies to application of nanomaterials for soil remediation: an integrated view of methods for decontamination of toxic metal (oid) s." *Environmental Science and Pollution Research* 27, no. 10 (2020): 10205-10227.

Sumalatha, J., P. V. Sivapullaiah, and R. Prabhakara. "Immobilization Remediation of a Heavy Metals Contaminated Soil: A Case Study of Dump Site at Bangalore, India." *Journal of The Institution of Engineers (India): Series A* (2022): 1-10.

Trakal, L., M. Neuberg, P. Tlustoš, J. Száková, V. Tejnecký, and O. Drábek. "Dolomite limestone application as a chemical immobilization of metal-contaminated soil." *Plant, Soil and Environment* 57, no. 4 (2011): 173-179.

Wuana, Raymond A., and Felix E. Okieimen. "Heavy metals in contaminated soils: a review of sources, chemistry, risks and best available strategies for remediation." *International Scholarly Research Notices* 2011 (2011).

Xie, Haijian, Hao Ding, Huaxiang Yan, Dandi Yang, Zhanghua Lou, ZhanhongQiu, and Yun Chen. "A semi-analytical solution to organic contaminants transport through composite liners considering a single crack in CCL." *Environmental Science and Pollution Research* (2022): 1-13.

Yadav, Ajar Nath. "Nanotechnology for agro-environmental sustainability." *Journal of Applied Biology and Biotechnology* 9, no. 4 (2021): i-i.

Yan, Huaxiang, Haijian Xie, Hao Ding, Dandi Yang, Cong Chen, Yunmin Chen, and Yun Chen. "Analytical solution for one-dimensional steady-state contaminant transport through a geomembrane layer (GMBL)/compacted clay layer (CCL)/attenuation layer (AL) composite liner considering consolidation." *International Journal for Numerical and Analytical Methods in Geomechanics* (2022).

Yasir, Arafat Md, Jie Ma, Xiaoxue Ouyang, Junying Zhao, Yujie Zhao, Liping Weng, Md Shafiqul Islam, Yali Chen, and Yongtao Li. "Effects of selected functional groups on nanoplastics transport in saturated media under diethylhexyl phthalate co-contamination conditions." *Chemosphere* 286 (2022): 131965.

Ye, Shujing, Guangming Zeng, Haipeng Wu, Jie Liang, Chang Zhang, Juan Dai, WeipingXiong, Biao Song, Shaohua Wu, and Jiangfang Yu. "The effects of activated biochar addition on remediation efficiency of co-composting with contaminated wetland soil." *Resources, Conservation and Recycling* 140 (2019): 278-285.

Yong, Raymond Nen, Abdel-Mohsen Onsy Mohamed, and Benno P. Warkentin. *Principles of contaminant transport in soils*. Elsevier, 1992.

Zeng, Xiaoxi, Hong Xu, Jijie Lu, Qimin Chen, Wen Li, Ling Wu, Jianxin Tang, and Liang Ma. "The immobilization of soil cadmium by the combined amendment of bacteria and hydroxyapatite." *Scientific Reports* 10, no. 1 (2020): 1-8.

Zhang, Ming, and Liheng Xu. "Transport of micro-and nanoplastics in the environment: Trojan-Horse effect for organic contaminants." *Critical Reviews in Environmental Science and Technology* 52, no. 5 (2022): 810-846.

Chapter 10

The Effectiveness of Arduino Ultrasonic and Passive Infrared Receivers on the Perception of Physical Distancing

Rita Dwi Pratiwi[1], Fenita Purnama Sari Indah[1], Sandeep Poddar[2,*], Tukimin Bin Sansuwito[2], Faridah Mohd Said[2] and Riris Andriati[1]

[1]STIKes Widya Dharma Husada Tangerang, South Tangerang, Indonesia
[2]Lincoln University College, Selangor, Malaysia

Abstract

The World Health Organization has started to use the term physical distancing, which means physical distance as a substitute for the term social distancing which is considered less effective in preventing COVID-19 (WHO, 2020). Covid necklace is one type of simple object worn by humans as a physical distancing alarm based on Arduino ultrasonic sensors and PIR passive infrared receiver sensors. Objective: to compare the effectiveness of ultrasonic arduino with passive infrared receiver (PIR) on the perception of physical distancing in the community of villa pamulang, Depok City. Methodology: This research is a quantitative research that uses primary data (questionnaire) and uses a quasi-experimental research method with a two group pree test posttest control group design. Data collected by using a questionnaire. The number of samples is 80, of which 40 are samples for ultrasonic sensors and 40 are for sensors (PIR). Results: the characteristics of the respondents of the Pamulang villa community were that most of the

* Corresponding Author's Email: sandeeppoddar@lincoln.edu.my

In: Computational Intelligence for Sustainable Development
Editor: Brojo Kishore Mishra
ISBN: 979-8-88697-198-9

respondents were 17-25 years old (50.0%), and (55.5%). Basedon the gender, the users of the ultrasonic arduino sensor necklace and passive infrared receiver are mostly female, namely 25 respondents (62.5%) and 24 respondents (60.0%). based on the characteristics of the health condition of respondents with ultrasonic arduino and passive infra red receivers, it was found that all respondents, 40 respondents from PIR sensors (100.0%), and 40 respondents from ultrasonic sensors (100.0%) were in good health. the description of physical distancing before being given treatment with the ultrasonic arduino sensor necklace and passive infrared receiver, the results obtained in the poor category, namely 20 respondents (50%), and 22 respondents (55%). While the description of physical distancing after being given treatment with an arduino ultrasonic sensor necklace and passive infrared receiver, the results obtained in good categories, namely 40 respondents (100%) and 36 respondents (90%). The results of statistical tests with T simples t test between ultrasonic arduino variables and passive infrared receiver (PIR) obtained p = 10,575 (p> 0.000) on ultrasonic sensors P = 3.975 (p> 01,000) on pir sensors. Arduino ultrasonic sensors have a higher effectiveness than passive infrared receiver sensor necklaces.

Keywords: physical distancing, sensors, ultrasonic arduino, passive infrared receiver (PIR)

Introduction

Corona Virus Disease (COVID-19) is a disease caused by severe acute respiratory syndrome coronavirus 2 (SARS-CoV-2). This virus is a new type of virus from the Coronavirus family that can cause various diseases of the respiratory system, ranging from mild symptoms to Severe SARS-CoV-2 is transmitted through direct contact with patients, droplets released by patients when coughing/sneezing and hands touching the mouth, nose, and eyes after touching objects contaminated with the virus (Qi et al. 2020; Dewiyuliana & Septiana, 2020).

According to the World Health Organization (WHO), COVID-19 is an extremely dangerous pandemic. So far, on January 21, 2021, there were 95,321,880 cases and 2,058.22 total deaths worldwide, while Indonesia reached 939,948 with 772,790 total of recovery and 26,857 deaths. Meanwhile, globally, as of July 26, 2021, there has been an increase in cases. 194,070,340 people have been infected since the first Covid-19 case was found in December 2019. The total number of deaths was 4,160,337, and 176,

157,715 recovered. Currently, on July 26, 2021, Indonesia is in position one with the highest number of deaths in the world, which is as many as 1000 cases per day. The possibility of this data will continue to grow if the community does not implement physical distancing. With the pandemic, it is necessary that the chain of transmission get broken while the population get protected from risk (Zhang, et al. 2020).

Indonesia with the number of Covid-19 exposures has reached 11,192 positive spread across 34 provinces, and 326 regencies/cities on May 13, 2020. Indonesia still represents an increase of positive-covid 19-cases due to the lack of public awareness in implementing health protocols and maintaining distance. The government has implemented various efforts to reduce the number of positive cases in Indonesia. One of them is the physical distancing policy, intending to break the chain of transmission in the community, such as the way of maintaining a physical distance of more than 1 meter in communicating in their social environment (Widyaningrum & Putri, 2020; Sulistiadi, 2020).

The term "physical distancing" means limiting physical distance or keeping a distance. This term is used instead of the term "social distancing," which is considered less effective in preventing COVID-19. The use of the term "social distancing" causes a misunderstanding that doing social distancing means limiting social interaction between people. Changing the term "social distancing" to physical distancing, hopefully, people will understand and be more orderly in carrying out physical distancing (Purnamasari & Raharyani, 2020).

Based on a trial survey conducted by the Indonesian Ministry of Health in March 2020, almost 60 percent of the Indonesian people, especially children, and adolescents, have not been able to implement physical distancing properly. The researcher surveyed to find out whether there are impacts and influences such as knowledge and prevention on the physical distancing implementation in Indonesian society. Based on data from the Depok City government's Covid-19 task force, it found that around (69%) with around (34%) spent time outside the house, and (35%) spent time out of the house. And based on the Covid task force, Bojongasri sub-district, Villa Pamulang, RW 11, Depok City, it was found that around (65%) of the community had not been able to implement physical distancing effectively.

Therefore, people need technology automatically detect the meter distance between people and other people. The rapid development of technology has made the field of robotics also increasingly developed, which ultimately led us to the era of technology in object detector sensors (Arsada,

2017; Muqorobin & Rais, 2020; Qi, et al. 2020; Riedel, et al. 2022). The World Health Organization recommends physical distancing or keeping this distance as an effective way to break the chain of the spread of COVID-19. 19. It considers the new coronavirus (Sars-CoV-2), which causes COVID-19 to spread quickly. It is necessary to have a device as an alarm/sign that the distance between one person and another must be at least 1-2 meters. It's all the background for the need for these devices (Joshua, 2020).

According to the Indonesian Ministry of Health, in 2020, the implementation of 3M, namely wearing masks, washing hands, and maintaining distance, is such important keywords in preventing the transmission of COVID-19. In the Afrianti & Rahmiati, 2021, there are research results on 163 respondents, namely, that 89.6% of the community has high adherence to the COVID-19 health protocol. This study is in line with the research of Yandi, et al. 2020, which describes the high level of community compliance in complying with the recommended health protocols during the COVID-19 pandemic. This is supported by other researcher that there are 74.19% of the community has high compliance with the use of masks. The same thing was conveyed through the results of a survey by the Central Statistics Agency (BPS) in 2020, which noted that Indonesia's level of compliance in implementing health protocols was good enough to prevent the COVID-19 virus (Afrianti & Rahmiati, 2021; Syapitri, et al. 2020; Hermawati, et al. 2021).

Necklaces are one of the earliest types of objects worn by humans in modern times, especially in this era of globalization. Necklaces are branding materials that are easy to carry and used by everyone. They use acrylic packaging as a synthetic polymer from methyl methacrylate, thermoplastic, and transparent. It's good, strong, flexible, and durable. People use Covid necklace as a physical distancing alarm (Setyowati, 2017).

The necklace contains an ultrasonic sensor and a pear sensor. Both of these sensors both have a function as distance detectors. The working principle of ultrasonic sensors is that it generates ultrasonic waves through a device called a certain frequency. This piezoelectric will produce ultrasonic waves (usually 40kHz frequency) when it applies an oscillator to the object. Generally, this tool will shoot ultrasonic waves towards an area or a target. After the wave hits the target surface, the target will reflect the wave (Gunawan, 2017). While the working principle of the PIR sensor is to capture passive infrared rays emitted by the human body, which have a different temperature from the environment, causing pyroelectric material (Ridwan, 2016). Government policies such as maintaining a distance of 1-2 meters, washing hands with soap, wearing masks, and using hand sanitizers. Hopefully many people who do not fully implement health protocols will implement them properly.

Problems Formulation

"Physical distancing" is the practice of staying at least 1-2 meters away from one another to ensure the disease does not spread (Aquarini 2020/in Faiz 2020). "Physical distancing" has also been implemented in various regions in Indonesia. However, "physical distancing" will not run easily, if people do not comply with what has been conveyed by the government. In this case, the role of technology greatly contributes to helping the community implement "physical distancing." Based on this description, the formulation of the problem in this study is how effective is the Arduino ultrasonic necklace with a passive infrared receiver in the perception of "physical distancing" in the community?."

Objectives and Benefits of Research

Many people still ignore the rules of physical distancing/keeping this distance when they have to be outside because of something urgent. When people have to leave their homes and be in crowded places, sometimes they do not pay attention to the safe distance between one person and another. Starting from this background, the aims and benefits of this research are:

Research Objectives

1. To identify the description of the characteristics of respondents in the application of physical distancing at Villa Pamulang
2. To analyze the effectiveness of using an Arduino ultrasonic sensor necklace on the perception of physical distancing
3. To analyze the effectiveness of the use of a passive infrared infrared receiver sensor necklaces on the perception of physical distancing in the community of Villa Pamulang RW 11 Depok City

4. To analyze the differences in the results of the effectiveness of Arduino with ultrasonic sensor necklaces and PIR sensors. Hopefully the public can use necklaces that are beneficial to health so that people can maintain a minimum distance of 1-2 meters to break the chain of spreading COVID-19.

Research Benefits

This research has some benefits for developing technology implementation of Arduino and PIR ultrasonic sensor necklaces to detect distance more effectively and break the chain of spreading the Covid-19 virus. The researcher expects this research will be a reference and contribution for further researchers in developing theory and technology more perfect and will be a better educational material.

Research Method

Research Design

The research design of this study used the True experiment with the method of Two groups of free test post-test control group design. The procedure of this research design carries a free test before treatment and a post-test after treatment in the experimental and control groups. The control group doesn't get treatment, while the experimental group will get treatment (Arikunto 2013).

An experimental design that only has treated groups is not enough. According to Taniredja in Arikunto (2013), an experiment can contain an attempt to compare the effects of a particular treatment with different treatment. The experimental group and control group at least have the same or close to the same characteristics. The control group is not given treatment and then observed to see or compare the differences or changes that occured in the experimental group.

The following is an observational research design conducted before the experiment (01) called free-test and observations after the experiment (02) called post-test (Arikunto 2013).

The study design of the *nonrandom free-test pos-test control grup desain*.

Location and Time of the Research

The researcher conducted this research at Villa Pamulang RW 11 Depok City. Many people did not apply physical distancing in this environment. According to the Covid-19 task force, Bojongsari Villa Pamulang sub-district, (69%) people did not maintain a distance of 1-2 meters, so they were vulnerable to the spread of COVID-19. The researcher carried out this research from May to June 2021.

Subject of the Research

The subjects in this study were the people in Pamulang villa RW 11, consisting of RT 01, 02, 03, and 04 of Depok City, totaling 303 families. In this study, researchers took many samples for simple research, namely as many as 80 persons. 40 persons for Arduino ultrasonic, 40 persons for PIR sensor as a general rule. A sample size of 10 to 20 could be effective, depending on the sampling method used and the research question used.

a) Inclusion
   1. The people living in the neighborhood of Villa Pamulang RW 11, Depok City
   2. Community of Villa Pamulang RW 11 Depok City who was willing to be respondents.

b) exclusion
   People who were not in the place when the research takes place.

Data Collection Instruments

The researcher used a side-by-side systematic random and a questionnaire as the data collection techniques. All methods and data collection were to compare the effectiveness of ultrasonic Arduino with passive infrared on the perception of physical distancing in the community at Villa Pamulang RW 11 Depok City. (Sugiyono, 2016). Of the 303 population of the community in Villa Pamulang RW 11 Depok City, the instrument or data collection technique that the researcher used was the systematic random sampling technique, which was the taking of sample members based on the order of the population members who had got serial numbers (Sugiyono 2014).

Literature Review

Ultrasonic Sensor and PIR Sensor Necklace

> PIR sensor is a sensor that detects the presence of infrared rays. This PIR sensor is passive, meaning that this sensor does not emit infrared light but only receives infrared radiation from the outside. This passive sensor only responds to energy from passive infrared rays of every object detected. This sensor usually detects objects, like the human body. (Alfazri, 2015)

According to Novi Lestari 2017, this PIR sensor works by capturing heat energy generated from passive infrared rays of every object with an object temperature above absolute zero. Like the human body, which has a body temperature of approximately 320 C, a typical heat temperature in the environment. Generally, PIR sensors have an effective reading range of up to 5 meters. PIR sensors have a range of reading distances and angles that vary depending on the characteristics of the sensor.

Source: Bagu srifqy alistia. Wordpress.Com.

Figure 1. Sensor diagram PIR.

The Principals and Procedure of Pir Sensor

PIR sensors are widely used in human life, one of which is automatic sliding doors. The way it works is when humans are in front of the PIR sensor in a silent condition, the PIR sensor will calculate the wavelength produced by the human body. This constant wavelength causes the production of heat energy almost the same as the surrounding environmental conditions when humans make movements (Maryanto, 2013 In Desmira, 2020).

The human body will produce passive infrared rays with varying wavelengths and produce different heat. The Pyroelectric sensor detects heat generated and converted into different currents. The current resulting is forwarded to the ADC (Analog to Digital Converter) and continued to the microcontroller (Prayogo, 2015 In Desmira, 2020).

According to Novi Lestari 2017, this PIR sensor works by capturing heat energy generated from passive infrared rays of every object with an object temperature above absolute zero. The human body has a body temperature of approximately 320 C, a typical heat temperature in the environment. The Pyroelectric sensor captures an infrared beam. The PIR sensor core causes the Pyroelectric sensor, consisting of gallium nitride, cesium nitrate, and lithium tantalate, to produce an electric current.

This passive infrared beam carries heat energy then it converted into electric current. The process is almost the same as the electric current formed when sunlight hits a solar cell. This electric current will cause a voltage and is read analogously by the sensor. Then this signal will be amplified by an amplifier and compared by a comparator with a particular reference voltage (the output is a 1-bit signal) (Tempongbuka, 2015 In Desmira, 2020).

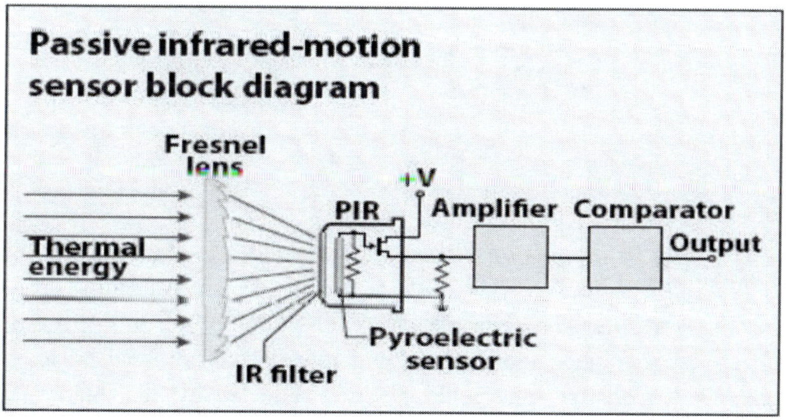

Source: Bagu srifqy alistia. Wordpress.Com.

Figure 2. Sensor PIR diagram.

Arduino Nano Ultrasonic Sensor

Ultrasonic Sensor

Ultrasonic waves are the waves that can estimate the distance between the sensor and an object. It means radar detecting the presence of an object. The workings of this sensor change the amount of sound (physical) into electrical quantities or vice versa. Based on the principle of reflected sound waves, this sensor is to measure the distance of an object with a particular frequency. It is called an Ultrasonic Sensor because this sensor uses ultrasonic waves (Robby Yuli Endra et al., 2019).

An ultrasonic sensor is a sensor to convert physical quantities (sound) into electrical quantities and vice versa. The way this sensor works is based on the principle of the reflection of a sound wave so that people can use it to interpret the existence (distance) of an object with a particular frequency (Kadir in Rudi Kurniawan 2018).

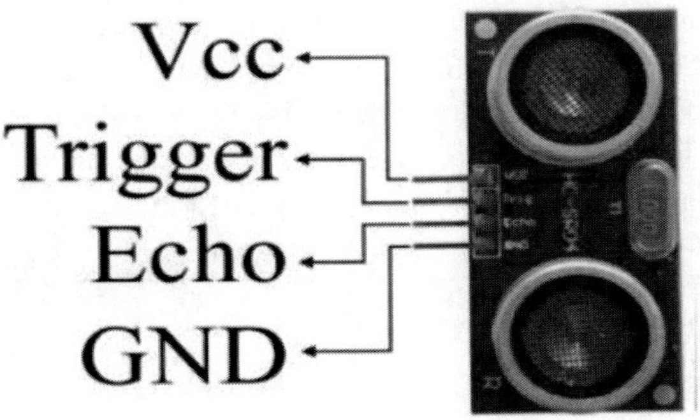

Source: Muzawi, et al. 2020.

Figure 3. Datasheet hc-sr04 ultrasonic.

The Working Principals of Ultrasonic Sensor

The working principle of HCSRF-04 is that the transmitter emits a beam of ultrasonic signal (20 kHz) in the form of pulse, then if in front of HCSRF-04 there is a solid object, the receiver will receive the ultrasonic signal. The receiver will read the pulse width (PWM) reflected by the object and the difference in broadcast time. Then it will recognize the object's distance in front of the sensor with these measurements (Abdulkhaleq, et al. 2020).

In ultrasonic sensors, a piezoelectric device generates ultrasonic waves with a particular frequency. This piezoelectric will produce ultrasonic waves (usually 40 kHz frequency) when the object applies an oscillator. Generally, this tool will shoot ultrasonic waves towards an area or a target. After the wave hits the target surface, the target will reflect back. The sensor will capture the reflected wave from the target then the sensor calculates the difference between the time of the wave sent and the time the reflected wave is received (Abdulkhaleq, et al. 2020).

Ultrasonic sensor works as an explosion signal transmitted for a short duration (emitted) by the transmitter. After that, there will be a period of silence. This period is called the "response time" and is the waiting time for the reflected wave. The acoustic emitted signal may or may not find an obstacle. If it finds an obstacle, the acoustic signal will bounce back from the obstacle. This bounced back signal is called an "echo." The receiving transducer receive an echo and convert it into an electrical signal. Usually, this signal is amplified, filtered, and can be converted into digital form. The elapsed time between transmission and reception, the distance between the Ultra Sonic system and the obstacle/object, can be calculated (Abdulkhaleq, et al. 2020).

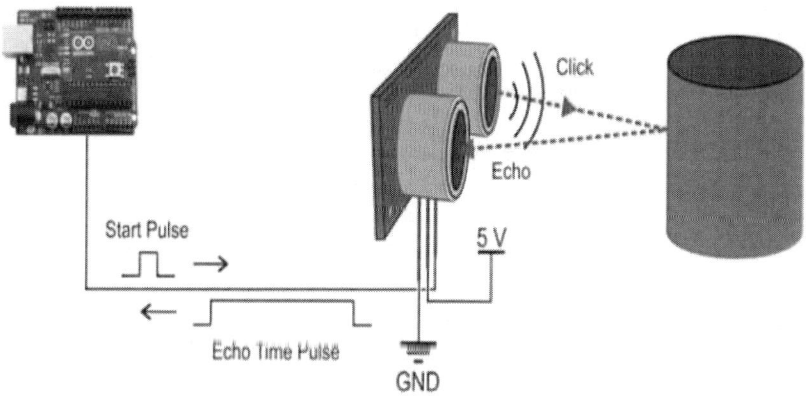

Source: Muzawi, et al. 2020.

Figure 4. Working Principles of *ultrasonic sensor*.

Arduino Nano

Arduino Nano is an Arduino often used by people, is easy to get, and is relatively cheap. The equipment of Arduino is an ATMEGA328P version R3 module and microcontroller, which is the last version to support the microcontroller so that it can work. Arduino nano is an electronic kit or open-source electronic circuit board in which the main component is a microcontroller chip with the AVR type from the Atmel Company (Zlatanov, 2015).

From some of the quotes above the researcher concluded that Arduino nano is an open-source physical computing platform. Arduino is not just a developer tool but is a combination of hardware, programming languages, and a sophisticated Integrated Development Environment (IDE). IDE is software whose role is to write programs, compile them into binary code and upload them to the microcontroller memory.

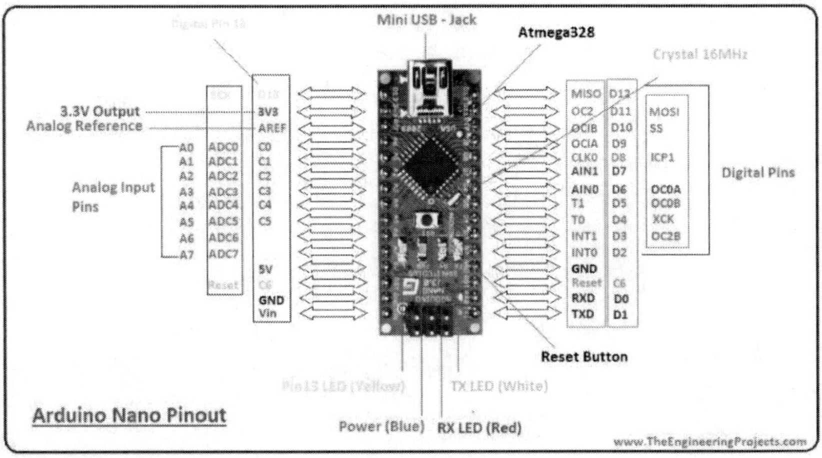

Source: Bagu srifqy alistia.

Figure 5. Arduino nano.

Components in Ultrasonic Sensors and Pir Sensors

Buzzer

An electric buzzer is an electronic component that can convert an electrical signal into sound vibrations. Usually the buzzer is an audio device often used in anti-theft circuits, alarms on watches, doorbells, reverse warnings on trucks, and other hazard warning devices. The type of buzzer that often found and used is the Piezoelectric type buzzer because the piezoelectric buzzer has various advantages such as being cheaper, relatively lighter, and easier to incorporate into other electronic circuits. Buzzers included in the transducer family are also called beepers. The coil is energized so that it becomes an electromagnet. The coil will pull in or out, until the direction of the current, the magnetic polarity, and the buzzer, become an indicator of an alarm device (Ismail, et al. 2016).

The buzzer is a component that works in converting electric current into sound, and basically, the working principle of a buzzer is almost the same as a speaker. The buzzer consists of a diaphragm that has a coil. When the coil is electrified and becomes an electromagnet, it will be attracted in or out according to the polarity of the magnet. Because the coil is mounted on the diaphragm, every diaphragm vibration will be back and forth so that the air vibrates and produces sound. This buzzer becomes an indicator when the motorcycle handlebar is forced straight and is locked (Ismail, et al. 2016).

Source: Teknikelektronika.com.

Figure 6. Buzzer.

LED

Light Emitting Diode or LED is an electronic component that can emit monochromatic light when a forward voltage is applied. LED is a family of diodes made of semiconductor materials. The colors of light emitted by LEDs depend on the type of semiconductor material used. LEDs can also emit infrared rays invisible to the eye while we often encounter TV Remote Control or Remote Control other electronic devices.

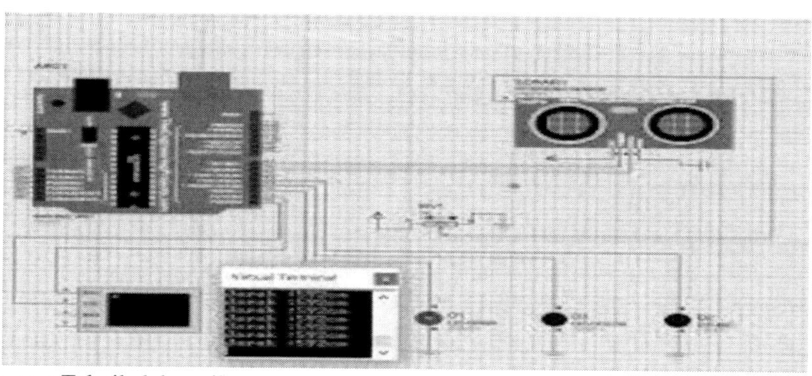

Source: Teknikelektronika.com.

Figure 7. Led Source TeknikElektronika.com.

Batteries

Batteries are electrical energy storage devices and are the most necessary part of portable electronic devices. Lithium-based batteries have the advantage because they are made of lithium, the lightest metal among other metals. So that it can produce high energy density and have a long life cycle (Tarascon & Armand, 2011).

The battery is an electric cell with a high efficiency reversible electrochemical. It means there is converting chemicals process into electric power in the electrochemical process of the battery.

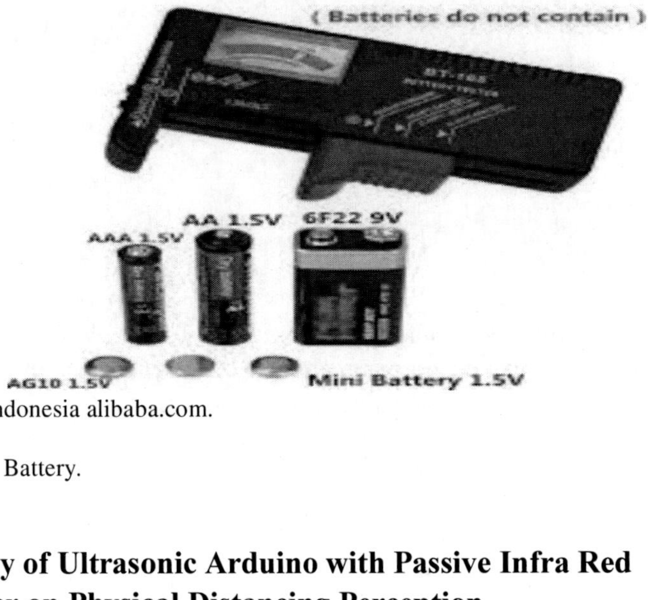

Source: Indonesia alibaba.com.

Figure 8. Battery.

Strategy of Ultrasonic Arduino with Passive Infra Red Receiver on Physical Distancing Perception

One of the precautions taken to break the chain of the spread of the Covid-19 virus (Coronavirus Disease) is: to promote social distancing rules, avoid crowds, no direct contact with other people, wear masks, wash hands using soap and running water or hand sanitizers, do not touch face area before washing hands and applying good coughing and sneezing etiquette (Indonesian Ministry of Health, 2020). Related to efforts to break the chain of virus spread, there are studies related to this. In order to deal with this situation, it is necessary to strive for innovation so that people obey to keep their distance. Therefore, it is necessary to develop a technology-based innovation using ultrasonic and PIR sensors.

There was a study result about the correlation between the effectiveness of Arduino Ultrasonic with Passive Infra Red Receiver on Perception of Physical Distancing in the Community at Villa Pamulang RW 11 Pondok Petir, Bojongsari Village, Depok City. The researcher conducted this research in May 2021, with 80 respondents. The researcher presented the results in general data (univariate) and special data (bivariate). Univariate general data contained characteristics of age, gender, and community conditions. While, the special data (Bivariate) consisted of comparative studies of effectiveness before and after wearing the ultrasonic arduino sensor necklace and Passive Infra Red Receiver (PIR). This study used the IBM SPSS program version 23 for windows. It presented in tabular and textual form based on univariate and bivariate analysis.

Based on the results of the Arduino Uno microcontroller circuit and passive infrared receiver, the researcher carried out a test to give a command program to the microcontroller by inputting data from the computer into the microcontroller as follows:

1. Software Test

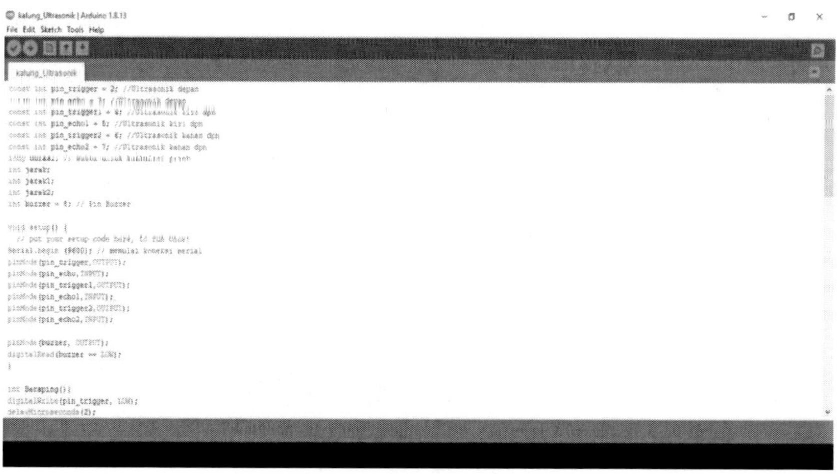

Figure 9. Ultrasonik arduino system software.

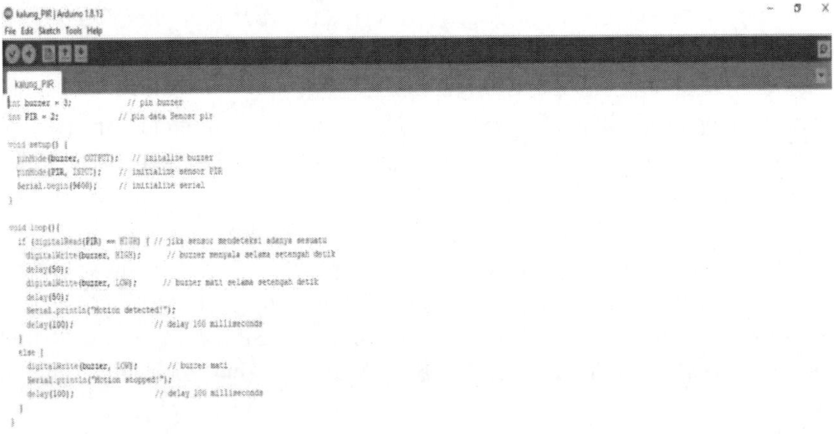

Figure 10. Device system (PIR).

2. Hardware Test

Based on the results of the Hardware test, the completed circuit is designed for a distance detector using a necklace based on Arduino ultrasonic sensors and passive infrared receivers. Then the unification of all devices is carried out. After the hardware device has been programmed with the microcontroller and executed using the downloader, the program automatically enters the microcontroller. Here is a picture of the results of the system design:

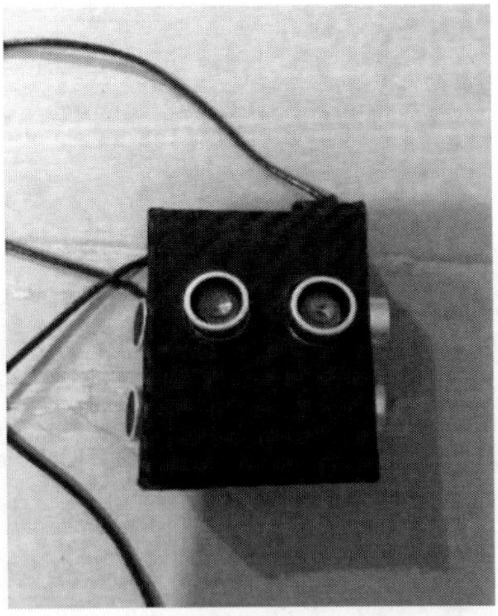

Figure 11. Ultrasonic sensor.

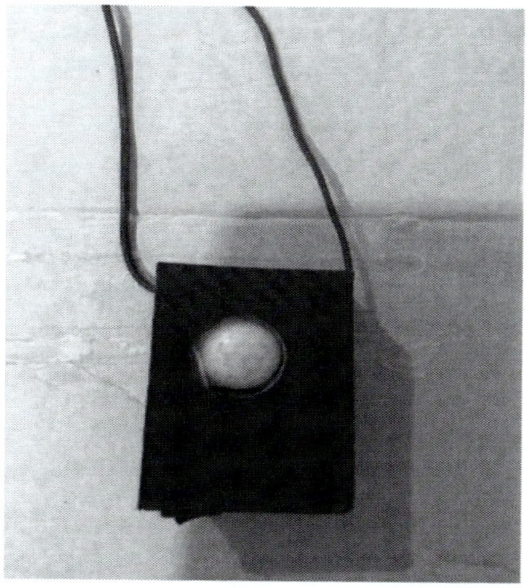

Figure 12. Pir sensor.

According to "The room security system uses the HC-SR501 PIR sensor, the HC-SR04 ultrasonic sensor, and the Arduino Mega-based camera VC0706," the results show that the security system can detect the presence of someone entering the room without permission and give a warning (Niyamaza, et al. 2020).

In line with the research, "Designing an Arduino Control System in A Trash Can Using A PIR Sensor and An Ultrasonic Sensor," the results obtained are: that the distance detected by the PIR sensor is 1-2 m, the distance of the person detected by the ultrasonic sensor is 25cm-50cm when the mechanism is closed, garbage at a distance of 25cm open and at a distance of 50cm closed.

Description of Respondents by Gender

The researcher found the research results regarding the frequency distribution of respondents using an Arduino ultrasonic sensor necklace and passive infrared receiver based on gender in the Villa Pamulang community, Depok City. The results showed that most of the respondents, namely 25 respondents (62.5%), were female, and most of the respondents were 24 respondents (60.0%), were female. According to Pertiwi & Budiono 2021, regarding "Community Physical Distancing Behavior in the COVID-19 Pandemic," the results of variables related to the behavior of implementing physical distancing, namely female gender, obtained results (46%) who did not apply physical distancing. This research is in line with Guspisa's research (2021) on "Analysis of Adolescent Physical Distancing Behavior in the city of Palembang during the covid-19 Pandemic." The univariate analysis results shows that good physical distancing behavior was carried out by 95 respondents, more than half of them, which was around 55.1%, who were female, couldn't apply physical distancing effectively. According to Budiono 2020, the researcher obtained the results of variables related to physical distancing, namely the female gender variable, with a percentage (95%). Preliminary results revealed that out of 115 respondents (80%) had used masks properly and correctly, (60%) had implemented handwashing with soap, and only (20%) had implemented physical distancing. The researcher concluded that increasing outreach activities, disseminating information, and restricting social activities, it shows that the community's physical distancing behavior is still low. The PIR sensor (100%) is in good health.

Descriptions of Respondent's Age

The researcher found the research results regarding the frequency distribution of respondents using an Arduino ultrasonic sensor necklace and passive infrared receiver based on gender in the Villa Pamulang community, Depok City. The results showed most of the respondents were 21 persons (52.5%) aged 17-25 years old, and a small proportion of respondents were 5 persons (12.5%) aged 46-55 years old. The users of passive infrared receiver necklaces found that most of the respondents were 24 persons (55.5%) aged 17-25 years old, and a small proportion of respondents were 4 persons (10.0%) aged 46-55 years old. According to Fatmawaty 2020, most the teenagers want and demand freedom. Because of this, many teenagers do not comply with the Covid-19 protocol.

Description of Respondents Based on Condition

The researcher found the research results regarding the frequency distribution of respondents using an Arduino ultrasonic sensor necklace and passive infrared receiver based on gender in the Villa Pamulang community, Depok City. The results obtained for all respondents, namely 40 ultrasonic sensor respondents (100%) in healthy condition. 40 respondents according to Firdaus 2020 on "Description of Attitudes and the Health Lifestyle of the Indonesian

People during the covid-19 Pandemic," the results obtained were multivariate respondents (66657). The results of the attitude and condition of the community regarding Covid-19 were (6,130) respondents agreed that the covid pandemic could be controlled, (6,038) respondents from the Indonesian community would continue to be vigilant, and (6,489) respondents consumed balanced nutritional foods.

Description of the Public's Perception of Physical Distancing before the Treatment (Arduino Ultrasonic Sensor Necklace)

The researcher found the research results regarding the frequency distribution of respondents using an Arduino ultrasonic sensor necklace and passive infrared receiver based on gender in the Villa Pamulang community, Depok City. The results obtained that most of the respondents were 20 respondents (50%) in the poor category. This research is in line with research conducted by Latif (2021), that half of the respondents with a percentage (50.0%), as many as 70 out of 140 respondents did not implement physical distancing.

Physical distancing has become an important tool in reducing infections and deaths during the COVID-19 pandemic. Physical distancing can reduce the risk of transmission and slow the spread of COVID-19. Public compliance with health measures and policies can have an important effect on the state of public health.

Based on Google's research in 2020, it ound that the mobility of Indonesians is still high compared to ASEAN countries. The data was obtained from 131 countries implementing Physical Distancing by reducing activities in various sectors such as shopping centers, parks, stations, terminals, offices, and housing. In Indonesia, Google noted that there was a decrease in activity in public places and an increase in activity at home or residence by 15%. The trend of mobility in restaurants, cafes, cinemas, libraries, museums, amusement parks, shopping centers decreased by 47%, markets, warehouses, food, drug stores, and pharmacies decreased by 27%, beach national parks, plazas, and public parks experienced a decline. A decrease of 52%, the trend of mobility in bus terminals, stations, or other public transportation areas has decreased by 54%, and in offices, it has decreased by 15%.

Description of the Public's Perception of Physical Distancing before Being Given Treatment (Passive Infrared Receiver Sensor Necklace)

The researcher found the research results regarding the frequency distribution of respondents using an Arduino ultrasonic sensor necklace and passive infrared receiver based on gender in the Villa Pamulang community, Depok City. The results obtained that most of the respondents were 22 respondents (55.0%) in the poor category. This research is in line with the research conducted by Mardhia (2020) on "The Iimplementation of Health Protocols and the Impact of covid-19 on the Community." It showed that (93%) of the public knows enough about the covid-19 pandemic, (84%) others are quite worried, (65%) do not apply physical distancing, (7%) are very worried and (9%) have no effect. According to Paul 2020 on "Physical Distancing and Emotional Closeness Amidst covid-19," the average reproduction number, namely the average number of people infected with the coronavirus and those who are vulnerable to help break the chain of infection and slow down, if not stop the spread of the epidemic on average. The average person infected by a covid-19 individual is estimated to be around two or three. It requires us to isolate at least 50-70% of the population to flatten the pandemic curve.

Description of the Public's Perception of Physical Distancing after Being Given Treatment (Arduino Ultrasonic Sensor Necklace)

After being given treatment with an Arduino ultrasonic sensor necklace in the community of Villa Pamulang, Depok City, it obtained results for all respondents, namely 40 respondents (100%) with good categories. It's in line with research conducted by Pindrayana (2020). It obtained the results of the Arduino ultrasonic sensor test, which can detect human movement at a distance of 1-5 meters. According to Bakhtiyar 2020, the results of sensor testing with a distance of (10-200cm) are the results of experiments from the ultrasonic HCSR sensor with a maximum distance of (60cm). The sensor did not experience an error. But when it exceeds the distance of (71 cm) the error showed on average of (142%), but it was still accurate due to the tolerance of the researcher. Similar to the research supported by Stefano's son (2020), the results of the ultrasonic sensor accuracy value with a percentage of (99%) and a precision value of (97%) with sensor characteristics of (97%).

This research is in line with research supported by Latif 2021 on "Analysis of How the Arduino Uno microcontroller and Ultrasonic Sensors work for Smart Jacket Design as the Implementation of Physical Distancing." The researcher carried out a test with the results of the system working where the output of all working conditions corresponds to actual conditions. The sensor could detect the distance to the object, and it serially showed the distance to the object with the provisions of a distance of 2-5m. An object < 100cm was detected in a condition that violates the rules of physical distancing, the LED is on, the buzzer sounds, and the results are appropriate. Detected objects more than 100 cm applied physical distancing rules. The led light did not turn on, the buzzer did not beep, and the results matched it.

Description of the Perception of Community Distancing after Treatment (Passive Infrared Receiver Sensor Necklace)

After being given treatment with a passive infrared receiver sensor necklace in the community of Villa Pamulang, Depok City, it showed that most of the respondents were 36 respondents (90.0%) with good categories. Based on research conducted by Zulhafandi (2020), most respondents had a good level of knowledge, namely 205 respondents, with a percentage result of (94.4%) of the majority being able to do physical distancing.

Survey on "Prevent transmission of Covid-19, Banyumas operates distance detection devices," the researcher operated 200 devices at the Banyumas public service mall, Central Java. The mall visitors got the devices, so that when they take care of licensing they can maintain a safe distance according to health protocols. This tool was developed by Bapedalibang with the Telkom Purwokerto Institute of Technology. With this sensor, it will detected whether someone is close or not. If the distance is less than 1-2 meters, the device will beep. It's in line with the research conducted by Yusup 2020, physical distancing is effective in preventing transmission of viral infections that transmitted through physical contact. It includes sexual contact and indirect physical contact, namely by touching a contaminated surface, or transmission through the air. It can also be transmitted through the air, contact with splashes, or droplets from coughing or sneezing. It concluded that the provision of a Passive Infra-Red Receiver (PIR) sensor necklace affects the public's perception of Physical Distancing because it had a value after being given treatment, namely with a percentage of (80%), the mean value was 4,800.

The Effect of the Effectiveness of the Arduino Ultrasonic Sensor Necklace on the Perception of Physical Distancing

The mean result before being given sensor treatment (Arduino ultrasonic) was 72.95, while the mean result after being given sensor treatment (Arduino ultrasound) was 83.53. There was an increase in the average before and after being given treatment with the ultrasonic Arduino sensor necklace. The mean result of 10,575 with a statistical test obtained a value of 0.00. It concluded that there is a significant difference between before and after being given treatment (Ultrasonic Arduino).

According to Arsada 2017 regarding "Ultrasonic Sensor Applications for Detecting Distance Positions in Space Using Arduino Uno" the results obtained were 3cm-60cm distance accuracy with an error value of 0%. The distance was 60cm to 200cm with an error accuracy rate of 1.78% with the original distance on the hcsr04 ultrasonic sensor user. It's in line with other research, that the ultrasonic sensor distance of 50cm in buzzer conditions was 1-2m distance accuracy could detect distances effectively (Arsada, 2017).

The research is in line with the research supported by Susilo et al. (2020) on "Car Parking Distance Controller Using Ultrasonic Sensors Based on Arduino Un." The car approached the sensor with a distance of 2-5cm valid. The sensor output sound emitted the best valid sound. The output sound emitted a valid forward sound if the car approaches the sensor exactly 2-5cm, the sound output made a valid sound. Thus the Arduino Ultrasonic sensor necklace affected the perception of physical distancing of the Pamulang community when the researcher gave them the treatment of the Arduino Ultrasonic sensor necklace to the community's perception of physical distancing in maintaining a more effective distance.

The Effect of the Passive Infrared Receiver (PIR) Sensor Necklace on the Perception of Physical Distancing in the Community

The mean result before being given sensor treatment (PIR) was 71.28, while the mean result after being given sensor treatment (PIR) was 75.25. There was an increase in the average before and after being given treatment with a passive infrared receiver sensor necklace. The mean result of 3.975 with a statistical test obtained a value of 0.00. So it was concluded that there is a significant difference between before and after being given treatment (PIR).

According to Muzawi et al. 2020 regarding "Prototypes of COVID-19 physical distancing tools using Passive infrared receivers." The study obtained results where the sensor module parameters (PIR) ranging from 50 cm - 4 meters can detect distances and the design of COVID-19 physical distancing tools designed to detect the distance between one person and another can work well in preventing and fighting COVID-19. The use of a Passive infrared receiver sensor to measure the distance and the Arduino Uno program will analyze the distance, when the distance is less than 1 meter (PIR) will turn on the buzzer (bell). With a passive infrared receiver technology prototype that is used safely and comfortably in the application of physical distancing. Thus, the passive infrared receiver (PIR) sensor necklace affects the public's perception of physical distancing. When given the Passive infrared receiver sensor necklace treatment, the public's perception of physical distancing in maintaining distance is more effective.

Comparison of the Effectiveness of Ultrasonic Arduino with Passive Infrared Receiver (PIR) on the Perception of Physical Distancing in the Pamulang Villa Community

Based on table 5.15, the Arduino ultrasonic sensor and passive infrared receiver had a test for Paired T samples. On the ultrasonic sensor, there is a P-value (2 tiles) of 0.000 (<0.05) with a T count of (14.245). And the PIR sensor had a significant result of 0.001 (<0.05) with a T count of (3.554). The T shows the paired test statistic where the T count must be greater than the T table, then compare the T count with df.

It showes the degrees of freedom from the test. Then the df results from these two sensors were (39). Then equate the T count with the T table in number (39). Then it got the results with a df value of (2.791), where the T count was greater than on the T table. It means that the results of these two sensor necklaces were significant. However, it concluded that the Arduino ultrasonic sensor necklace has a higher level of effectiveness than the passive infrared receiver sensor (Sugiyono 2018). The results of research conducted by Hardayani 2020 on health technology journals and social sciences related to "Detection Devices for the Presence of Children in the room using a PIR sensor based on an Arduino microcontroller, the fuzzy logic method." The experimental results obtained if the distance is 5-100 cm, the speaker would turn on with a loud sound to notify that there were children in the room. If the distance was 75 cm-399 cm, the speaker would be on with a medium sound to notify the room that there was a child. If the distance was 300 cm-600 cm, the speaker would be off, or no sound at all, to let you know that there were no children in the room.

Strategy of Ultrasonik Arduino with Passive Infra Red Receiver on the Percepion of Physical Distancing

Conclusion

1. Characteristics of respondents identified using an Arduino ultrasonic sensor necklace and a passive infrared receiver sensor necklace (gender, age, and condition) of the community at Villa Pamulang RW 11, Depok City. Most of the respondents were female, as many as 25 respondents (62.5%) and 24 respondents (60%), most of the respondents aged 17-25 years old were 21 respondents (50.0%) and 22 respondents (55.5%), all ultrasonic sensor respondents with health conditions were 40 respondents (100%), and 40 respondents (100%) PIR sensors.
2. Identified the results of the effectiveness of using an Arduino ultrasonic sensor necklace on the perception of physical distancing in the community of Villa Pamulang RW 11 Depok City before being given treatment with an average value of the level of difference (Mean) of 72.92 and after being given the treatment with an average value of 83, 53.
3. Identified the results of the effectiveness of using a passive infrared sensor necklace on the perception of physical distancing in the people of Villa Pamulang RW 11, Depok City, before being given the treatment with an average value of the level of difference (Mean) of 71.28 and after being given the treatment with an average value average 75.25.
4. Analyzed the comparison of the effectiveness of the ultrasonic Arduino sensor necklace with the passive infra-red receiver (PIR). The results obtained are ultrasonic Arduino with a P-Value of 0.00 (<0.05) and a passive infrared receiver of 0.01 (<0.05). It concluded that H0 was accepted and Ha was rejected. It means that there is a comparison of the effectiveness of the Arduino ultrasonic sensor necklace with the passive infrared receiver sensor necklace.

References

Abdulkhaleq, N. I., Hasan, I. J., & Salih, N. A. J. (2020, February). Investigating the resolution ability of the HC-SRO4 ultrasonic sensor. In *IOP Conference Series: Materials Science and Engineering* (Vol. 745, No. 1, p. 012043). IOP Publishing.

Afrianti, N., & Rahmiati, C. (2021). Factors affecting public compliance with the COVID-19 health protocol. *Scientific Journal of Permas: Scientific Journal of STIKES Kendal, 11*(1), 113-124.

Aquarini. (2020). *The Influence of Political Policies on Physical Distancing Compliance Preventing the Spreading of COVID-19.* (2)19,88.

Arsada, B. (2017). Ultrasonic sensor application for distance position detection in space using Arduino Uno. *Journal of Electrical Engineering,* 6(2).

Bakhtiyar A, (2020) *Ultrasonic Sensor Application for Detecting Distance Positions in Space Using Arduino Uno.* 2020 (1) 2-3.

Budiono. (2020) *Community physical distancing behavior during the covid-19 pandemic.* 1(1)2-3. http;//Journalunes.ac.id/sju/index.php/1/PHN.

Dan Sensor Ultrasonik Untuk Perancangan Smart Jacket Sebagai Penerapan Physical Distancing. *OSF Preprints.* January, 2.

Dewiyuliana, D., & Septiana, N. (2020). The Effect of Spreading Covid-19 Information Through Whatsapp on Community Preparedness in Facing Covid-19. *Scientific Journal of Permas: Scientific Journal of STIKES Kendal,* 11(1), 103–112.

Gupisa, C., & Fajar, N. A. (2021). *Analisis Perilaku Physical Distancing Remaja Di Kota Palembang Pada Masa Pandemi Covid-19* (Doctoral dissertation, Sriwijaya University).

Hardayani, E. (2021). Detection Devices for the Presence of Children in the room using a PIR sensor based on an Arduino microcontroller, the fuzzy logic method. *Journal of Health and Social Sciences 3*(1), 11-21.

Hermawati, B., Handayani, O. W. K., Mahendrasari, D. S., Mukti, F. A., & Wijayanti, A. (2021). Health Education Regarding Handwashing Practices for Teachers as an Effort to Prevent Covid-19. Journal of Obsession: *Journal of Early Childhood Education. Journal of Obsession: Journal of Early Childhood Education*, 6(2), 894-902.

Ismail, R., Omar, Z., & Suaibun, S. (2016). Obstacle-avoiding robot with IR and PIR motion sensors. In *IOP Conference Series: Materials Science and Engineering* (Vol. 152, No. 1, p. 012064). IOP Publishing.

Latif, A. A. (2021). Analisis Cara Kerja Mikrokontroler Arduino Uno Dan Sensor Ultrasonik Untuk Perancangan Smart Jacket Sebagai Penerapan Physical Distancinguino Uno.

Ministry of Health of the Republic of Indonesia. (2020). Guidelines for Adapting New Habits during the Covid-19 Pandemic. Jakarta. Ministry of Health of the Republic of Indonesia. (2020). Regulation of the Minister of Health of the Republic of Indonesia Number 9 of 2020 Concerning Guidelines for Large-Scale Social Restrictions for Acceleration of Corona Virus Disease 2019 (Covid-19). Jakarta.

Muqorobin, M., & Rais, N. A. R. (2020). Analysis of the Role of Information System Technology in Lecture Learning During the Corona Virus Pandemic. In *Proceedings of the National Seminar & Call for Paper STIE AAS* (pp. 157-168).

Muzawi, R., Efendi, Y., & Rio, U. (2020). Prototype Alat Physical Distancing Covid-19 Menggunakan Arduino Uno. *JOISIE (Journal of Information Systems and Informatics Engineering)*, *4*(2), 121-127.

Niyamaza, M., Ubaya, H., & Sembiring, S. (2020). *Perancangan Sistem Keamanan Kendaraan Menggunakan RFID (RADIO FREQUENCY IDENTIFICATION) BERBASIS IoT (INTERNET of THINGS)* (Doctoral dissertation, Sriwijaya University).

Pertiwi, G. S., & Budiono, I. (2021). Perilaku Physical Distancing Masyarakat pada Masa Pandemi Covid-19. *Indonesian Journal of Public Health and Nutrition*, *1*(1), 90-100.

Purnamasari, I., & Raharyani, A. E. (2020). The level of knowledge and behavior of the people of Wonosobo Regency about Covid-19. *Scientific Journal of Health,* 10(1), 33-42.

Qi, L., Yang, Y., Jiang, D., Tu, C., Wan, L., Chen, X., & Li, Z. (2020). Factors associated with the duration of viral shedding in adults with COVID-19 outside of Wuhan, China: a retrospective cohort study. *International journal of infectious diseases*, *96*, 531-537.

Riedel, S., Morse, S. A., Mietzner, T. A., & Miller, S. (2022). *Microbiologia Médica de Jawetz, Melnick & Adelberg-28*. McGraw Hill Brasil.Romedzo M, Des (2020) *Covid-19 Physical Distancing Tool Protoytype Using Arduino Uno*.2020.4(2)121-127.

Sulistiadi, W., Slamet, S. R., & Harmani, N. (2020). Handling of public stigma on covid-19 in Indonesian society. *Kesmas: Jurnal Kesehatan Masyarakat Nasional (National Public Health Journal).*

Susilo, A., Rumende, C. M., Pitoyo, C. W., Santoso, W. D., Yulianti, M., Herikurniawan, H., ... & Yunihastuti, E. (2020). Coronavirus disease 2019: A review of the current literature. *Indonesian Journal of Internal Medicine,* 7(1), 45-67.

Syapitri, H., Siregar, L. M., & Saragih, F. L. (2020). Prevention of Covid-19 Transmission Through Socialization and Distribution of Masks at Medan's Pringgan Market. *Journal of Community Service Creativity* (PKM). *3*(2), 274–282.

Tarascon, J. M., & Armand, M. (2011). Issues and challenges facing rechargeable lithium batteries. In *Materials for sustainable energy: a collection of peer-reviewed research and review articles from Nature Publishing Group* (pp. 171-179).

Widyaningrum, N., & Putri, Y. D. (2020). Gambaran Penerapan Physical Distancing Sebagai Upaya Menekan Persebaran Covid-19 di Provinsi Daerah Istimewa Yogyakarta. *NUSANTARA: Jurnal Ilmu Pengetahuan Sosial*, *7*(2), 470-481.

World Health Organization. (2020). *Clinical management of severe acute respiratory infection when novel coronavirus (nCoV) infection is suspected: interim guidance, 25 January 2020* (No. WHO/nCoV/Clinical/2020.2). World Health Organization.

Zhang, T., Wu, Q., & Zhang, Z. (2020). Probable pangolin origin of SARS-CoV-2 associated with the COVID-19 outbreak. *Current biology*, *30*(7), 1346-1351.

Zlatanov, N. (2015). Arduino and open source computer hardware and software. *J. Water, Sanit. Hyg. Dev*, *10*(11), 1-8.

Zulhafandi, Z., & Ariyanti, R. (2020). The relationship between knowledge about Covid-19 and compliance with physical distancing in Tarakan. *Mutiara Mahakam Midwifery Journal,* 8(2), 102-111.

Chapter 11

Analytical Model for Orifice as a Flow Measurement Device

Santosh Kumar Panda[1,*], Kali Charan Rath[1,†] and Balaji Kumar Choudhury[2,‡]

[1]Department of Mechanical Engineering, GIET University, Odisha, India
[2]Parala Maharaja College of Engineering, Berhampur, Odisha, India

Abstract

Flow measuring devices play an important role in various industrial applications. Among the flow measuring devices, orifice meters are one type of flow controlling and measurement devices. The Orifice meter carries utmost importance because of its simplicity in design, low cost of manufacturing and reliability for a wide range of fluid flow measurement applications. The applications of the orifice are in the oil and gas industry, Research and testing Laboratories, power plants, refrigeration industries, chemical and processing industries, etc. The orifice meter replaced the wellhead chokes to overcome the difficulty of petroleum production industries with better control and accuracy in the mass flow rate measurement. Single and multi-phase fluid flow measurement reading has been carried out with the estimation of pressure drop (Δp) along the flow, and coefficient of discharge Cd with the applications of Bernoulli's principle. The present chapter considers an orifice as a flow measuring device for the various applications on the variation of shapes and sizes with respect to its classifications, design, and operating parameters for better understating and designing of orifice meters. The

* Corresponding Author's Email: santosh.kumarpanda@giet.edu
† Corresponding Author's Email: krath@giet.edu
‡ Corresponding Author's Email: balajichoudhury@gmail.com.

In: Computational Intelligence for Sustainable Development
Editor: Brojo Kishore Mishra
ISBN: 979-8-88697-198-9

study also presents the flow characteristics output as p and C_d of the orifice on the effects of area ratio, space ratio, Reynolds number along the upstream and downstream of the flow. The flow dynamics behavior has been carried out with the use of numerical simulation based on Finite volume method (FVM) and Computational Fluid Dynamics (CFD) technique. ANSYS Fluent is a commercial software used to determine flow parameters. The article gives an overview of the numerical analysis of the orifice for better understanding of the flow characteristics.

Keywords: flow measuring device, orifice, coefficient of discharge, pressure drop, area ratio, space ratio

Abbreviations

P	Pressure, kPa
V	Velocity, m/s
A	Flow area, m^2
A_c	Area of contraction, m^2
D	Diameter of pipe, m
s	Orifice thickness, m
S	Slip
x	Quality, kg/kg
C_d	Coefficient of discharge
m	Mass flow rate, kg/s
z	Length, m
Δ	Difference
Φ	Two phase multiplexer
q	Primary phase
p	Secondary phase
ρ	Density, kg/m^3
σ	Area ratio
σc	Contraction coefficient
l	Liquid
g	Gas

1. Introduction

A wide range of flow measuring devices is used such as turbine meter, vortex shedding meter, Coriolis meter, and magnetic resonance meter for different fields of flow measurement applications. An orifice meter is another type of device like a venturiimeter, venturi tube, nozzle, and orifice and density meter to measure the flow rate with the use of fluid behavior. The orifice works on the theory of Bernoulli's equation which gives the relation between pressure and velocity as an energy head of a fluid flow. The pressure drops (through the flow is predicted as mass flow rate or volume flow rate. The principle of orifice meters is applied in the mining, chemical processes, oil, energy, nuclear, refrigerant devices, electricity production systems, oil pits pipeline, and food industries, where the fluid flows as a single-phase and multi-phase flow, which plays an essential role. Wellhead chokes are a flow measurement device used in the fuel and gas industry. Wellhead chokes regulate the production rates of wells. Due to pressure shock and turbulence in the downstream wellhead chocks not estimating a gap the flow rate accurately also creates internal damage to the surface and the fluid reservoir, Gassan (1991). Minimize the above disturbance effect orifice used as a flow measuring device on the above field.

1.1. Flow Measuring Device

Flow measurement devices play an important role for the determination of flow profile for a fluid. Types of flow measuring devices are described as follows

a. *Orifice Plates*

   The most common and simple flow measuring device is orifice which will be discussed in detail during this book chapter.

b. *Weir*

   A weir is an obstruction used as an orifice for open fluid flow like reservoirs, dams, lakes, ponds and canals. Normally water (incompressible fluid) is used for weir flow for the irrigation and drainage application.

c. *Nozzle*

A nozzle is a flow passage that has varying cross section area used to increase the flow velocity with the expense of pressure drop. The types of nozzles are convergent, divergent and convergent – divergent used as per the required applications. The other application for the nozzle is to control the mass flow rate, speed, velocity, change of flow direction, etc. of a flowing stream.

d. *Venturimeter and Venturi Tubes*

Venturimeter is an obstruction type flow meter similar to a nozzle having a convergence and divergence part used to measure the flow with the help of Δp. The Δp is less in the venturimeter compared to the orifice meter but has a more complicated design structure.

e. *Pitot Tubes*

A pitot tube is a flow (velocity) measuring device used to measure the difference between static and kinetic pressures at a single (stagnation) point of a flowing fluid. Bernoulli's theorem is used to calculate the velocity of the flow for small and larger ducts. After calculating the velocity with the multiplication area, mass and volume flow rate is determined.

f. *Rotating Meters*

Rotating meters are classified into two types: Turbine meter and paddle wheel meter. Both the meter uses the linear velocity of a rotating element to measure the flow with the impact of fluid flowing velocity or actual flow rate of flow. This blade or impeller is mounted on a shaft with the support of a bearing at the center and locked with a key. When the fluid affects the tip of the blade, the impeller and shaft etc., with the fluid force. The rotating blades create angular velocity and are converting to flow quantity with the calculation. This device is more accurate for gas, liquid and gas-liquid flow.

g. *Vortex Shedding Flow Meters*

Bluff body with vortex, vortices frequency and wake, concept is used for measurement of flow velocity. The vortices are directly proportional to velocity of flow and mass flow rate.

h. *Magneto Flow Meters*

Magnetic resonance meters operate on the principle of induction. A conductive liquid surrounded with a magnetic field along the flow to measurable signal which is proportional to the flow velocity.

i. *Coriolis effect Meters*

This type of flow meters is affected with the use of the Coriolis effects (means motion mechanic of the fluid) which is directly related to mass flow rate applied to measure the mass and volume flow rate.

j. *Hot Wire/Film probes*

The low magnitude flow velocity of air, liquid and gas flow rate is measured with the help of hot wire or hot film anemometer. The principle of this type of flow device is that a resistance wire is placed along the flow direction with a constant current flow and the velocity will be measured with the resistance of the wire. Based on the property of the material the velocity of the flow will be measured. 2 – Wire and 3 – wire anemometer is a more popular measuring device for the velocity measurement.

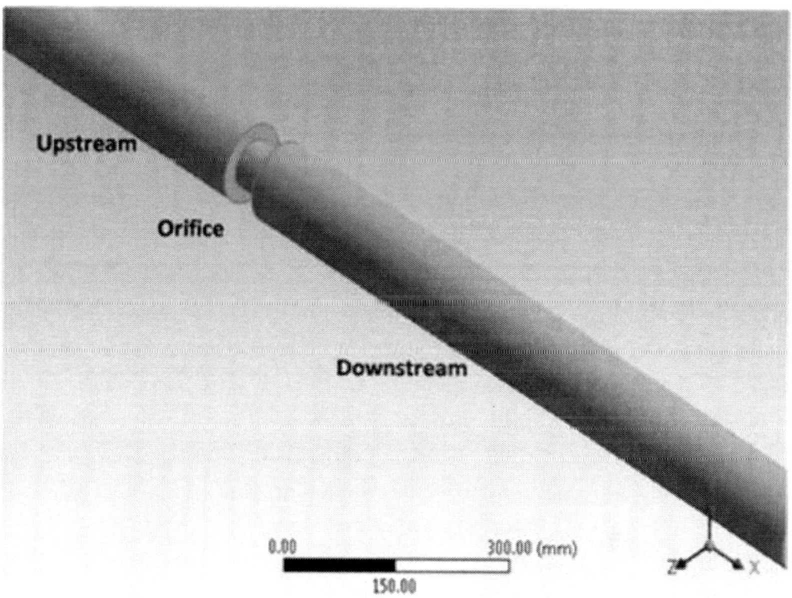

Figure 1. Measuring system used with orifice.

Figure 1 shows the geometric profile of the orifice in a pipe. In a fluid flow, the fluid passes upstream of the flow along the orifice, after a certain level of filling the fluid passes downstream. During the flow the pressure drops suddenly after the orifice and the velocity increases. The pressure recovers after a certain distance pass from the orifice plate known as the expansion zone. Differential manometer is used as the measuring device for the measurement of pressure difference by manometric fluid height. The tapping point kept a 2D distance on the upstream side and a 0.5D distance apart from the orifice plate in the downstream. The smallest cross-sectional area (A) of the orifice meter is called vena contractor. The major advantage of the orifice meter over the other flow-measuring device is accurately determining flow parameters that the reason acts as a calibration device. Orifice meters have simple devices to operate, construct, are able and cost less with no moving parts. Due to the simple structure, an orifice meter doesn't require calibration for fluid flow in an extensive series of temperature and pressure conditions. The drawback of the orifice is the output results are nonlinear in nature due to turbulence in the downstream, also in the longer run, the edge of the orifice erodes which affects the flow profile. The flow coefficient is also affected by pipe layout, nature of flow and types of fluids which creates interest in the research field.

1.2. Introduction to Orifice Plate

The operational parameters of fluid flow are classified into two types such as positive displacement and Inferential. Positive displacement fluid flow meters which consist of parts, such as the primary component, which has direct contact with the fluid interface and the secondary part through which fluid passed gives an output signal as volume flow rate. Positive displacement flow meters are very accurate instruments used for low flow rates. These devices are not affected by the fluid flow properties such as density, viscosity and floating nature. Reciprocating piston, rotating piston, rotating disk, sliding and rotating vanes, and gear and lobed impeller are the categories of positive displacement type appropriate for smaller flow rates. The advantages of such a flow meter are that it is used for high viscous and corrosive types of fluids. Orifice Plates, Flow Nozzles and Venturi Tubes, Pitot Tubes, Rotating meters, Vortex Shedding Meters, Magnetic Meters, Coriolis Meters are known as Inferential types. The flow meter works with the principle of flow phenomenon. Nozzle and venurimeter is the primary interfacial device with 60% higher flow rate than the orifice meter for a particular flow condition. For a particular pressure drop, a smaller diameter nozzle or venturimeter gives a better flow rate but the devices are complicated and more expensive than orifice meters. In the comparison of the percentage of p of different interfacial type flow meters along the orifice flow for a particular diameter ratio, and found orifice plate has utmost pressure drop and guides better results for flow accuracy (Emerson, 2017) over other flow measuring devices. The discharge of the mass flow rate through an orifice is presented in equation (1) which describes Bernoulli's equation considering with no energy loss during the flow.

Orifice devices have more reliability, range-ability and repeatability for wider applications. The measurement of flow coefficients of the orifice meter is having uncertainty of ± 0. 5 wide range of orifice diameter ratio. This uncertainty further enhances if the diameter ratio is below 0.2 and higher than 0.7. Similarly, the uncertainty gives minimum within a Re range of 4000 to 100000, which is the wide range and reliability for the flow and geometrical application. The operational life time of orifice is 5000 to 17000 m³/hour with maintaining ± 1% error. Due to simple construction, orifice plates are giving the same reading for same flow conditions, which gives more repeatability capability for the flow measurement.

$$C_d = \frac{\dot{Q}}{\frac{A_1 A_2 \sqrt{2g\frac{\Delta p}{\rho}}}{\sqrt{A_1^2 - A_2^2}}} \quad \text{or} \quad C_d = \frac{\dot{Q}}{\frac{A_2 \sqrt{2g\frac{\Delta p}{\rho}}}{\sqrt{1-\beta^2}}} \qquad \ldots (1)$$

where,

C_d = Coefficient of discharge

Q = Volume flow rate, m^3/s

A_1= Area of the pipe, m^2

A_2= Area of the orifice, m^2

G z = Acceleration due to gravity, m/s^2

Δp= Pressure difference, Pa

ρ = Density of fluid, kg/m^3

x = Manometric height difference, m

β = Area ratio (a/A)

1.3. History of Orifice

ISO 5167 series used for the design of orifice with the standard procedure and standard dimensions. The ISO 5167 gives information, which doesn't require any calibration for any fluid flow through the orifice. The first general concept for a flow measuring device was generated by Sextus Julius Frontinus, former Governor of Britain to solve the water crisis problem. He got the concept during his tenure as Water Commissioner in the city of Rome with the use of pressure difference along the fluid flow. In 18th century Daniel Bernoulli (1700 – 1782) a Professor of Mathematics in St. Petersburg derived the Bernoulli's theorem in 1738. After Bernoulli's theorem, the first flow measuring device was invented, the venturimeter, by Giovanni B. Venturi (1746 – 1822) at the University of Modena. The scientist carried out the experiment with the venturimeter by flowing water and found that the fluid gained velocity in the convergent part but decreased a large amount of static Δp after the throat of the ventuimter. The above work concludes that a flow measuring device can be designed with the concept of differential pressure drop. An orifice device for fluid measurement was introduced for the first time in 1797 by Giovanni B. Venturi, but used in practice by Clemons Herschel (1842 – 1930) in Holyoke Water Power Company, 1887 for a 20" diameter pipe to get stability in the water flow problems. Herschel has no information regarding the use of the orifice device but later it is reported. In 1890, Forrest M. Towl observed the application of orifice meters to measure mass flow rate of natural gas in Columbus, Ohio. Around 1904 Thomas R. Weymouth used orifice with flange tapping with an orifice plate and a Pitot tube to get a comparison study. The above study was not published anywhere because of unsuccessful results but in the year 1910 Alkali Company came up with useful results. Orifice meter invented by Robinson University of Ohio Professor S.W. implemented the gas pressure behavior in Columbus, Ohio in 1990. T. Pennsylvanian experiments lead to the formation of coefficients for flange taps to produce flange taps flow analysis. American Gas Association (AGA) and the American Society of Mechanical Engineers (ASME) carried out lot of research during 20th century and reported in articles "History of Orifice Meters and The Calibration, Construction, and Operation of Orifices for Metering" which gives brief about details thing about the orifice research.

1.4. Types of Orifice Plate

Figure 2 demonstrates a range of orifice plates for different flow measurement appliances for different ranges of flow rate. The different type's orifice plates according to the shape are thin sharp edges, concentric, Eccentric, Segmental, Quadrant, orifice plates are used. Sharp edge type orifice is commonly used for a wide range application because of accuracy of measurement for low viscous fluids such as water, air and gas. Eccentric orifice plates are used for non-abrasive, solids and small amounts of liquids flows, like chemical manufacturing, nuclear plants, steel and paper industries. Segmental or sector orifice plates are used for liquid and gas, non-abrasive impurities, light slurries and dirty gases like sewage application. The quadrant edge and conic orifice plate is used for high viscosity fluids or low flow Re. The inlet tip of the conic orifice is bevel type shape and follows a counter-bored passage along the thickness. These types of orifices play an impact on the a/A of the orifice and other applications as per the requirements. The wake created downstream of the orifice also impacts on the shape of the orifice on the determination of Δp and the flow coefficient.

According to the thickness of the orifice, it is categorized as a thin and thick orifice. Different researchers such as Fossa (2002), Fossa et al. (2006), Roual and Das (2010) classify the orifice based on the geometrical parameters thickness to diameter ratio (s/d). If the s/d ratio is less than 0.5, then the orifice is thin and if the s/d ratio is more than 0.5 then it is thick, as shown in Figures 3(a) and 3(b). The expansion length of orifice is also considered for the above categorization of the orifice. If the length of the orifice is less than the expansion length of the orifice then it is called a thin orifice, otherwise, the orifice is a thick orifice. This classification plays an important parameter in the stability and accuracy of the determination of the flow rate on two-phase flows and highly turbulent flows. The orifice is further classified as per the interaction edge of flowing fluids. The classifications are Square, square with 45°, Sharp, streamlined approach, Slopping and Quadrant edges. The material properties required for orifice plates have good rigidity, non-corrosion, and non-erode against flowing fluid. Erosion can damage the edges of the orifice which affects flow coefficient, pressure drop, smoothness of flow, accuracy in measurement and contraction coefficient. The expansion of fluid in the downstream affects the flow expansion which causes loss of pressure energy. To overcome the problems high thermal expansion materials, have to be chosen. Different types of materials are used like stainless steel, carbon steel, and titanium, for improving strength Inconel, brass, copper, and Aluminum can be considered two phase flow where heat transfer phenomena take place. Glass, ceramic, different plastic also used as orifice material in visualization, corrosive free reactive fluids.

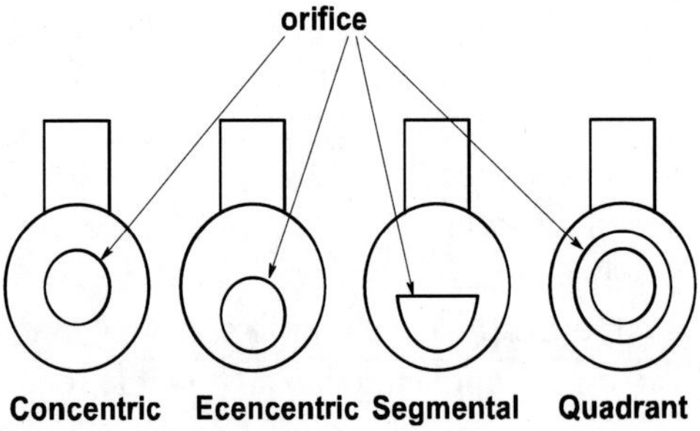

Figure 2. Types of orifice plate according to the shape of orifice.

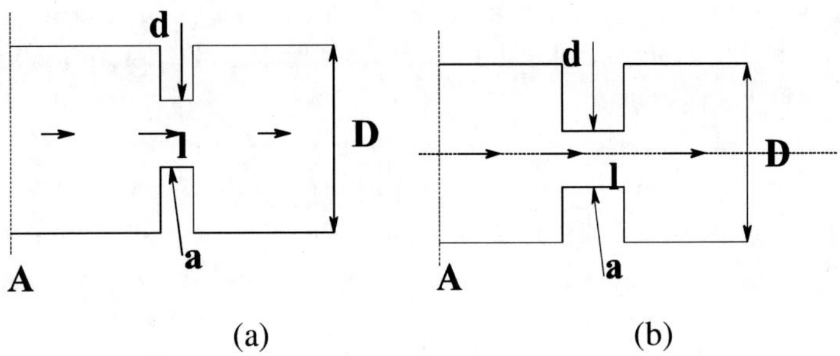

Figure 3. Types of orifice plate according to the thickness of the orifice.

1.5. Orifice Design

The design specifications of an orifice have been documented in ISO 5167-1:2003 and ISO 5167-2:2003. The ISO standards also provide the specification details such as diameter, diameter ratio to pipe, edge shape, ratio of pressure in upstream and downstream, roughness etc. The design parameters such as shape and size of the circularity of orifice holes, flatness, and roughness of orifice surface and parallelism of the two faces affects flow profile. Similarly, for the pipe parameters such as tapings shape, size and position, pipe roughness, concentricity of the orifice circularity and cylindricity of the pipe also considered for determining the pressure loss. Diameter ratio (β= d/D) of the orifice provides better accuracy to the measurement of coefficient of discharge on a range of $0.2 \leq \beta \leq 0.6$. The uncertainty increases if β value is within the range of $0.5 \leq \beta \leq 0.6$. Due to roughness and eccentricity in the pipe surface, Δp increases. The orifice plate should be thin but the thickness has less effect on C_d, but thick orifice gives stability to the flow to minimize the disturbance of the forces imposed to Δp, Mincks (2002). Parallelism of orifice surface does not affect on the smaller diameter pipe up to 200 mm, but if the diameter of pipe is more than 200 mm then the fluid pressure creates negligible impact on the plate surface on the C_d Michael (2015). An inclination of the orifice plate orientation gives errors up to ± 0.2%, Jepson and Chipchase (1975), roughness produces 0.28% Reader-Harris (2010), edge sharpness gives up to 0.33% to 0.51%, for thin orifice, 0.10% to 0.20% for thick orifice Botros (1992), and pipe tapping gives 0.26% on the C_d Gallagher (1968). As per the different literature review reports concluded that the diameter ratio and thickness ratio influence the C_d and Δp measurement with the help of an orifice.

1.6. Orifice Taping or Pressure Tapping

The orifice ISO manual, ISO 5167 reported three different types of pressure tapping like flange tapping, D and D/2 tapping and corner tapping

1.6.1. Flange Tapping

Generally, Flange tapping is used for oil and gas flow measurement for distance of 1 inch (2.54 cm) apart from orifice plate in both upstream and downstream face. The distance will be provided to hold the orifice in proper location during the flow with the help of proper packaging. Flange tapping is more suitable because of its simple design, easy to manufacture. The diameter ratio of inner and outer diameter of flange tapping is so high, which is giving advantage of less or no pressure expansion at the tapping during the fluid flow. That's the reason no correction factor was added for determination of C_d. Similarly, the roundness, sharpness and depth of tapping also do not affect the flow for a wider range of Re. The drawbacks of introducing orifice plates with flange tapping make the flow measuring device more complex. The size of 50 mm or lesser diameter flange tapping affects the measurement of C_d because of pressure energy recovery in the downstream zone and vena contracta region.

1.6.2. D and D/2 Tapping

This type of tapping is positioned 1D distance in the upstream and D/2 distance downstream apart from the orifice plate. The 1D distance minimizes the effect of flow disturbance that occurs in the upstream and downstream of orifice because of wake created due to sudden expansion in the flow passage. The fabrication and joining process such as welding creates protuberance of the flange, pipe and orifice disturbed the neighborhood flow velocity. Proper care has to be taken to welding protuberance during fabrication with careful and skillful machining of flange. The flow disturbance can be avoided with correctly placed tapping as mentioned in different databases for a wide range of specifications. The specifications are mentioned as follows: ISO 5167-2: 2003 provides the flange taping diameter is 0.13D and 1.3 cm. Equivalent API standard, informs the tapping diameters should be 0.95 cm (3/8 inch) for 2 inch and 3-inch pipes and 1.27 cm (1/2 inch) for 4-inch pipes and bigger API (2000). Similarly, European data and the American data for tubes of API specification are 0.04D means0.95 cm (3/8 inch). Flange tapping spacing for upstream and downstream are

$$(2.54 \text{ cm} \pm 0.05) \text{ cm when } \beta > 0.6 \text{ or } D < 15 \text{ cm}$$
$$\text{and } (2.54 \text{ cm} \pm 0.1) \text{ cm } \beta \leq 0.6, \beta > 0.6 \text{ or } 15 \text{ cm} \leq D \leq 100 \text{ cm}.$$

1.6.3. Corner Tappings

Corner tappings are another parameter concern for flow along the orifice. Corner tappings mean the location between the orifice plate and the pipe wall. As per the guideline of ISO 5167-2: 2003 single tapping is recommendable for clean fluids and gas flow, but for sewage or high viscous fluids the more rapid corner tapping is recommendable. The tapping size of 2 mm is better for orifice design up to diameter less than 100 mm. Multiple tapping is used for the orifice and flange but each tapping should connect with each other with some arrangement. Fluid flow like hazards, reactive or toxic flows the tapping should give at the top half of the pipe to avoid damages.

1.7. Location of Tapping in the Orifice

The flow area of the fluid stream continuously reduces after flowing past the orifice throat or smallest diameter section. The static pressure will fall to the minimum level at the vena contract following the sudden fall at the throat of the orifice during the fluid flow. The pressure energy has been recovered after passing through the venacontracta. The differential pressure profile has been found on an orifice flow with the location of the pressure taps. The positions of the tapping should be maintained on a single plane, and the location will be where the minimum pressure gradient is to be found, which will affect less error to the measurement. The locations of pressure tap for the differential pressure taps recommended in five points based on the different studies as mentioned: (a) Flange taps (b) Venacontracta taps (c) Radius taps (d) Full flow or pipe taps and (e) Corner taps.

a. *Flange taps.* Flange tap is suitable for pipe diameters up to 5 cm and more or the centerline is 2.5 cm apart from the orifice wall plate surface. The flange taps are not used for lesser pipe size of 35 cm diameter or the venacontracta may be closer than 2.5 cm from the orifice plate.
b. *Venacontracta taps.* Venacontracta taps placed in the upstream of the pipe of orifice plate. The tap is also placed downstream where the minimum pressure point occurs in the orifice plate. The Venacontracta taps are used if the pipe diameter is more than 15 cm.
c. *Radius taps.* Radius taps are equivalent to venacontracta type taps and are located at a distance of half diameter of pipe in the downstream side. These orifices are better than venacontracta taps and do not affect variation of orifice diameter ratio.
d. *Pipe taps.* These type taps are pointed at a distance of 2.5 times of pipe diameters in the upstream and 8 times the pipe diameter distance in the downstream located from the orifice plate. The pipe tap's locations are effects with the irregular in the pipe surface, detonation of the tip of the orifice and dimensional irregularities.
e. *Corner taps.* Corner taping has the same application as flange taps, where the pressure is measured at the Corner of the orifice plate and the wall surface of the pipe. The corner tap is suitable for the diameter is less than 5 cm.

2. Literature Review and Methodology

The literature reviews more concern about the shape and size of the orifice with its wider application for various fluids and fields. High sediment and viscous fluid with gravity dominated flow are significant applications for a sediment flow having specific problems to concern during flow measurement.

The circular orifice has limitations for the above problem, to evade that a segmental or sector orifice is considered for sediment flow to find out flow constraints such as Δp, pressure profile and the C_d. The opening angle of the sector varies the flow area positioned at the bottom of the pipe which allows moving the segmental material freely to avoid material accumulation in the upstream side. Sector orifices are considered as an alternative to the concentric plate for flow measurement. The diameter of the pipe is 40 mm with the consideration of the diameter ratio of 0.2 to 0.8, length to diameter ratio 0.2 to 0.9, the angle of sector having value 72º to 288º to determine the C_d. Based on the flow parameter a correlation developed as a function of Re for a particular a/A ratio, Yadav (1995). Similar type applications, an elliptical orifice plate is another type of orifice shape for semimetal flow. An elliptical orifice plate with horizontal, angular and vertical position is considered in a 38.1 mm circular pipe which maintains diameter ratio range of 0.2 - 0.8 with experimental study conducted to evaluate the C_d. Water is used as a working fluid. The distance of the vena contract from orifice in the downstream or the expansion length is determined based on the a/A ratio, l/d ratio, elliptical, displacement angle in the downstream. The operating parameters like non-dimensional pressure, discharge and limiting discharge coefficient considered as a function of length to diameter ratio along the downstream, mass flow rate, Re, a/A, ellipticity of the orifice, Athar (2012).

The energy phenomenon of the flow depends on the velocity, pressure etc. Similarly conventional techniques such as eddy formation, separation, wake, cavitations, friction losses etc. are used to determine the energy losses as a sudden drop in Δp along the flow. The sudden drop in Δp causes the flow pressure below vapour pressure which causes evaporation in multi-square-hole orifice passage, causing energy losses. Similarly, the above flow phenomenon is creating utmost energy losses along the orifice plate. 3-D Particle Image Velocimetry (PIV) and High-Speed Photography used in a flow through multi-square-hole orifice in a hydrodynamic cavitation reactor. The photography and images studies the velocity and pressure distribution for each multi-square-hole orifice plates for flow analysis, Dong (2013). Flow analyses of a slotted orifice are carried with area ratio (a/A) as an input parameter with wet gas (air and water mixture) as flowing fluid. Single and multi-phase flow simulation is applied with CFD (Computational Fluid Dynamics) analysis tool for slotted orifice, which has multi-passage to determine pressure profile and Δp. The outputs of the studies are estimated as static pressure, velocity and density contour along the flow domain for the slotted orifice. Compared to multi-hole-orifice, slotted orifice provides better pressure recovery on upstream and downstream flow regimes and gives uniform and steady velocity in the flow field. Further the experiment studies were carried out with two series connections of slotted orifices to improve the flow measurement accuracy of the gas and liquid flow of wet gas measurement along the pipeline. The advantages of slotted orifices give better flow control in wet gas measurement and flow pattern which enhances the design of orifice over the standard orifice, Geng (2006). Slotted orifices are utilized as a flow measurement appliance for the two-phase mixture and the natural gas industry. The orifice is used to avoid the uncertainty in the measurement of the upstream and downstream flowing condition for a diameter ratio of 0.5. Among the holes in the slotted orifice four large holes provide for the flange fastening and small four holes provide for decreasing the upstream gas disturbance effect along the flow passage. After the performance comparison the slotted orifice is replaced by the turbine flow meter and orifice meter for better control and avoids the uncertainty in the measurement, Morrison (2001).

A combination of orifices such as two-stage orifices is a new type of orifice configuration for the stability of flow measurement. Theoretical, CFD and experimental analysis consider a two-stage orifice for a comparison study with a single orifice on the determination of C_d and Δp analysis. Mineral oil was used as fluid for the two-stage orifice with range of Re 900 to 1700 having diameter of 3 mm orifice and found better design concept on the flow analysis compared to the single orifice. The study also provides better design for micro-orifice and micro-channel flow fields. The authors concluded that for a particular diameter of the two stage orifice gives better C_d on the comparison of single orifice, Gao and Wu (2019). Two-phase flow pressure analysis based on reversible and irreversible losses are carried out on sudden expansion and contraction of thick and thin orifice plates. R-113 is used as a working substance to evaluate the pressure drop. The volume-averaged methods used for solving the governing equation with the variation parameters such as quality of the fluid, liquid-to-vapour density and viscosity ratios. Local slip ratios play an important role to evaluate the Δp for the numerical analysis, Kojasoy et al. (1997). A number of flows measuring devices such as small orifices and nozzles are considered for an experimental study, carried out on subsonic and critical-flow regions. The study also considers a number of compressible gases such as air, CO_2, Ar, and He and a mixture of Ar - He to determine the C_d. Knife-edge, straight-bore, and nozzles are rounded-entry, and elliptical-entry of series of convergent shape orifice considered with diameters range of 0.9 to 1.9 mm in this study. Based on the number of studies with the variation of number of fluid and number of edges orifice correlations for Δp and C_d are developed in the different flow regions. The study concluded that at a particular free-area ratio perforated plate orifice produces better flow coefficient than circular single orifice and increase of plate thickness of 1.5 times Δp reduces significantly, Gan (1997).

A correlation-based comparison analysis on a perforated plate orifice having the same a/A with straight edged orifice in a single-phase flow on the variation of Re and geometrical parameters reported, Shanfang (2013). Comparison between numerical and experimental analysis of cylindrical, square edged and perforated orifice have been presented on the evaluation of the pressure loss coefficient. The input variation is considered as plate thickness, a/A and Re on the comparison study. The comparison study originates in the variation of Re for a scrupulous a/A for a perforated plate gives the improved pressure loss coefficient over others orifice, Bullen (1997). The relationship between the volumetric flow rate and Δp of a perforated orifice determines for the geometric structures of orifice such as thickness, porosity, hole distribution and upstream disturbance flow. The diameter of the pipe measured 2.9 cm with water as working fluid. The study reported an enhancement of 22.5 to 25.6% on the C_d in the perforated orifice plate over the standard orifice plate, Idelchik (1994).

A pipe design enhanced study conducted on the evaluation of Δp as a function of pressure loss contraction along the flow of the pipe. The experimental work reported within a limit of Re (4×10^4 to 2×10^5), a/A (0.13 to 0.7) and contraction sharpness as varying input parameters. The static pressure distribution profile of small and large diameter pipe is used to determine the pressure loss coefficient and report that the pressure loss coefficient is increasing with decreasing Re, Morris (1996). Analysis for determining discharge rate and flow coefficient of a safety valve is described on the effect of diameter ratio, and size of the valve. The static and total pressure distribution at the pipe inlet and orifice region is considered along the flow passage for the improvement of the safety valve design. The work concluded the flow stability of the value depends on entry pipe loss, diameter ratio and gravitational head during the flow, Dugdale (1997). Molasses have been chosen as viscous fluids for a thin orifice plate flow inside a fluid chamber. The fluid chamber has a larger orifice diameter ratio to analyze flow characteristics. The orifice size is 0.1469m, thickness 0.0075 cm and the chamber pipe diameter 2.85 cm. The molasses viscosity varies with the varying composition of glycerin and water. The Δp was evaluated using pipe flow-based Hagen-Poiseuille governing equation over the pipe length, Chisholm (1967).

The orifice is classified as a sharp edge, Rounded edge, short tube, and Borda edge as per the discharge point. Among these rounded orifices have better flow coefficients, but sharp-edge orifices produce more curate flow measurement devices. A two-phase theoretical study with sharp-edged orifices considered for a gas-liquid mixture to evaluate the flow coefficient with the consideration of interface shear forces by solving governing equations. The fluid mixture works as an incompressible fluid with phase velocities, Δp and weight ratio of fluid mixture as the input parameter, Lin (1982). Similarly, type of orifice considered for flow measurement of a phase separated on different vapour-liquid density ratios of R-113 refrigerant on the evaluation of Δp, C_d. A series of correlations for sharp edge orifice developed as a function of mass flow rate, quality of vapour in the mixture, vapour-liquid density ratio, diameter of the orifice and pipe based on experimental and theoretical study by Akshay (2018). Due to turbulence effect sound waves, disturbance, vibration, noise and cavitations generate in the macroscopic level of orifice flow. The flow disturbance degrades the energy transmission along the flow which prevents the losses of the acoustic determined on the study of flow through orifice. The acoustic behavior has been evaluated with variation of sound power spectra and frequency spectra. The geometrical shapes of orifice are considered as circular, square and rectangular duct types for large eddy simulation turbulent model in the above study, Hasan (2021).

A conical entry orifice used for better flow measurement devices to the Sharp edge orifice according to stability view. An experimental and numerical study conducted by the conical entry orifice with variation of conical angle for a wide range of Re to optimize the conical angle. Conical angles (15° – 45°), diameter ratio (0.4-0.63) as geometrical and Re (5000- 240000) as flow input parameter, with water as a working fluid used for this conical orifice. The studies concluded that a 30° conical angle gives better results for Δp, a value of 37-52% less than sharp-edge orifice. The flow analysis also produces a length of axial characteristics, recirculation region, and flow recovery with numerical correlations along the flow domain by Ho (1985). A conical entrance orifice considered for low Re (100 - 1000), diameter ratio (0.247 - 0.448) for a concentric and fully eccentric shape to determine C_d. The conclusion of the studies is C_d increased with diameter ratio, but eccentricity has no effects by Sanghani (2016). The comparison of different type geometrical orifice edges studies reported that for a fixed Re sharp edge geometry produces more Δp and streamlined approached orifice produces less Δp.

CFD based approaches are used for a fluid flow analysis of different types of orifices according to the edges. The study was conducted with six different edges of orifices such as square, square-edged, square-edged with 45° back-cut, sharp-edged, streamlined approach, sloping approach and quadrant-edged for water flow. Among the orifices, square edged with 45° back cut gives maximum pressure loss and a streamlined approach produced less pressure loss on particular flow input by Bagade (2016). A review paper reported on multi-hole orifice on different types of studies such as CFD, FEM, numerical methods, analytical and Experimental methods along the flow through an orifice which gives ideas about the flow analysis topic by Grace (1951). Based on the above literature review a chapter is proposed for different types of orifices for the flow through an orifice which gives a basic idea about the design of orifice on different flow input parameters. This chapter also gives better ideas on the orifice flow for betterment, accuracy, and reliability for a wide range of applications.

3. Calculation of the Parameter through Analytical Method

The Δp in a single-phase flow for a thin orifice is reported as a function of flow a/A or σ and contraction ratio (σ_c) suggested by Chisholm (1967) as eq. (2). Where ρ is the density of the fluid and V is the average flow velocity.

$$\Delta P_{sp} = \frac{\rho V^2}{2}\left[\left(\frac{1}{\sigma\sigma_c} - 1\right)\right]^2 \tag{2}$$

If the orifice is thick, the loss of pressure energy due to sudden expansion in the flow passage is reported as eq. (3).

$$\Delta P_{sp} = \frac{\rho V^2}{2}\left[\left(\frac{1}{\sigma\sigma_c}\right)^2 - 1 - \frac{2}{\sigma^2}\left(\frac{1}{\sigma_c} - 1\right) - 2\left(\frac{1}{\sigma} - 1\right)\right] \tag{3}$$

The local Δp also reported as a function of C_d as follows as reported Lin (1982) and Grace and Lapple, (1951) in eq. (4).

$$\Delta P_{sp} = \frac{\rho V^2}{2}\left[\left(\frac{1}{\sigma}\right)^2 - 1\right]\frac{1}{C_d^2} \tag{4}$$

Similarly, the contraction ratio (σ_c) for thin and thick orifice as eq. (5) and eq. (6)

$$\sigma_{c,thin} = \frac{1}{\sigma + \sqrt{\frac{1-\sigma^2}{C_d}}} \quad \text{... (5)}$$

$$\sigma_{c,thick} = \frac{1}{\sigma + \sqrt{\left[\frac{1-\sigma^2}{C_d^2}\right] - 1 + 2\sigma - \sigma^2}} \quad \text{... (6)}$$

The relation between flow a/A (σ) and contraction ratio (σ_c) are as follows in eq. (7), Chisholm [24], Bullen et al. (1987).

$$\sigma_c = \frac{1}{\left[0.639(1-\sigma)^{0.5} + 1\right]} \quad \text{... (7)}$$

Slip ratio is defined as the ratio of phase velocity of gas to phase velocity of liquid, Chisholm (1967) and Morris (1996). Due to difficulty in phase velocity measurement the slip ratio is defined with a function of the dryness fraction of the gas phase and the fluid density ratio as in eq. (8)

$$S = \left[1 + x\left(\frac{\rho_l}{\rho_g} - 1\right)\right]^{0.5} \quad \text{... (8)}$$

The dryness function x is defined as the ratio of gas mass flux to mixture mass flux as in eq. (9)

$$x = \frac{\rho_g V_{sg}}{\rho_g V_{sg} + \rho_l V_{sl}} \quad \text{... (9)}$$

Kojasoy et al. (1997) modified the Chisholm (1967) expression for S with eq. (10)

$$S=\left[\left\{1+x\left(\frac{\rho_l}{\rho_g}-1\right)\right\}^{0.5}\right]^n \qquad \dots (10)$$

The two-phase multipliers defined as the ratio of the two-phase Δp through the orifice to the single-phase Δp in eq. (11)

$$\Phi_{lo}^2=\frac{\rho_l}{\rho_g}x+(1-x) \qquad \dots (11)$$

Chisholm (1967) developed the following expression eq. (12)

$$\Phi_{lo}^2=1+\left(\frac{\rho_l}{\rho_g}-1\right)\left[Bx(1-x)+x^2\right] \qquad \dots (12)$$

Where B = 0.5 for thin orifice
= 1.5 for thick orifice.

4. Result Analysis of Virtual Orifice Model

A number of literatures are reported for the flow through an orifice in single and multi-phase analysis. The result section presented a few common results for a better understanding of the flow through the orifice on different input parameters. Thin and thick orifices consider as per the thickness ratio for two-phase experimental analysis on the evaluation of determining local Δp along the flow passage of pipe. Air-water is used as a two-phase fluid mixture. The diameter of the pipe considered is 60 and 40 mm with the parameters ranges are Re 3 x 10^4 to 2 x 10^5, the area contraction ratio in the orifice is 0.73 to 0.54, and the orifice thickness ratio is 0.025 to 0.59. The two-phase flow parameters are varied such as void fraction, gas and liquid superficial velocity ranging from 0.3 to 4 m/s and 0.6 to 2 m/s respectively. The predicted result from the experimental analyses has been measured up with the available literature according to the parameters, Fossa and Guglielmini (2002). An experimental study for two-phase horizontal intermittent flow along the orifice contractions has been conducted based on an impedance method to determine the void fraction at different pipe locations down the flow with the air-water fluid combination. A two-phase flow structure is also plotted downstream of the orifice in 40 mm diameter pipes with different orifice plates, plate thicknesses, and restriction ratio of pipe to the orifice. The predicted results are the void fraction, with a peak value at a space of 1D from the orifice on the downstream flow, but for fully developed flow the peak achieves at 2D space for a particular slip ratio. In the stratified two-phase region, the liquid void fraction is lower in the pipe flow region and decreases with the decreasing area ratio, Fossa et al. (2006).

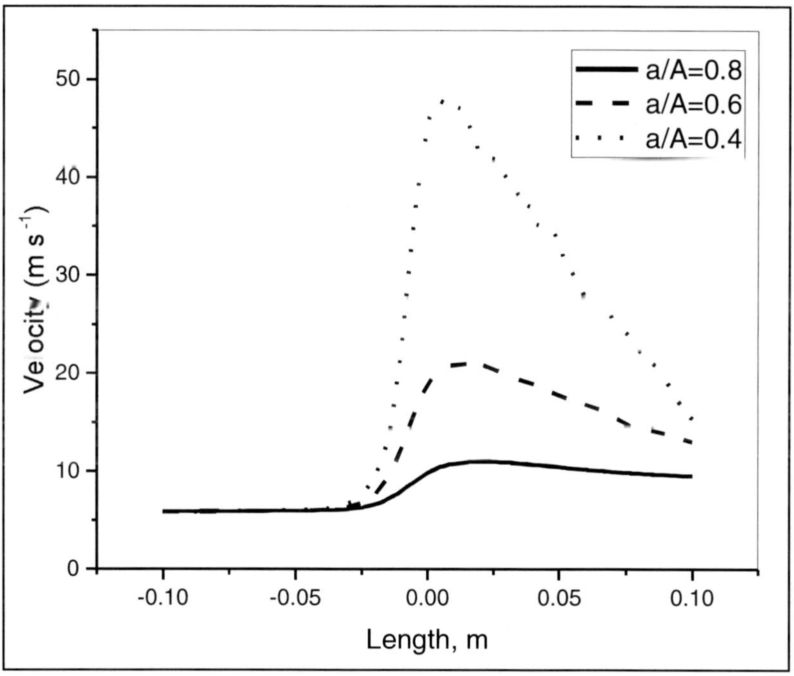

Figure 4. Analysis between velocity profile vs Reynolds number.

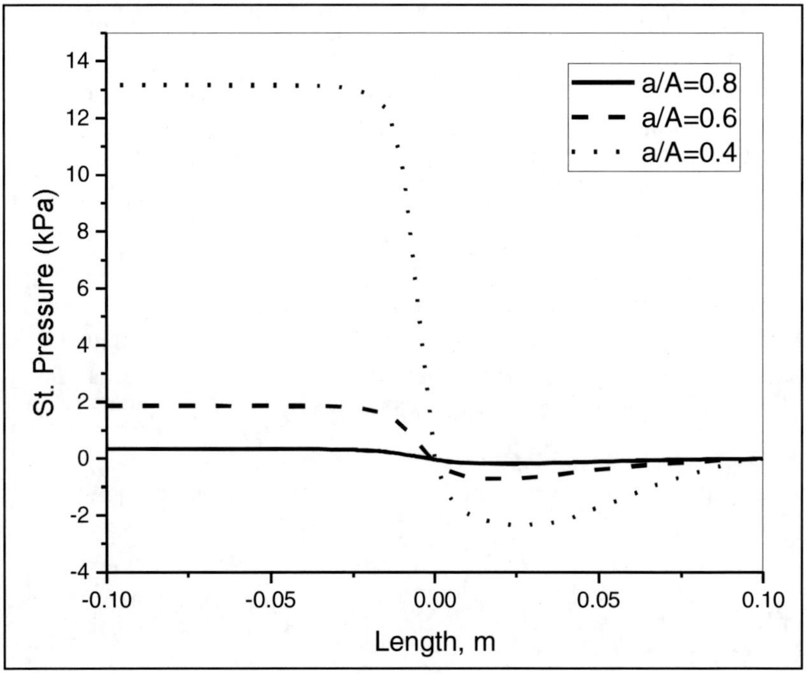

Figure 5. Δp on the variation of length to diameter ratio.

A CFD based numerical analysis was carried out with Eulerian - Eulerian multi-phase model to know the effect of individual fluids on an orifice flow. The local Δp, C_d and Φ_{lo}^2 (multi-phase pressure drop) are the outcome of the analysis with inlet parameters as velocity and the geometric of the orifice. The study found that on a single flow analysis contraction coefficient has less effect with Re, but Φ_{lo}^2 affected with Re. Based on two-phase results a correlation developed which showed good agreement with existing correlation. The effect of void fraction along the flow evaluated with maximum and minimum in the downstream of the orifice flow as slip ratio, Roul and Das (2012). The study was carried out for a diameter of 40 mm, Re (up to 200000), a/A (0.54 and 0.73). l/d also varies for thin and thick orifice for a range of 0.025-0.59. The predicted result shows maximum Δp occurs at maximum a/A and maximum l/d. A velocity and static Δp profile along the flow through orifice presented in Figure 4 and Figure 5 which gives on the variation of diameter ratio. The static pressure drops at orifice and increases steadily to get a stable profile in the downstream, from which C_d is calculated.

Similarly, the velocity profile increases at the orifice and drops to stable condition later in the downstream.

Similarly type numerical study carried out in the present work for thin and thick orifice having a/A = 0.4 - 0.8, l/d= 0.125 - 0.5, and Re 5000 - 50000 with air, water as single-phase fluid and water liquid and water vapour as multi-phase fluid to know the effects on flow parameters. The studies determine Δp, static pressure, velocity profile and contraction coefficient for the variation of a/A, l/d, Re and different fluids. Figure 6 (a) suggests that the local Δp decreases with the area ratio because of more flow passages. The Δp increases on the increases of flow velocity (Re) along the flow passages. Similarly, Figure 6 (b) suggests that a less space ratio provides more p compared to high pressure ratio because of less flow stability for lesser thickness. The other factors considered for the orifice studied are such as coefficient of contraction, C_d, flow coefficient, Δp by various authors, with the variation along the flow length, shape and size of orifice and pipe, etc. Figure 7 shows the velocity profile along the flow passage of the orifice including downstream and upstream along the space ratio. The velocity is predicted to have a higher value on the upstream side for a higher space ratio but on the downstream the profile is reverse. The phenomenon occurs due to the dynamic nature of the fluid after passing the orifice plate which guides the dust along the flow.

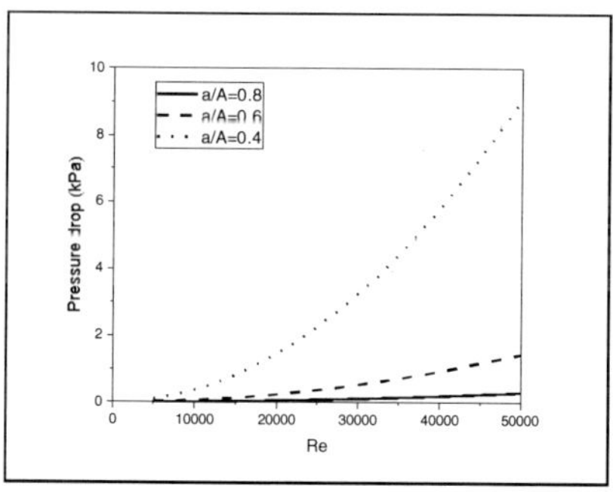

(a) Area ratio (a/A)

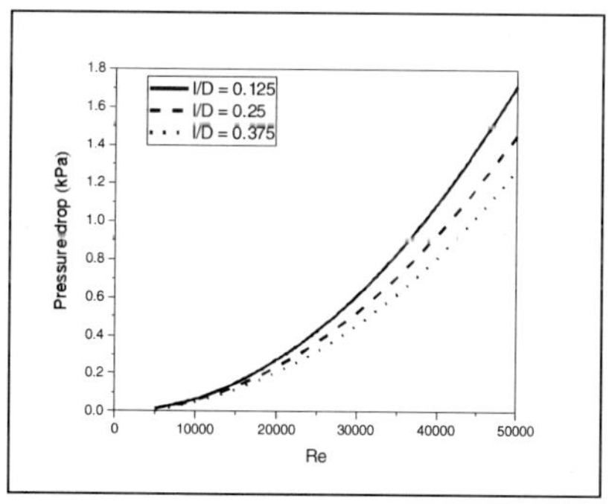

(b) Space ratio (l/d)

Figure 6. Local Δp along the flow through orifice with varying (a) a/A (b) l/d.

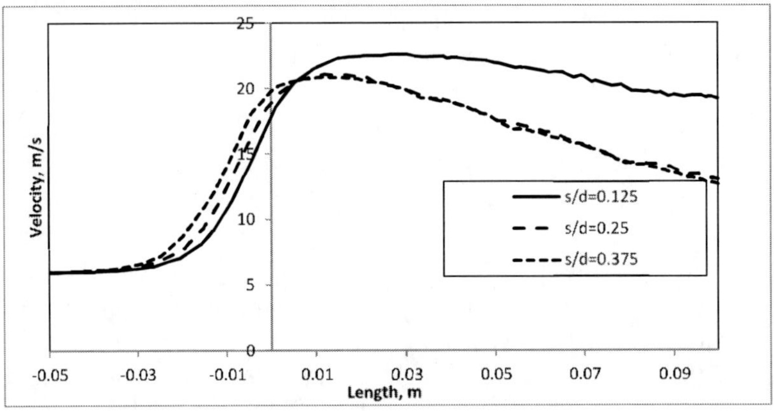

Figure 7. Velocity profile of the flow along the length.

Conclusion

The present chapter gives brief details about an orifice as a flow measurement device. The chapter has discussed different types of orifices according to the shape, thickness and edges of their application. The various shapes of an orifice such as elliptical, perforated, slotted, crescent, segmental, two-stage, conical entry, etc. are studied. The orifice flow studied with single and multi-phase types with the output of Δp, flow coefficient and C_d. The number of holes also varies (single, multi-hole) with single and multi-phase flow along the variation of geometrics for various applications, as a flow measurement device. The book chapter also describes the variation with thickness (thin and thick) of orifice with a/A and l/d. The other classification also varies according to edges of the orifice such as square, square edge with 45° back cut, sharp, conical with angle variation, Quadrant edge, etc. are discussed. This analysis also carried out a comparison study for the flow parameters (Reynolds number). The results conducted for Δp, velocity profile and contraction coefficient on the variation of Re, a/A and l/d. The present chapter gives a broad range of orifice in spite of the future design, better understanding, accurate in the measurement, stability of flow, minimize error, utmost life time for the flow measurement point of view.

References

Akshay R. Kadam, Mangesh B. Chaudhuri, "Acoustic study and behaviour of flow through orifices, "*Procedia Manufacturing,* 20, pp. 154 - 159 (2018).

Athar M., ISH M. A. Ansari M. & Khan M. A., "Flow characteristics of elliptical orifice plates," *ISH Journal of Hydraulic Engineering,* 9:2, pp. 22 - 35 (2003).

Bagade V.S, Suryawanshi P.M., Nalavade S.M., "A review of multi-hole orifice plate," *International Journal for Research in Applied Science & Engineering Technology,* 7 (4), 3197 -3208 (2016).

Botros K. K., Studzinski W, Barg P, "Results of NOVA'gas metering research. Canadian Gas Association," *Measurement School,* Vancouver, May 26 – 29, (1992).

Bullen, P. R., Cheeseman, D. J., Hussain, L. A., and Ruffel, A. E., "The Determination of Pipe Contraction Coefficients for Incompressible Turbulent Flow," *Int. J. Heat Fluid Flow*, 8, pp. 111 – 118 (1987).

Chisholm D., "Flow of incompressible two-phase mixtures through sharped edged orifices," *Mechanical Engine metering Science* 9, pp. 72 – 78 (1967).

Dong Zhiyong, Yang Yong Gang, Chen Qiqi and Shi Bin, "A study of hydraulic characteristics of multi-square-hole orifice plates," *Applied mechanics and Material,* 256, pp. 2470 - 2473 (2013).

Dugdale D. S., "Viscous flow through a sharp-edged orifice," *Int. J. Engg. Science,* 35 (8), pp. 725 - 729 (1997).

Emerson, "*Fundamentals of orifice meter measurement,* White paper," https://www.emerson.com/documents/automation/white-paper-fundamentals-of-orifice-meter-measurement-daniel-en-us-188730.pdf (2017).

Fossa M., and Guglielmini G., "Pressure Drop and Void Fraction Profiles during Horizontal Flow through Thin and Thick Orifices," *Exp. Thermal Fluid Sci.,* 26, pp. 513 – 523 (2002).

Fossa, M., Guglielmini, G., and Marchitto, A., "Two-Phase Flow Structure Close to Orifice Contractions during Horizontal Intermittent Flows," *Int. Commun. Heat Mass Transfer,* 33, pp. 698 – 708 (2006).

Gallagher G R, "Measuring edge sharpness of orifice plates." *The Engineer,* (1968).

Gan Guohui, Riffat Saffa B. "Pressure loss characteristics of orifice and perforated plates," *Experimental Thermal and Fluid Science*, 14, pp. 160 - 165 (1997).

Gao Junxia, Wu Fenghe, "Investigation of flow through the two-stage orifice," *Engineering Applications of Computational Fluid Mechanics*, 13 (1), pp. 117 - 127, (2019).

Gassan H. Abdul-Majeed A., Maha R. A., "Correlations developed to predict two phase flow through wellhead chokes," *The Journal of Canadian Petroleum Technology* 30, pp. 47 –55 (1991).

Geng Yanfeng, Zheng Jinwu, Shi Tianming, "Study on the metering characteristics of a slotted orifice for wet gas flow." *Flow Measurement and Instrumentation* 17, pp. 123 – 128 (2006).

Grace H. P., and Lapple C. E., "Discharge Coefficient of Small Diameter Orifices and Flow Nozzles," *Trans. ASME,* 73, pp. 639 – 647 (1951).

Hasan Duz, "Effect of conical angle on the hydraulic properties of plate flows: A numerical approach," *Flow Measurement and Instrumentation*, 81, 102026 (2021).

Ho Y. S., Leung, T. P., "Performance of conical entrance orifice plates at low Reynolds numbers," *International journal of Heat and Fluid Flow*, 6 (2), pp. 122 - 125 (1985).

Idelchik I. E., Malyavskaya, G. R., Martynenko, O. G., and Fried, E., "*Handbook of Hydraulic Resistances,*" CRC Press, Boca Raton (1994).

Jepson P, Chipchase R, "Effect of plate buckling on orifice meter accuracy." *J. Mech Eng. Sci.* 17 (6): 330 – 337, (1975).

Kayser John C., Shambaugh Robert L., "Discharge coefficients for compressible flow through small-diameter orifices and convergent nozzles," *Chemical Engineering Science,* 46 (7), pp. 1697 - 1711 (1991).

Kojasoy G., Kwame M. P., and Chang C. T., "Two-Phase Pressure drop in Multiple Thick and Thin Orifices Plates," *Exp. Thermal Fluid Sci.,* 15, pp. 347 – 358 (1997).

Lin Z. H., "Two Phase Flow Measurement with Sharp Edge Orifices," *Int. J. Multiphase Flow,* 8, pp. 683 – 693 (1982).

Michael Reader-Harris, Experimental fluid mechanics, "*Orifice plates and venture tubes,*" Springer, ISBN 978-3-319-16880-7 DOI 10.1007/978-3-319-16880-7 (2015).

Mincks Leo Michael, "*Pressure drop charteristics of viscous flow across orifices,*" A Master Science thesis, Iowa state university, Ames, Iowa, (2002).

Morris S. D., "Liquid flow through safety valves: diameter ratio effects on discharge coefficients, sizing and stability," *Journal of Loss Prevention in the Process Industries,* 9 (3), pp. 217 – 224 (1996).

Morrison Gerald L., Terracina Dwayne, Brewer Carl, Hall K. R., "Response of a slotted orifice flow meter to an air/water mixture," *Flow Measurement and Instrumentation* 12, pp. 175 – 180 (2001).

Reader-Harris M J, Barton N, Hodges D, "The effect of contaminated orifice plates on the discharge coefficient". In: *Proceedings of 15th FLOMEKO,* Taipei (2010).

Roul Manmatha K., Dash Sukanta K., "Single-Phase and Two-Phase Flow through Thin and Thick Orifices in Horizontal Pipes," *Journal of Fluids Engineering, ASME,* 134, 091301-1 (2012).

Sanghani C. R., Jayani D. C., Dobariya C. R., Sabhadiya G. K., Gohil J. R., "Comparative Analysis of different profile geometrics for pressure drop," *International Journal of Science Technology & Engineering, 2* (10), pp. 494 - 498 (2016).

Shanfang Huang, Taiyi Ma, Dong Wang, Zhong Lin, "Study on discharge coefficient of perforated orifices as a new kind of flow meter," *Experimental Thermal and Fluid Science,* 46, pp. 74 - 83 (2013).

Yadav H. S., ISH M., Athar M., ISH M., Kaushal D. R., "Flow Characteristics of Sector Orifice Plates," *Ish Journal of Hydraulic Engineering,* 1 (2), pp. 84 - 94, 1995.

Chapter 12

Research Hotspots in Quantum Error Correction Using Artificial Intelligence Techniques

Seema Verma[1,*], Rakesh Kumar Sheoran[1,†] and Savita Kumari Sheoran[2,‡]

[1]Banasthali Vidyapith, Banasthali, Rajasthan, India

[2]Indira Gandhi University Meerpur, Rewari, Haryana, India

Abstract

Quantum computing is a nascent field of research jousting on principles of quantum physics and information theory. It has vast potential to solve the complex computational problems in polynomial time which are otherwise intractable in traditional computing. The main challenges in harnessing the full potential of this technology are the faults occurring due to interaction of system with environment. Various methods have been developed for fault-tolerant quantum computation for correcting the quantum errors but none of them can be considered as a full-proof solution to this issue. Recently artificial intelligence (AI) has emerged as a smart technique to mimic human intelligence in machines and other computational systems. The algorithms based on AI techniques *viz.* machine leaning, reinforcement learning, deep learning, artificial neural network, genetic algorithms, support vector machine, and natural language processing are found effective in mitigating the quantum error which is an entirely different paradigm from classical error correction. This chapter intends to analyse the research hotspots in quantum error

* Corresponding Author's Email: seemaverma3@gmail.com.

† Corresponding Author's Email: sheoran.rk@gmail.com

‡ Corresponding Author's Email: savita.sheoran@igu.ac.in

In: Computational Intelligence for Sustainable Development

Editor: Brojo Kishore Mishra

ISBN: 979-8-88697-198-9

correction along with possible applications of artificial intelligence techniques to gaze the present status and future research possibilities for making the fault-tolerant quantum computation practically feasible.

Keywords: quantum error correction, fault-tolerant quantum computing, AI for QEC, and quantum machine learning

1. Introduction

Quantum physics and information theory are the two most revolutionary developments of the 20th century which have profoundly influenced society by solving long-standing questions in science and helping humanity to confront various challenges. After a century-long affluent intensification, these powerful technologies converged as "quantum computing" to harness unique features of quantum physics to perform computations and store data. The disruptive fusion of these two technologies possessing weird potential manifested its prospective technology in the realm of high energy physics, molecular dynamics, DNA decoding, drugs discovery, etc granting it unparallel business value never witnessed by any technology in past. According to an estimate by the Boston Consultancy Group (BCG), quantum computing is all set to transform multiple industries and fetch a revenue of US$850 billion by 2040 [1]. Its current market size of US$ 472 million is expected to grow by a Compound Annual Growth Rate (CAGR) of 30.2% to reach US$ 1765 in 2026. Apart from governments across the world; giant players like IBM, Micro-Soft, D-Wave, Google, Toshiba, Intel, Hewlett Packard, Regitti, IonQ, Honeywell, Baidu, Alibaba, etc are surging their investment in this nascent technology affirming it as central importance. In India, the government has leap forwarded this race by announcing a National Mission on Quantum Technologies & Applications (NM-QTA) in 2018 with a budgetary outlay of INR 8000 Crores (approximate US$ 10.75 billion) to be implemented in the next five years [2].

Quantum computing is an entirely different paradigm of computing that uses principles of quantum physics such as interference, superposition, entanglement, and teleportation to carry out computational tasks [3,4]. Despite huge opportunities associated with it, the practical implementation of quantum computer poses enormous challenges and its developments is still confined between the most versatile versions of quantum annealing, noisy intermediate-scale quantum computing, and fault-tolerant universal quantum computing

[5]. The realization of a versatile quantum computer requires the implementation of a universal set of gates (a set of unitary transformations) at the logical level which is highly prone to error due to its interaction with the environment, noise, and other undesired interference. The temporal drifting of error in the quantum circuit operations may cause a higher error rate at the later stage and destroy the computing environment. To sustain a scalable and robust quantum computing environment and achieve high fidelity in operations it is necessary to keep the error below the desired threshold. The process of detecting and mitigating such undesired errors is known as *Quantum Error Correction (QEC).* Classically the error correction code manages to produce multiple copies of information but the no-cloning theorem puts a bar on copying of quantum information because it is not possible to copy the exact quantum state. Therefore, QEC essentially requires a surreal mechanism for error correction which relies on the characteristics of the noise of the particular physical system. Such a computational system needs to follow a mechanism to calibrate, code, decode, and characterization based upon the learning from its past error [6]. Artificial intelligence (AI) techniques such as machine learning (ML), reinforcement learning (RL), deep learning (DL), artificial neural network (ANN), and genetic algorithms (GA) are strong tools to learn the system from the data [7, 8]. Therefore, AI techniques may be helpful in developing an error correction system based on the experiences of real-time quantum computers. This chapter intends to study quantum error correction using AI techniques.

1.1. Quantum Computing

Since Moore's Law was propounded in 1965 by Intel co-founder Gordon E. Moore; the number of transistors on a dense integrated circuit increase, the size of electronic components on it shrinks exponentially [9]. Presently this size of electronic components shrank below a nano-scale, where only the laws of quantum physics can be applied. Such computing concatenating non-intuitive subjects of physics and computer science based on the properties of matter below the atomic scale, having no classical counterpart, is known as quantum computing [10]. The classical computing fulcrum around bits having values either '0' or '1', contrary to which the quantum system is based on the operation of qubit (quantum bit) using linear algebra. A qubit is a two-state mechanical system which can be realized as a complex wave function representing a coherent superposition of two orthogonal quantum states as

$|\psi\rangle = \alpha|0\rangle + \beta|1\rangle$ where co-efficient 'α' and 'β' are complex numbers specifying the quantum states of qubit such that $|\alpha|^2 + |\beta|^2 = 1$. The qubit states can be geometrically represented on a unitary Bloch sphere with $|0\rangle \rightarrow (0,0,1)$ and $|1\rangle \rightarrow (0,0,-1)$, as shown in Figure 1.

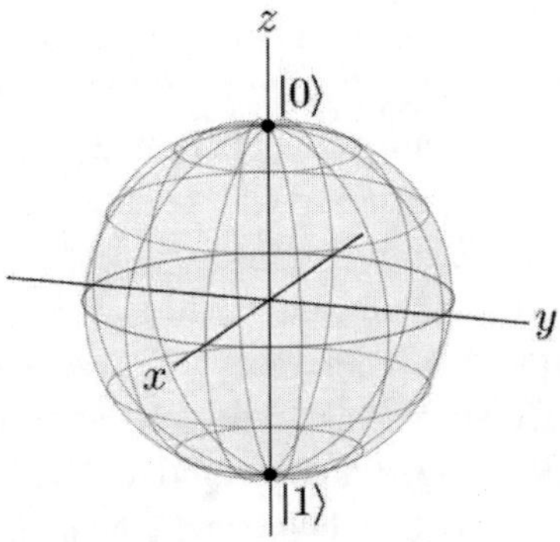

Figure 1. Bloch Sphere Representation of Qubit.

There can be many other states of $|0\rangle$ and $|1\rangle$ on the equator of the Bloch sphere which can be reached by changing the relative phase of the corresponding quantum state as summarized in equation (1) and can be calculated by the coefficients (amplitude) of $|0\rangle$ and $|1\rangle$ by applying conditions of Heisenberg uncertainty principle.

$$|+\rangle = \frac{1}{\sqrt{2}}(|0\rangle + |1\rangle)$$
$$|-\rangle = \frac{1}{\sqrt{2}}(|0\rangle - |1\rangle)$$
$$|i\rangle = \frac{1}{\sqrt{2}}(|0\rangle + i|1\rangle)$$
$$|-i\rangle = \frac{1}{\sqrt{2}}(|0\rangle - i|1\rangle) \quad (1)$$

The quantum computer carries out an exponentially faster computation, even faster than supercomputers, because the superposition of $|0\rangle$ and $|1\rangle$ enhance information density by providing '2^n' computational states in comparison to '$2n$' states in classical computing. The best example of a qubit is spin–½ particle in '$\mathbb{C}^2$' normalized vector space whose energy can be represented as unitary time evolution operator $U|\psi\rangle = e^{-iHt/\hbar}|\psi\rangle$ in Hilbert space (H) taken as the computational basis. Such a system can be realized by performing unitary operations on an array of qubits and can be further sped up by using a d-dimensional state system $(d > 2)$ called qudits. These entangled states can be labeled in 2^n-dimensional Hilbert Space using tensor product between $|0\rangle \otimes |0\rangle \otimes \ldots \otimes |0\rangle = |00\ldots0\rangle$ and $|1\rangle \otimes |1\rangle \otimes \ldots \otimes |1\rangle = |11\ldots1\rangle$. The unitary operations on qubits form quantum gates (quantum counterpart of classical gates) which further concatenate to form quantum circuits [11]. The qubits can be physically realized through physical systems such as a photon, trapped ion, cold atom, Nuclear Magnetic Resonance (NMP), quantum dots, diamond defect qubit, superconductor, and Josephson junction and neutral atom in the lattice by manipulating quantum information globally so that it remains unaffected from global properties. To practically realize quantum computing the quantum gates must be linear, reversible, universal, and normalized. The unitary operation of simplest quantum gate on a qubit can be represented as $U|\Psi\rangle = A(|0\rangle + B|1\rangle) + C(|0\rangle + D|1\rangle)$, where quantum gate '$U$' is a unitary operator such as $UU^* = 1$ which can transform the basis to achieve the desired operation. Table 1 depicts the common quantum gates along with their computational actions.

Table 1. Computational actions of some common quantum gates

Quantum Gate	Computational Action
Identity Gate	$I|0\rangle = |0\rangle, I|1\rangle = |1\rangle$
Pauli-X Gate	$X|0\rangle = |1\rangle, X|1\rangle = |0\rangle$
Pauli-Y Gate	$Y|0\rangle = i|0\rangle, Y|1\rangle = -i|1\rangle$
Pauli-Z Gate	$Z|0\rangle = |0\rangle, Z|1\rangle = -|1\rangle$
Phase-S Gate	$S|0\rangle = |0\rangle, S|1\rangle = i|1\rangle$
Toffoli (T) Gate	$T|0\rangle = |0\rangle, T|1\rangle = e^{i\pi/4}|1\rangle$
Hadamard (H) Gate	$H|0\rangle = \frac{1}{\sqrt{2}}(|0\rangle + |1\rangle), H|1\rangle = \frac{1}{\sqrt{2}}(|0\rangle - |1\rangle)$

1.1.1. Historical Developments

The earlier testaments of quantum mechanics were accumulated during 1895-1925 by efforts of various scientists while the modern information theory can be traced back to the Turing Machine of Alan Turing proposed in 1936 followed by the paper 'A Mathematical Theory of Communication' of Claude Shannon's in 1948. Meanwhile, in 1935 the EPR Paradox (Einstein-Podolsky-Rosen, based on locality and reality assumptions, discarded the completeness of quantum mechanics. In 1964, John Stewart Bell discarded this paradox by developing the inequality popularly known as Bell's inequality [5,6]. First of all, a theoretical model of quantum computing was proposed by Yuri Manin in 1980 [12] The concept of quantum computing in its present form was propounded by great physicist R.P. Feynman in 1982 on the consideration that quantum computer is required to efficiently simulate the equations of quantum physics [13, 14]. In the paper of 1982, Paul Benioff described the theoretical feasibility of quantum computing in the reversible Turing machine. In 1985, David Deutsch was the first to present the concrete concept of a universal quantum computer having the potential to outperform a classical computer [15]. He suggested a stronger version of the 'Church-Turing thesis' stating that every finitely realizable physical system can be perfectly simulated by a universal model computing machine operating by finite means. The real potential of quantum computing was adorned ten years later, in 1994, when Peter Shor developed his famous Shor algorithm, enabling efficient prime factorization of an integer 'N' within a polynomial time '$logN$' which was NP-hard in classical computing [16]. The nascent paradigm of computing was further strengthened by Luv Grover's algorithm in 1996 for searching an unsorted database with 'N' entries in '$ON^{1/2}$' time and using '$OlogN$' storage space [17]. In 2000 theoretical physicist David P. DiVincenzo formulated the five-point DiVincenzo criteria which essentially to be followed by every quantum computer, paved the way for the practical realization of a quantum computer [18]. The Di Vincenco criteria essentially to be followed by every quantum computer is based on the following criteria:

- A well charectrized and scalabile physical system with well-characterised qubits
- The ability to initialising qubits to a simple fiducial state
- Long relevant decoherence times
- A universal set of quantum gates
- A qubit-specific measurement capability

The quantum computational journey took a bizarre leap in 2017 when IBM presented a commercial model of a practical quantum computer [19]. In 2019 Google demonstrate a sycamore chip composed of 54-qubits superconducting loops and claimed quantum supremacy [20] while china achieved the quantum supremacy milestone in 2020 for Gaussian-Boson sampling [21]. In 2021 IBM achieved another milestone by breaking a barrier of 100-qubits and unveiled the 'Eagle', a 127-qubits quantum processor [22]. Earlier IBM was having processors of 65-qubit 'Hummingbird' and 27-qubit 'Falcon' and planning to launch 433-qubit 'Sprey' in 2023 and 1121-qubit another processor by 2026. To enhance the research in this domain various companies have developed quantum simulators such as IBM-Q, Google Ciraq, Microsoft-Azure and extended the same to research communities through cloud-based services. India too has joined the quantum race with the announcement of the National Mission on Quantum Technology and Application (NM-QTA) with a budget outlay of Rs. 8000 crores in the 2020 budget. To address the common challenges in advancing towards quantum computing research India has launched its indigenous quantum simulator 'QSim' developed by IISc Bengaluru, IIT Roorkee, and C-DAC in 2021. Similarly, the Ministry of Electronics and IT (MeitY) has signed an MoU with Amazon Web Services (AWS) for setting up of quantum research lab in India.

1.1.2. Essential Quantum Computing Properties

The 20th century science peeped into the fundamental nature of the microscopic world known as 'quantum mechanics'. This study of the miniaturized matter revealed unparallel benefits in the form of 'quantum supremacy' or 'quantum advantage' vacuuming around the realization of qubits through a physical system in two eigenstates of field *viz.* +½ and -½. A quantum state is the characterization of embodiment containing all the possible outcomes of a measurement of a physical system. The spin can acquire any of the two states up-spin ($|\uparrow\rangle = |1\rangle$) or down-spin ($|\downarrow\rangle = |0\rangle$) or superposition of these two states as ($\alpha|\uparrow\rangle + \beta|\downarrow\rangle) = (\alpha|1\rangle + \beta|0\rangle)$ but the uncertainty principle hinders the accurate measurement of system states because measurement may collapse the states between $|\uparrow\rangle$ or $|\downarrow\rangle$. Apart from it, the interaction of states of the computational system with the outer environment may cause decoherence in the system resulting in the partial or complete destruction of computation. The following quantum properties are essential to understand the quantum computing system:

- *Superposition:* Quantum physics depart from its classical counterpart on fundamental principles of superposition according to which a state $|\Psi\rangle$ can be a superposition of two states if $|\Psi_A\rangle$ and $|\Psi_B\rangle$ and every measurement of $|\Psi\rangle$ is also the possible measurement of $|\Psi_A\rangle$ and $|\Psi_B\rangle$ [23]. In general, an n-qubit system can be represented in Hilbert Space by wave function $|\Psi_B\rangle = \sum_{i=1}^{2n-1} c_i\,|i\rangle$ in 2^n-dimensional computational space. The constant c_i must satisfy the normalization condition $\sum_{i=1}^{2n-1}|c_i^2| = 1$. The superposition principle can express wave-particle duality as $|\Psi_B\rangle = \sum_i a_i|W_i\rangle + b_i|P_i\rangle$; $|W_i\rangle$ and $|P_i\rangle$ are the wave and particle aspects of the state.

- *Entanglement:* Quantum entanglement is a bizarre property where the quantum state of paired or interactive particles can't be expressed independently i.e., *the* states can't be factored into product states [24]. The concept was initially expressed in EPR-Paradox and doesn't have a counterpart in classical physics. Initially, it was rejected by Einstein in his famous paradox of 'spooky action at a distance' but later Bell's inequality affirmed it based on the theory of local hidden variables. A pure quantum state $|\Psi\rangle$ with basis $\{|u_1\rangle, \dots\dots\dots |u_n\rangle$ can be expressed as $|\Psi\rangle = c_1|u_1\rangle + c_2|u_2\rangle + \dots\dots\dots + c_n|u_n\rangle$. The entanglement of two pure states in n-dimensional Hilbert space (H_n) and m–dimensional Hilbert space (H_m) with orthogonal basis $\{|u_i\rangle; i = 1,2, \dots\dots, n\}$ and $\{|v_j\rangle; j = 1,2, \dots\dots, m$ respectively can be expressed as $H_{nm} = H_n \otimes H_m$ in nm-dimensional Hilbert space with orthogonal basis. $\{|u_i\rangle \otimes |u_j\rangle$. Since the pure state can be identified as density operator $\rho = |\Psi\rangle\langle\Psi|$. For a statistical mixture of pure states $\{\rho_i = |\Psi_i\rangle\langle\Psi_i|; i = 1,2, \dots\dots\dots, n\}$ with probabilities $\{p_i = i = 1,2, \dots\dots\dots, n\}$ then its density operator $\rho = \sum_{i=1}^{n} p_i \rho_i$ will be positive Hermitian of unit trace with Neumann entropy equal to $S(\rho) = -tr\ \rho\log\rho$ which can be measured using local operations performed locally on a sub-system: $T = T^n \otimes T^m$. The final measurement of entanglement 'E' between two pure states in Hilbert Space $H_n \otimes H_m$ is given by the formula, $E(\rho) = -tr(\rho^m \log \rho^m$ where $\rho^m = tr_n(\rho)$.
- *Quantum Parallelism:* It is the fundamental property of a quantum system to heuristically encode multiple computational results into a quantum state to simplify the task of quantum computing to simultaneously evaluate a function '$f(x)$' for different values of 'x' by carrying out a large number of parallel computations [26, 27]. Evaluation of all the possible permutation of a binary function $f : \{0,1\}^n \rightarrow \{0,1\}$ require 2^n operations but quantum parallelism

may use a reversible function $U|x\rangle|y\rangle \rightarrow |x\rangle|y \oplus f(x)\rangle$; where '$\oplus$' represent operation module 2 addition; to compute the a same in single move. Quantum parallelism forms the basis of many quantum algorithms to enhance their computational power in comparison to classical computers.

1.2. Fault-Tolerant Quantum Computing

The quantum computer promises to solve complex computational problems in real-time but noise and de-coherence are its most formidable enemies. The art and science of building an error-free computing system in the presence of environmental fault are known as fault-tolerant computing. Both 'no-cloning theorem' and EPR Paradox discards the possibility of quantum error correction but Bell's inequalities open a way forward to distribute the information over multiple entangled qubits based on which Shor's Code and Streane's codes are developed to study fault-tolerant quantum computing which can be possible only if the error probability of the individual qubits is observed to be lower than a threshold of 10^{-12} per qubit prescribed by 'quantum threshold theorem'. The condition under which QEC code 'C' with orthogonal basis $|\Psi_i\rangle$ can correct the error $|E_i\rangle$ can be defined as $\langle \Psi_i|E_a^{\dagger} E_b|\Psi_i\rangle = C_{ab}\delta_{ij}$; where $C_{ab} \in C$.

1.3. Quantum Error Correction (QEC)

The quantum systems are highly error-prone and QEC requires mapping of 'k' (2^k - dimensional Hilbert Space) qubit into 'n' qubits (2^n - dimensional Hilbert Space) such that '$n > k$'. The 'k' qubits are known as physical qubits and 'n' qubits are known as logical qubits obtained by concatenating the former. The quantum system is inevitably entangled with the larger non-local environment system. Let $|0\rangle$ be the state of environment then the unitary evolution (U) will interact with the environment as $U|\psi\rangle \otimes |0\rangle \rightarrow \alpha|0\rangle \otimes |e_{00}\rangle + \alpha|1\rangle \otimes |e_{01}\rangle + \beta|0\rangle \otimes |e_{10}\rangle + \beta|1\rangle \otimes |e_{11}\rangle$

Therefore,

$$U|\psi\rangle \otimes |0\rangle \rightarrow (\alpha|0\rangle + \beta|1\rangle) \otimes \frac{1}{2}(|e_{00}\rangle + |e_{11}\rangle) + (\alpha|0\rangle + \beta|1\rangle) \otimes \frac{1}{2}(|e_{00}\rangle - |e_{11}\rangle) + (\alpha|0\rangle + \beta|1\rangle) \otimes \frac{1}{2}(|e_{01}\rangle + |e_{10}\rangle) + (\alpha|0\rangle + \beta|1\rangle) \otimes \frac{1}{2}(|e_{01}\rangle - |e_{10}\rangle)$$

In symbolic form, it can be written as equation (2),

$$U|\psi\rangle \otimes |0\rangle \rightarrow I|\psi\rangle \otimes |e_I\rangle + Z|\psi\rangle \otimes |e_Z\rangle + X|\psi\rangle \otimes |e_X\rangle + Y|\psi\rangle \otimes |e_Y\rangle \quad (2)$$

Where $I|\psi\rangle \rightarrow$ no error, $Z|\psi\rangle \rightarrow$ bit − flip error, $X|\psi\rangle \rightarrow$ phase − flip error and
$Y|\psi\rangle \rightarrow$ both bit − flip, and phase − flip errors simultenously
The error of n-qubit can be described as, $E_a = M_1 \otimes M_2 \otimes \ldots\ldots\ldots\ldots\ldots\ldots \otimes M_n$. with $M_i \rightarrow$ Pauli operators (I, X, Y, Z). The necessary and sufficient condition to detect an error in two code-words $|i\rangle$ and $|j\rangle$ can be written as $\langle i|E_a|j\rangle = C_a\delta_{ij}$. The measurement and correction of error involve introducing a syndrome qubit to the quantum system which represents the error state. The unitary operations are applied on the syndrome qubit to reverse the quantum error states. The effective QEC code concatenates the physical qubit into logical qubit(s) of minimum overhead such that bit-flip error is reduced exponentially and causing only a linear expansion in phase-flip error through a process called error-biasing. The repetition codes are applied on error-biasing for extremely fault-tolerant computation. There can be various types of QEC codes based upon the system's hardware and other requirements representing error state *viz.* stabilizer, non-stabilizer, CSS, topological codes, etc which can be further classified in two classes of *Additive* and *Non-additive codes*.

1.4. Quantum Error Correction Using Artificial Intelligence Techniques

Quantum error correction using AI techniques is a nascent and innovative idea to devise an error correcting system by learning from experimental data.

However, not much work has been done in this regime but fewer researchers have explored it to mitigate issues that arise out of imperfect quantum operations or de-coherence due to interactions with the environment [28, 29, 30]. Consider a two-state quantum system $|\psi\rangle = \alpha|0\rangle + \beta|1\rangle$. Physically a QEC system is that which can make it fault-tolerant by protecting it from environmental errors and noise which can be achieved by coding its qubits in the sub-space of higher-dimensional Hilbert space. The simplest code is the 3-qubit code which encodes the information as shown in equation (3).

$$|000\rangle \dashrightarrow |0\rangle\otimes|0\rangle\otimes|0\rangle \tag{3}$$

$$|111\rangle \dashrightarrow |1\rangle\otimes|1\rangle\otimes|1\rangle$$

This code can correct a single Pauli-X error 'σ_x' in the encoded qubit. Similarly, Pauli-Y and Pauli-Z error can be mitigated as $\sigma_x \dashrightarrow X, \sigma_y \dashrightarrow Y, \text{and } \sigma_z \dashrightarrow Z$. Classical machine learning is slow and can't solve complex problems while quantum machine learning (QML) recognizes a pattern in the quantum error data and predict the futuristic error within the real-time by training the quantum circuits through two approaches *viz. simulator-based* and *hardware-based approach.* Various artificial intelligence algorithms have been developed as learning protocols for such quantum systems *viz.* quantum Boltzmann machines (QBM), hybrid Helmholtz machine, hybrid quantum auto-encoders (HQA), quantum reservoir computing (QRC), quantum multiclass classifier (QMCC), support vector machines with a quantum kernel estimator (QSVM-Kernel), quantum neural networks QRNN, hybrid k-neighbors-nearby model (HKNN), quantum generative adversarial networks (QGAN), recurrent quantum neural networks (RQNN) and quantum recurrent encoding-decoding neural networks (QREDNN), etc [31].

2. Literature Review

The development of quantum computing as a discipline of technology does not happen all of a sudden, rather it is the result of an amalgamation of knowledge acquired over a century. Quantum computing originated from the work of Feynman, Benioff and Deutsch span a long untiring journey till climb to the pinnacle of present-day fault-tolerant quantum computer [11, 13, 14,

15]. Owing to its potentially disruptive applications across all the fields of engineering, technology to social life, the research in this regime has increased leap and bound. Therefore, the publications on quantum computing, quantum error correction and proliferation of artificial intelligence in QEC have increased steeply during the last four decades since the 1980s. Despite huge research spending by industry and culminating efforts of scientific community the practical quantum computer is still a long distance away. To gaze the issue popular search engine Google and eminent research database if IEEE Xplore is explored on four key-words *viz.* 'Quantum Computing', 'Quantum Computer', 'Quantum Error Correction', 'Fault-Tolerant Quantum Computing and AI' and the results are shown in Figures 2 and 3 respectively.

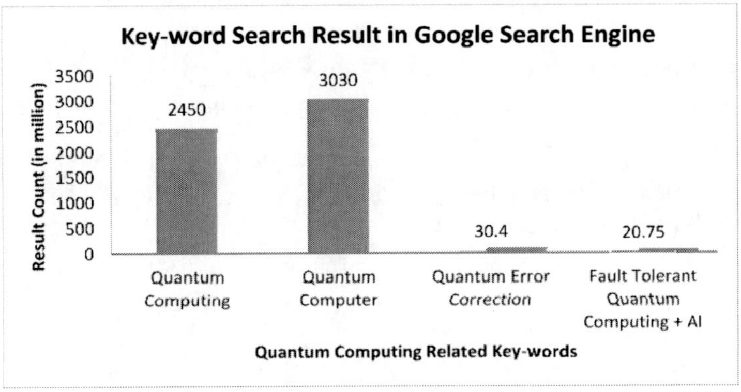

Figure 2. Quantum Computing Key-words Search in Google Search Engine.

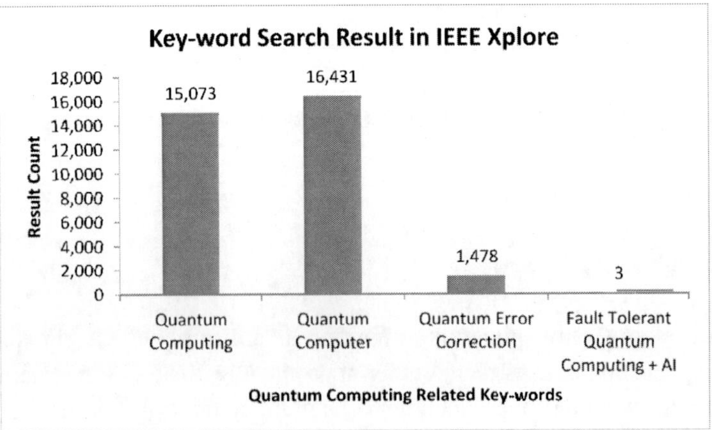

Figure 3. Quantum Computing Key-words Search in IEEE Xplore.

2.1. Review of Quantum Computing

Andrew Chi-Chih Yao in 1993 explained the complexity of quantum circuits which is much easier than a quantum Turing machine. The Bernstein-Vazirani algorithm has further kick-started the quantum complexity theory by propounding the universal quantum Turing machine [36]. This introductory work of quantum computer could not gain much limelight until the Shor algorithm and Grover algorithms theoretically effectuated the concept in their work [16, 17] which was experimentally implemented by Chuang et al. reported for fast quantum search [37]. The field of quantum computing rapidly shifts from classical to quantum components to formulate the noisy intermediate-scale quantum computer based on a special class of algorithms called 'variation quantum algorithms' [38, 39]. However, the full potential of quantum computing can be realized only through large-scale fault-tolerant universal quantum computers. Artificial intelligence techniques and machine learning can further enhance computational abilities [40, 41]. The use of logical gate-based quantum computers for various applications was studied by Bravyi et al. [42]. Bremner et al. used a quantum array of quantum states to construct a stable quantum memory [43]. Processing units are the central part of any computational system. The Quantum Processing Unit (QPU) using the quantum bus for intra-component communication was developed by Meter in 2006 [44]. The quantum control and measurement can be achieved by manipulations of qubits [45]. Detection and characterization of error is essential for developing a fault-tolerant quantum computing system. Nagayama et al. have designed an error correction system using the amplitude and phase of a quantum state [46]. Mukai et al. have studied the approaches of quantum computation by classifying into analog and digital systems [47].

2.2. Review of Quantum Error Correction

The main challenge before the fault-tolerant quantum computer is to develop an effective, sustainable and stable quantum error correction schema to correct the error caused by environmental noise and de-coherence. A good quantum error correction code converts the unmanageable continuum of errors into a manageable discrete set [48]. The first quantum QEC of Shor was by entangling 9-qubits and Khalifa et al. have evaluated the performance of such QEC [49]. Saeed et al. have designed a simulation system to investigate the effect of system load on quantum error correction for 5-qubit, 7-qubit, and 9-

qubit systems [50]. Babar et al. have developed a quantum feedback control for correcting the quantum error in such a system [48]. Grassl et al. investigated the quantum error correction during the natural decay process of multilevel atoms and qudit amplitude dumping of the bosonic system which uses the principle of mathematical induction to correct the error in 'A_{ij}' using '$j = i + 1$' for known code 'A_0' [51, 52]. The quantum computing operations can be carried out using quantum gates which are unitary mathematical operations of linear algebra. Apart from error correction from qubit, the error correction can be used due to faulty gates is investigated by Ahn et al. [53]. Majumdar Sur-Kolay has explored that redundancy control can be used for error correction in the classical system but can't be directly applied due to the no-cloning theorem [54]. De Brito et al. have developed a bit-flit model to diagnose the error in encoded code [55].

2.3. Review of QEC Using AI Techniques

Recently artificial intelligence techniques have been identified as a strong tool to train the system on past data and perform the task in a fast and adaptable manner. Initially, Maria et al. explored classical machine learning on a quantum computing system for solving computationally difficult subroutines in a faster and more robust manner [56]. The artificial intelligence techniques were applied for optimizing the QEC in near-term fault-tolerant quantum computing in [57]. Decoding is an important strategy in QEC and ANN has been deployed for the same by Varsamopoulos et al. for solving such problem while Sweke et al. has utilized Reinforcement Learning for such issues which were further developing a general framework for fast and near-optimal machine-learning-based decoder for topological stabilizer codes [58, 59, 60]. The work of Andreasson et al. is based on the application of deep reinforcement learning to study QEC for toric codes while Chinni et al. used the same for checking parity in such code [61, 62]. In [63] a stochastic neural network called a Boltzmann machine has been designed for QEC in topological codes. The work of [64] is the first of its kind to initiative to explore the application of a genetic algorithm for QEC while [65] accomplished the same using the natural language processing (NLP) approach. With the AI techniques are gaining traction in QEC various giants have come forward to launch specific end-to-end quantum computing platforms /tools out of which Google's TensorFlow [66], Xanadu's PennyLane, IBM Qiskit QML

Library, Amazon AWS QML Library, and MS-QML are few such popular platforms. The summary of research work in the area of QEC using AI techniques in tabulated in table 2.

Table 2. Summary of literature review on 'QEC using AI Techniques'

Research Work	Main achievements / research output
Maria et al. [2014) [56]	Utilized classical machine learning on a quantum computing system for solving computationally difficult subroutines in a faster and more robust manner.
Dunjko and Briegel (2018) [57]	Applied AI techniques for optimizing the QEC in near-term fault-tolerant quantum computing environment.
Varsamopoulos et al. (2017) [58]	Deployed ANN for solving decoding problem in quantum computing.
Sweke et al. (2018) and Devaasuren (2020) [59, 60]	Utilized Reinforcement Learning for developing a general framework for fast and near-optimal machine-learning-based decoder for topological stabilizer codes.
Andreasson et al. (2019) [61]	Applied deep reinforcement learning to study QEC for toric codes.
Chinni et al. (2019) [62]	Used deep reinforcement learning for checking parity in toric codes.
Torlai and Melko (2017) [63]	Designed a stochastic neural network called a Boltzmann machine for QEC in topological codes.
Acampora et al. (2021) [64]	Developed first of its kind model to explore the application of a genetic algorithm for QEC
Valiron et al. (2021) [65]	Developed QEC using the natural language processing (NLP) approach
Quantum AI Platforms [66]	Popular Quantum AI platforms are Google's TensorFlow, Xanadu's PennyLane, IBM Qiskit QML Library, Amazon AWS QML Library, and MS-QML

3. Research Hotspots in Quantum Error Correction Using Artificial Intelligence Techniques

3.1. Current Status

With unprecedented developments in the quantum computing world is at the brink of another revolution but the realization of a fully functional quantum computer is still far away. The stakeholders in Information Technology (IT) and computing regime are toiling hard to convert quantum computing into a working chip to form a commercially viable quantum computer. All the countries across the world, research institutes, universities, and IT companies are spending huge funds on quantum computing [67]. The details of major quantum systems developed so far are tabulated in Table 3.

Table 3. Summary of major quantum systems

Platform	Company	Country	Type	Qubit implementation	QC Cloud offering	Provides access to
Large multi-platform offering QPU, CPU and GPU in the Cloud	Amazon	USA	Multiple	Multiple	Yes	D-Wave Systems, IonQ, Rigetti
	Microsoft Azure	USA	Multiple	Multiple	Yes	Honeywell, IonQ, QCI
Quantum Processing Units (QPUs)	Alibaba Quantum Lab	China	Gate	Superconducting	Yes	Own system
	Alpine Quantum Technologies	Austria	Gate	Trapped Ion	Yes	Own system only (Pine, Ibex)
	D-Wave Systems	Canada	Annealer	Superconducting	Yes	Own system
	Google Quantum AI	USA	Gate	Superconducting	Yes	Own system
	Honeywell Quantum Solutions	USA	Gate	Trapped Ion	Yes	Own system only (HØ and H1)
	IBM	USA	Gate	Superconducting	Yes	Own system
	IonQ	USA	Gate	Trapped Ion	Yes	Own system
	Origin Quantum	China	Gate	Superconducting	Yes	Own system
	Rigetti Computing	USA	Gate	Superconducting	Yes	Own system
	Xanadu	Canada	Gate	Photonics	Yes	Own system

The work of giants like Google, IBM, Microsoft, Honeywell, D-Wave System, AWS, IonQ, and Rigetti are already in at advanced stages. In 2017, D-Wave Systems released its first quantum computer 2000Q and in 2019, Google has proved its supremacy while China has proved the same in 2020. Thereafter quantum supremacy has become an issue of strategic importance for developed and developing economies across the world and all of them have planned a road map to build their own full-fledged quantum computer. IBM Research and Google AI are the major research groups having influential research in this field. Google is planning to release its first practical quantum computer by 2029. To harness the benefits of the weird potential of artificial intelligence and machine learning in quantum computing, Google has released its 'TensorFlow'; a simulator harnessing state-of-the-art machine learning and quantum computing algorithms.

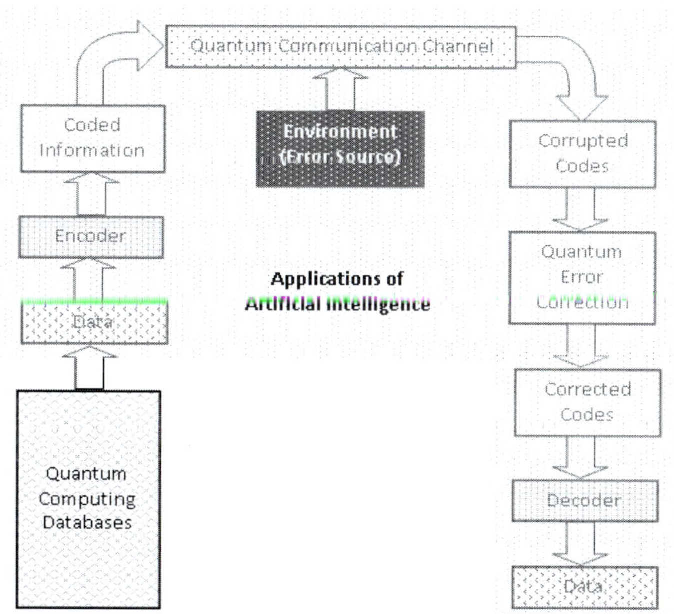

Figure 4. Application regime of artificial intelligence on quantum computing.

3.2. Research Concern

All these national and international efforts are trying to develop a fault-tolerant quantum system that can't be possible without minimizing the quantum error below a threshold prescribed by the 'quantum threshold theorem'. The error in the quantum system generally arises due to sensitivity in interaction with environment, reliable matrix transformations, constraints on state preparation, uncertainty and entropy of quantum gates, constraints on hardware, low temperature, etc. Artificial intelligence can reciprocate technological benefits with quantum computing by learning the system to optimize it as shown in Figure 4.

The AI agent learns from the environment (quantum computer) and decides the next action to maximize cumulative rewards while interacting with the environment. The main challenge in this regime is that quantum computers suffer from the issue of memorization instead of understanding. Therefore, quantum-inspired learning algorithms which can facilitate adaptive and dynamic learning are required for effective quantum computing.

4. Implementation of a Quantum Computer Using AI

Presently, various Python-based quantum computing simulators such as TensorFlow, PennyLane, IBM-Q, Azure, and QSim are available to study the quantum systems. These quantum simulators can be accessed via cloud-based services. Apart from it, the dataset for quantum machine learning is available on Mendeley Data [68], GitHub QDataSet [69], QM9 [70], and IEEE Data Port [71]. Various artificial intelligence algorithms have been developed as learning protocols for quantum systems. Following are the major AI algorithms used in implementation of quantum error correction:

- Quantum Boltzmann machines (QBM)
- Hybrid Helmholtz machine
- Variable depth quantum circuits (vVQC)
- Hybrid quantum autoencoders (HQA)
- Quantum reservoir computing (QRC)
- Quantum multiclass classifier (QMCC)
- Support vector machines with a quantum kernel estimator (QSVM-Kernel)
- Quantum neural networks QRNN
- Hybrid k-neighbours-nearby model (HKNN)
- Orthogonal neural networks
- Quantum Hopfield Neural Network

- Quantum fully self-supervised neural networks (QFS-Net)
- CNOT neural networks (CMN)
- Quantum convolutional deep convolutional neural networks (QDCNN)
- Quantum backpropagation neural networks (QBP)
- Feed foward neural networks (ffNN)
- Long short-term memory neural networks LTSM-QNN
- Quantum generative adversarial networks (QGAN)
- Recurrent quantum neural networks (RQNN)
- Quantum recurrent encoding-decoding neural networks (QREDNN)

Conclusion

Quantum computing is a nascent field of research jousting on the two most influential technologies of the 20th century, viz. quantum physics and computer science. However, the field of quantum computing has taken a big leap during the last few decades but the practical desktop quantum computer is still a distant away from reality. Governments, research institutes, universities, IT companies are investing a huge amount to design a practical quantum computer but QEC to minimize de-coherence and quantum noise up to the level of fault-tolerance is still not achieved. The Government of India has also roll out ambitious projects to compete in quantum race for quantum advantage. Various AI-based techniques have been developed which are capable of learning the system optimally and behaving smartly. This chapter summarized the research hotspots in quantum error correction using AI techniques which reveals that AI may be an exhaustive solution for quantum error correction. These research hotspots require immediate attention of the scientific community involved in this regime for making the quantum computers practically feasible and viable.

Acknowledgments

This chapter is based on the research proposal submitted by the corresponding author under supervision of Prof. Seema Verma at Banasthali Vidyapith, Rajasthan. He is grateful to co-authors and Vidyapith and for help and support.

References

[1] "*BCG*," [Online]. Available: https://www.bcg.com/en-in/press/21july2021-quantum-computing-transform-multiple-industries-create-850-billion-annual-value. [Accessed 20 Nov 2021].

[2] "*Budget 2018, GoI*," [Online]. Available: https://www.indiabudget.gov.in. [Accessed 20 Nov 2021].

[3] Rieffel, E. and W. Polak, *Quantum Computing: A Gentle Introduction*, London: The MIT Press, 2011.

[4] LaPierre, R. *Introduction to Quantum Computing*, Springer Link, 2021.

[5] Bernhardt, C. *Quantum Computing for Everyone*, Cambridge: The MIT Press, 2019.

[6] Gaitan, F. *Quantum Error Correction and Fault Tolerant Quantum Computing*, Boca Ranton: CRC Press, 2008.

[7] Andreasson, P., J. Joel, S. Liljestrand and M. Granath, "Quantum Error Correction for the Toric Code using Deep Reinforcement Learning," *Quantum*, vol. 3, p. 183, 2019.

[8] Watkins, C. J. C. H. and P. Dayan, "Q-learning," *Machine Learning*, vol. 8, p. 279–292, 1992.

[9] Moore, G. and G. Moore, "Cramming More Components Onto Integrated Circuits," *Electronics Magazine*, vol. 38, no. 8, pp. 114-117, 1965.

[10] Nielsen, M. A. and I. L. Chuang, *Quantum Computation and Quantum Information*, Cambridge: Cambridge University Press, 2000.

[11] Benioff, P. "Quantum Mechanical Hamiltonian Models of Turing Machines," *Journal of Statistical Physics*, vol. 29, pp. 515-546, 1982.

[12] Manin, Y. "*Computable and Uncomputable* (in Russian)," Sovetskoye Radioo, Moscow, p. 128, 1980.

[13] Feynman, R. P. *Lectures on Computation*, New York: Addison-Wesley Publishing Company Inc., 1996.

[14] Feynman, R. P. "Quantum Mechanical Computers," *Foundations of Physics*, vol. 16, p. 507–531, 1986.

[15] Deutsch, D. "Quantum Theory, the Church-Turing Principle and the Universal Quantum Compute," Proceedings of the Royal Society of London, Series A, *Mathematical and Physical Sciences*, vol. 400, no. 1818, pp. 97-117, 1985.

[16] Shor, P. W. "Polynomial-Time Algorithms for Prime Factorization and Discrete Logarithms on a Quantum Computer," *SIAM Journal on Computing*, vol. 26, no. 5, p. 1484–1509, 1995.

[17] Grover, L. K. "A Fast Quantum Mechanical Algorithm for Database Search," in STOC-1996: *Proceedings of the twenty-eighth annual ACM symposium on Theory of Computing*, Philadelphia, 1996.

[18] DiVincenzo, D. P. "The Physical Implementation of Quantum Computation," *Fortschritte der Physik*, vol. 48, no. 9-11, p. 771–783, 2000.

[19] "*IBM Quantum Computing*," [Online]. Available: https://www.ibm.com/quantum-computing/. [Accessed 30 Oct 2021].

[20] Arute, F., K. Arya, R. Babbush, D. Bacon, J. C. Bardin, R. Barends, R. Biswas, S. Boixo, F. G. S. L. Brandao, D. A. Buell, B. Burkett, Y. Chen, Z. Chen, B. Chiaro, R. Collins, W. Courtney, A. Dunsworth, E. Farhi, B. Foxen, A. Fowler, C. Gidney, M. Giustina, R. Graff, K. Guerin, S. Habegger, M. P. Harrigan, M. J. Hartmann, A. Ho, M. Hoffmann, T. Huang, T. S. Humble, S. V. Isakov, E. Jeffrey, Z. Jiang, D. Kafri, K. Kechedzhi, J. Kelly, P. V. Klimov, S. Knysh, A. Korotkov, F. Kostritsa, D. Landhuis, M. Lindmark, E. Lucero, D. Lyakh, S. Mandrà, J. R. McClean, M. McEwen, A. Megrant, X. Mi, K. Michielsen, M. Mohseni, J. Mutus, O. Naaman, M. Neeley, C. Neill, M. Y. Niu, E. Ostby, A. Petukhov, J. C. Platt, C. Quintana, E. G. Rieffel, P. Roushan, N. C. Rubin, D. Sank, K. J. Satzinger, V. Smelyanskiy, K. J. Sung, M. D. Trevithick, A. Vainsencher, B. Villalonga, T. White, Z. J. Yao, P. Yeh, A. Zalcman, H. Neven and J. M. Martinis, "Quantum Supremacy using a Programmable Superconducting Processor," *Nature*, vol. 574, pp. 505-510, 2019.

[21] Ball, P. "Physicist in China Challenge Goofle's 'Quantum Advantage'," *Nature*, vol. 588, p. 380, 2020.

[22] "IBM's Eagle — 127-Qubit Quantum Processor — Takes Flight, Another Step Toward 'Frictionless' Quantum in 2025," *The Quantum Insider*, [Online]. Availabfranle: https://thequantuminsider.com/2021/11/15/ibms-eagle-127-qubit-quantum-processor-takes-flight-another-step-toward-frictionless-quantum-in-2025/. [Accessed 18 Nov 2021].

[23] Wheeler, J. A. and W. H. Zurek, *Quantum Theory and Measurement,* Princeton: Princeton University Press, 2016.

[24] Bengtsson, I. and K. Zyczkowski, *Geometry of Quantum States: An Introduction to Quantum Entanglement*, Cambridge: Cambridge University Press, 2016.

[25] Moghadam, M. Z., M. Houshmand and M. Houshma, "Optimizing Teleportation Cost in Distributed Quantum Circuits," *International Journal of Theoretical Physics*, vol. 57, no. 3, pp. 848-861, 2018.

[26] Van Meter, R., K. Nemoto and W. Munro, "Communication links for distributed quantum computation," *IEEE Transactions on Computers*, vol. 56, no. 12, p. 1643–1653, 2007.

[27] Preskill, J. "Quantum Computing in the NISQ Era and Beyond," *Quantum*, vol. 2, p. 79, 2018.

[28] Criger, B. and B. M. Terhal, "Noise Thresholds for the [[4,2,2]]-Concatenated Toric Code," *Quantum Information and Computation*, vol. 15, no. 16, pp. 1261-1281, 2016.

[29] Kandala, A., A. Mezzacapo, K. Temme and M. Takita, "Hardware-efficient Variational Quantum Eigensolver for Small Molecules and Quantum Magnets," *Nature*, vol. 549, pp. 242-246, 2017.

[30] Li, Y. and S. C. Benjamin, "Efficient Variational Quantum Simulator Incorporating Active Error Minimization," *Physical Review* X, vol. 7, no. 021050, pp. 1-14, 2017.

[31] Mishra, N., M. Kapil, H. Rakesh, A. Anand, N. Mishra, A. Warke, S. Sarkar, S. Dutta, S. Gupta, A. P. Dash, R. Gharat, Y. Chatterjee, S. Roy, S. Raj, V. K. Jain, S. Bagaria, S. Chaudhary, V. Singh, R. Maji, P. Dalei, B. K. Behera, S. Mukhopadhyay and P. K. Panigrahi, "Quantum Machine Learning: A Review and

Current Status," in *Data Management, Analytics and Innovation*, Singapore, Springer, 2020, pp. 101-145.

[32] Shannon, C. E. "A Mathematical Theory of Communication," *The Bell System Technical Journal*, vol. 27, p. 623–656, 1948.

[33] Landauer, R. "Minimal energy requirements in communication," in P*hyscomp-1992: Proceedings of the Workshop on Physics and Computation*, Dallas, 1993.

[34] Pathak, A. *Elements of Quantum Computation and Quantum Communication*, Boca Raton: CRC Press, 2013.

[35] Yao, A. C.-C. "Quantum Circuit Complexity," in *Proceedings of 1993 IEEE 34th Annual Foundations of Computer Science*, Palo Alto, CA, 1993.

[36] Bernstein, E. and U. Vazirani, "Quantum Complexity Theory," in STOC-1993: *Proceedings of the Twenty-Fifth Annual ACM Symposium on Theory of Computing*, San Diego, 1993.

[37] Chuang, I., N. Gershenfeld and M. Kubinec, "Experimental Implementation of Fast Quantum Searching," *Physical Review Letters*, vol. 80, no. 15, pp. 3408-3411, 1998.

[38] Preskill, J. "Quantum Computing in the NISQ Era and Beyond," *Quantum*, vol. 2, p. 79, 2018.

[39] Huggins, W. J., J. Lee, U. Baek, B. O'Gorman and K. B. Whaley, "A Non-orthogonal Variational Quantum Eigensolver," *New Journal of Physics*, vol. 22, p. 073009, 2020.

[40] Devitt, S. J. "Classical Control of Large-Scale Quantum Computers," in In *International Conference on Reversible Computation*, Kyoto, 2014.

[41] Imamog, A., D. Awschalom, G. Burkard and D., "Quantum information processing using quantum dot spins and cavity QED.," *Physical Review Letters*, vol. 83, no. 20, p. 4204, 1999.

[42] Bravyi, S., D. Gosset and R. König, "Quantum Advantage with Shallow Circuits," *Science*, vol. 362, no. 6412, pp. 308-311, 2018.

[43] Bremner, M. J., A. Montanaro and D. J. Shepherd, "Achieving Quantum Supremacy with Sparse and Noisy Commuting Quantum Computations," *Quantum*, vol. 1, pp. 1-8, 2017.

[44] Meter, R. D. V. "Architecture of a Quantum Multicomputer Optimized for Shor's Factoring Algorithm.," Dissertation: Keio University, Keio, Japan, 2008.

[45] Meter, R. D. V. and S. Devitt, "The Path to Scalable Distributed Quantum Computing," *Computer* (Long. Beach. Calif), vol. 49, no. 9, pp. 31-42, 2016.

[46] Shota, N., A. G. Fowler and D. Horsman, "Surface Code Error Correction on a Defective Lattice," *New Journal of Physics*, vol. 19, no. 2, p. 023050, 2017.

[47] Sakata, M. H. K., D. Simon and W. Rui, "*Packaging Large-scale Superconducting Quantum Computer with Airbridge*," in APS 2019, 2019.

[48] Babar, Z., D. Chandra, P. Botsinis and S. X. Ng, "Duality of Quantum and Classical Error Correction Codes: Design Principles and Examples," *IEEE Communications Surveys & Tutorials*, vol. 21, no. 1, pp. 970-1010, 2019.

[49] Khalifa, O. O., Z. Ahmed, E. An, R. A. Saeed and A. Abdalla, "Performance Analysis of LTE Codes System using Various Modulation Techniques," in

Proceedings of the 2021 National Computing College Conference (NCCC), Taif, Saudi Arabia, 2021.

[50] Saeed, M. M., M. K. Hasan, R. Hassan, R. Mokhtar, R. A. Saeed, E. Saeid and M. Gupta, "Preserving Privacy of User Identity Based on Pseudonym Variable in 5G," *Computers, Materials & Continua*, vol. 70, no. 3, pp. 5551-5568, 2022.

[51] Grassl, M., L. Kong, Z. Wei, Z.-Q. Yin and B. Zeng, "Quantum Error-Correcting Codes for Qudit Amplitude Damping," *IEEE transactions on Information Theory*, vol. 64, p. 4674, 2018.

[52] Kubler, J. M., A. Arrasmith, L. Cincio and P. J. Coles, "An Adaptive Optimizer for Measurement-Frugal Variational Algorithms," Quantum, vol. 4, p. 263, 2020.

[53] Ahn, C., H. Wiseman and G. Milburn, "Quantum Error Correction for Continuously Detected Errors," *Physical Review Journals*, vol. 70, no. 2, pp. 834-839, 2003.

[54] Majumdar, R. and S. Sur-Kolay, "Approximate Ternary Quantum Error Correcting Code with Low Circuit Cost," in *Proceedings of the 2020 IEEE 50th International Symposium on Multiple-Valued Logic* (ISMVL), Miyazaki, Japan, 2020.

[55] de Brito, D. B., J. C. do Nascimento and R. V. Ram, "Quantum Communication with Polarization-encoded Qubit using Quantum Error Correction," *IEEE Journal of Quantum Electronics*, vol. 44, no. 2, pp. 113-118, 2008.

[56] Maria, S., I. Sinayskiy and P. Franc, "An Introduction to Quantum Machine Learning," *Contemporary Physics*, vol. 56, no. 2, pp. 172-185, 2014.

[57] Dunjko, V. and H. J. Briegel, "Machine Learning & Artificial Intelligence in the Quantum Domain: A Review of Recent Progress," *Rep. Prog. Phys.*, vol. 81, p. 074001, 2018.

[58] Varsamopoulos, S., B. Criger and K. Bertels, "Decoding Small Surface Codes with Feedforward Neural Networks," *Quantum Science Technology*, vol. 3, p. 015004, 2017.

[59] Sweke, R., M. S. Kesselring, E. P. L. v. and J. Eisert, "Reinforcement Learning Decoders for Fault-tolerant Quantum Computation," *Machine Learning: Science and Technology*, vol. 2, p. 025005, 2018.

[60] Davaasuren, A., Y. Suzuki, K. Fujii and M. Koashi, "General Framework for Constructing Fast and Near-Optimal Machine-Learning-Based Decoder of the Topological Stabilizer Codes," *Physical Review Research*, vol. 2, no. 3, p. 033399, 2020

[61] Andreasson, P., J. Johansson, S. Liljestrand and M. Granath, "Quantum Error Correction for the Toric Code using Deep Reinforcement Learning," *Quantum,* vol. 3, pp. 183-195, 2019.

[62] Chinni, C., A. Kulkami, D. MPai, K. Mitra and P. K. Sarvepalli, "Neural Decoder for Topological Codes using Pseudo-Inverse of Parity Check Matrix," in ITW 2019: *IEEE Information Theory Workshop*, Visby, Gotland, Sweden, 2019.

[63] Torlai G. and R. G. Melko, "Neural Decoder for Topological Codes," *Physical Review Letters*, vol. 119, p. 030501, 2017.

[64] Acampora, G., M. Grossi and A. Vitiello, "Genetic Algorithms for Error Mitigation in Quantum Measurement," in *IEEE Congress on Evolutionary Computation* (CEC), Kraków, Poland, 2021.

[65] Valiron, B., S. Mansfield, P. Arrighi and P. Panang, "*Quantum Physics and Logic*," EPTCS, vol. 340, p. 213–229, 2021.

[66] "Tensor Flow," [Online]. Available: https://www.tensorflow.org/. [Accessed 17 Oct 2021].

[67] "*Quantum Computing Report: Universities*," [Online]. Available: https:// quantumcomputingreport.com/universities/amp/. Accessed on Dec 17th 2021. [Accessed 03 Nov 2021].

[68] "*Mendeley Data*," [Online]. Available: https://data.mendeley.com/datasets/ pmycgb2bt7/1/files/999cfaf8-0964-4b8d-bd9b-63f77492781a. [Accessed 01 Dec 2021].

[69] "*QDataSet*," [Online]. Available: https://github.com/eperrier/QDataSet. [Accessed 11 Dec 2021].

[70] "*QM9 Dataset*," [Online]. Available: https://paperswithcode.com/dataset/qm9. [Accessed 5 Dec 2021].

[71] "*IEEE Data Portal*," [Online]. Available: https://paperswithcode.com/dataset/ qm9. [Accessed 19 Dec 2021].

Chapter 13

Crop Disease Detection Using Image Processing and Convolutional Neural Network

Goutam Sahu, Rutuparnna Mishra, Sujata Chakravarty* and Mamata Garanayak
Centurion University of Technology and Management, Bhubaneswar, India

Abstract

Cultivation is the backbone of the Indian wealth. About majority of population depends on agriculture by harvesting various types of food crops. Out of which Paddy is one of the primary food batch of India as it feeds more than 50 percent of the world's population. But currently, agronomists are cladding forfeiture in the producing of crops due to several causes. For the above issue, one of the most important causes is the diseases of the crops. This is due to insufficient awareness about the infection, pesticides and insecticides at hand to control the infections. To get the solution of the above discussed problem a deep learning prototype is developed using Convolutional Neural Network (CNN) procedure to predict the rice grain infection from the input image and take appropriate action to save the crops. The proposed model classifies into four classes i.e., Healthy leaf, Bacterial Leaf Blight, Brown Spot, and Hispa among which some are disease or pest. The recommended model has achieved an overall accuracy of 98.81% where the performance of leaf disease in paddy class like bacterial leaf blight and brown spot have achieved an overall 100% accuracy.

Keywords: deep learning, brown spot, hispa, bacterial leaf blight, convolutional neural network

* Corresponding Author's Email: sujata.chakravarty@cutm.ac.in.

In: Computational Intelligence for Sustainable Development
Editor: Brojo Kishore Mishra
ISBN: 979-8-88697-198-9

Introduction

Agriculture serves as livelihood for major population. Agronomists have a broad variety of crops to select from. The production of crops for full benefit and regular manufacturing is normally scientific. It could be produced with the assist of technical assistance. The overseeing of continuously growing crops necessitates supreme power, especially for management of diseases, which may have a major impact on the factors of production to create a profit (Devaraj, Abirami, et al., 2019; Kurniawati, Nunik Noviana, et al., 2009).

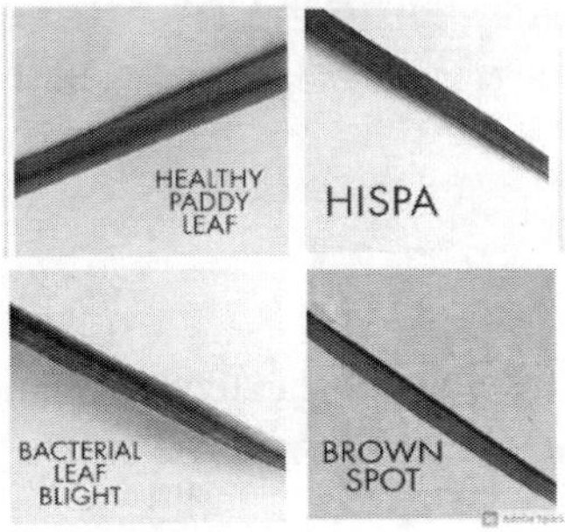

Figure 1. Healthy and disease affected leaf.

In agricultural countries, plantation of paddy is one of the important agricultural occupations. Paddy is a cereal crop and a staple food for many people around the world. A disease is an unusual disorder that harms or causes the plant to act inappropriately. Related noticeable changes in the plant make diseases easy to spot. There are several different types of diseases of paddy with similar symptoms: Blast Disease (BD), Brown Spot Disease (BSD), and Narrow Brown Spot Disease (NBSD) (Ramesh, Shima, et al, 2018; Narmadha, R. P. et. al., 2017). The disease comparison between the diseased and healthy leaf is pictorially represented in Figure 1. A variety of new applied science has come out to reduce processing of postharvest, heighten the agricultural sustainability, and increase the agricultural productivity. Different laboratory relied methods for disease detection have been used, including polymerase chain reaction, gas chromatography, mass spectrometry, thermography, and hyperspectral techniques. These methods, on the other hand, are inefficient and time-consuming. To improve the rate of recognition and overall accuracy of the outcomes, modern techniques like machine learning and deep learning procedures were utilized. Several studies have also been carried out in the area of machine learning for the identification and diagnosis of plant diseases, along with conventional machine learning methods such as artificial neural networks, support vector machines, and convolutional neural networks being among them (Ramesh, Shima, et al., 2018; Ramesh, S., and D. Vydeki, 2020).

Farmers currently rely on their expertise to detect diseases. Farmers use pesticides in large amounts because they don't recognize plant diseases. These pesticides don't help prevent disease, but they harm plants. While not too many rice infections may cause likely lesions or spots, different spots can result from the same disease due to different rice varieties and native weather conditions. As a result, their misclassification can harm production which leads assistance of a rice disease expert. Disease specialists in remote or rural areas are unable to provide timely solutions or guidance to farmers, and they often need expensive equipment and a significant amount of time to manually diagnose and classify diseases. Manual data analysis, on the other hand, necessitates the use of more eyes to check again and triple-check data for overall accuracy. An automated method, on the other hand, recognizes and classifies disease-affected images more reliably than a manual process (Hasan, Md Jahid, et al., 2019).

An automated framework based on machine learning techniques is proposed as a solution to this issue. Convolutional Neural Networks (CNN) and SoftMax are utilized in the proposed model to identify paddy diseases. CNN has been commonly used in recent years to investigate infection of the tomato, apple, corn, pear, tea, and various leaf diseases (Kurniawati, Nunik Noviana, et al., 2009). To achieve higher accuracy, a deep convolutional neural network is often used. The suggested method represented a system for classifying paddy plant disease and assisting farmers in making precise calculations as well as increasing production. The picture method is the most effective technique for agricultural application functions that do not require paying work. Image of plants can be used to detect illness. Aid for agricultural production will help to mitigate this impact. As most of the primary signs are microscopic, human being vision limits the disease's detection. This approach is tedious and time-consuming (Deshmukh, Radhika, and Manjusha Deshmukh, 2015).

In this paper, the Convolutional Neural Network (CNN) procedure is used to find paddy crop infection using an image of the crop, so that the crop can be saved by taking the right action at the right time. The proposed model has the best overall accuracy of 98.81 percent, with 100% accuracy for each paddy leaf infection class, such as bacterial leaf blight and brown spot, and 95% accuracy for Hispa.

Paddy Crop Diseases

Paddy Leaf is frequently affected by some diseases and pests like Brown Spot, Bacterial Leaf Blight, and Hispa, etc. which leads to a decrease in paddy yield. The main aim of this work is to identify the paddy leaf infections at an earlier stage so that appropriate action can be taken at right time to get a healthy paddy crop that is disease-free. Some of the images of Healthy Paddy leaf that are used in this research are shown in Figure 2.

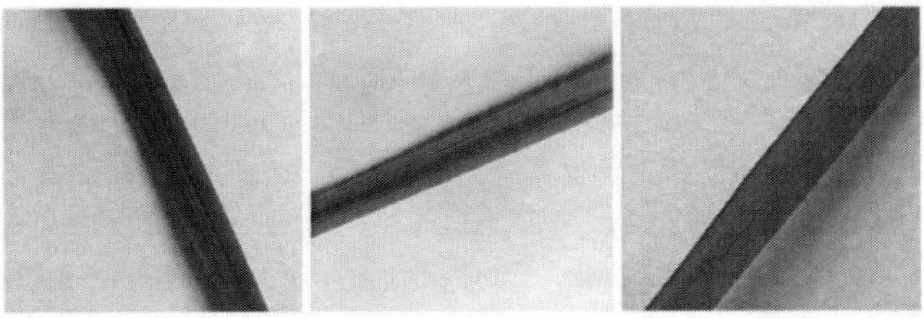

Figure 2. Healthy paddy leaf.

Brown Spot

Brown spot is also known as fungal blight or sesame leaf spot. The fungus threatens the seedling of the crop. The infection initiates as a little brown spot, then gradually increases to cylindrical, oval, and circular shapes. Spots with a width of 0.5 to 2.0 mm coalesce to form wide patches. The leaf then dries up as many spots coalesce. Infection can also be seen on the panicle and spine, which has a brown color. Glumes may also have dark brown or black spots. Crop infection results in seed germination failure, seedling mortality, and a reduction in grain quality and weight.

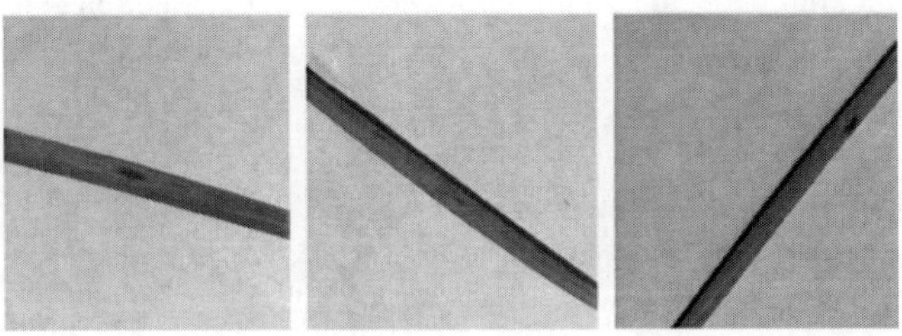

Figure 3. Brown spot.

In extreme cases, yields are reduced by 50%. It is possible to ensure that the paddy plant does not contract this infection by supplying the crop with adequate nutrients and avoiding stress of water. Seeds that have been treated with the chemical are less likely to become infected. Brown Spot paddy leaf disease photos are shown in Figure 3.

Hispa

Rice Hispa (also known as Dicladispaarmigera) is a major rice insect pest. Rice crops are harmed by both adults and grubs. They primarily damage the leaf blades of rice plants. Photosynthesis is hindered as a result, and yield is reduced. Some of the symptoms of damage are as follows: grubs' mining would be visible on the leaves, The upper surface of the leaf blade is scraped away, leaving only the lower epidermis as white lines parallel to the midrib, if larvae tunnel is detected through the leaf tissue, they leave behind irregular translucent white patches that run parallel to the veins and when a rice field is heavily infested, it appears burned. Figure 4 depicts some of Hispa Infected Paddy Leaf Disease images.

Figure 4. Hispa.

Bacterial Leaf Blight

Rice bacterial blight, also termed as bacterial leaf blight, is a bacterial infection that is one of the important destructive infections of agricutural paddy. Crop reduction can extend up to 75% in extreme epidemics, and millions of areas of paddy are pollute every year. Some of the harmful signs and symptoms are described are:-Yellowish streaks on leaf blades or beginning at the leaf tips with a wavy border, soaked in water, The main signs are undulated yellowish-white or golden-yellow marginal necrosis, drying of leaves back from tip and curling, and leaving the midrib intact. Early in the morning, bacterial ooze that looks like a milky or opaque dewdrop appears on young lesions, Leaves that have been severely contaminated appear to dry out easily and reduction in grain production may be as high as 75%. Figure 5 shows some of the Paddy Leaf Disease photos of Bacterial Leaf Blight.

Figure 5. Bacterial leaf blight.

Related Work

In (Devaraj, Abirami, et al., 2019), a software framework that mechanically detects and categorizes disease illness detection is implemented by steps such as loading of an image, image pre-processing, image segmentation, image extraction, and finally the image classification. The leaves images are utilized to detect plant infection diseases. As a result, image processing techniques can be used to get and diagnose infections in agricultural areas. Alternaria Alternata, Antracnose, Bacterial Blight, and Cercospora Leaf are all covered. By utilizing image processing methods in MATLAB software these automatic illnesses are detected. The development of automated detection systems using advanced technologies such as processing of an image aids agronomists in identifying infections at the very begining stage and giving the important required data for the management of diseases.

In (Kurniawati, NunikNoviana, et al., 2009), proposed a method for detecting infection sin paddy crops such as BD, BSD, and NBSD, which was primarily based on MATLAB application. The classification scheme was created using image processing techniques. The classification system was developed using 4 characteristics: lesion form, of the leaf colour boundary, colour of the spot, and the colour of the damage leaf of the paddy. The ratio of the lesion spot's height and width given a specific characteristic of the shape for calculating the forms of the lesions. To obtain the good outcome in detecting 94 paddy leaf images, 2 thresholding procedures were utilized. The best accuracy of the two local entropy threshold methods is about 94.6 percent. Since several values of intensity are less susceptible to illumination, the Otsu procedure is disabled for accurate segmentation.

In (Ramesh, Shima, et al., 2018), the algorithm aims to detect abnormalities on plants in their green houses or their environment. To avoid occlusion, the picture is normally considered with a clear backdrop. For precision, the algorithm was compared to other machine learning prototypes. The prototype was trained by utilizing 161 pictures of papaya leaves and a RF classifier. The model had a classification accuracy of around 70%. When using a large number of pictures and other characteristics in conjunction with global characteristics like SIFT (Scale Invariant Feature Transform), SURF (Speed Up Robust Features), and DENSE, along with BOVW, the accuracy can be improved (Bag of Visual Word).

In (Ramesh, S., and D. Vydeki., 20202), proposed a method for infections like bacterial blight, brown coloured spot, sheath rot, and blast, photographs of paddy leaves are directly taken from the agricultural area. To eliminate the background in pre-processing, RGB pictures are transformed into HSV pictures and hue component masking is performed. The segmentation of the unwelled and normal portions is done using a clustering process. Disease classification is carried out by utilizing the proposed DNN JOA system, with the JOA selecting the best weights. To fine-tune the stability of our system, we built a feedback loop. The overall accuracy, precision, F1-score, TNR, TPR, FPR, FNR, FDR, and NPV of the experimental results are evaluated and compared to ANN, DAE, and DNN. When compared to other classifiers, the DNN JOA method had a high accuracy of 98.8% for blast affected, 94.7 percent for bacterial blight, 92 percent for sheath rot, 93 percent for brown spot, and 90.57 percent for normal leaf picture. When comparing the training and testing accuracy, the DNN JOA classifier get the highest testing accuracy of 97 percent, followed by the ANN classifier at 83 percent, the DAE classifier at 90 percent, and the DNN classifier at 93.5%. In the future, any improved approach can be used to improve the identification and classification of plant diseases to produce the best results by reducing the false positives.

In (Hasan, Md Jahid, et al., 2019), with the use of Artificial Intelligence (AI) and powerful computer vision techniques, a novel and fruitful approach to detecting rice diseases have been introduced. In this article, an SVM classifier is used in conjunction with a deep CNN model. The proposed model has been improved using the transfer learning methodology. After that, the proposed model is re-trained using 1080 images from the dataset of nine distinct rice diseases. The features extracted from the DCNN model are then used to train the SVM classifier. The proposed model correctly described and categorized nine distinct types of rice diseases with a 97.5 percent accuracy.

In (Deshmukh, Radhika, and Manjusha Deshmukh., 2015), Picture segmentation, attribute extraction, and image classification are the three steps of the proposed methods. K-means clustering is used as an image segmentation technique to detect contaminated sections of the leaf. The paddy leaf picture is used to extract features in the feature extraction process. These characteristics are fed into the classifier for classification. The classifier in this experiment is an artificial neural network. For several years, many researchers have been focusing on real-time plant leaf diseases. This project will be used to detect leaf disease in real-time in the future. Farmers will benefit greatly from this project because it will enable them to detect paddy diseases at an early stage.

In (Narmadha, R. P., and G. Arulvadivu, 2017), this article aims to detect paddy infections. Blast Disease (BD), Brown Spot Disease (BPD), and Narrow Brown Spot Disease (NBSD) are the paddy diseases that stop the growth and defense of the paddy. Paddy may be infected at various stages of growth and in all parts of the plant, including the leaf neck and node. Bacteria, fungus, and other organisms may cause paddy disease. The technique was created to remove noise automatically, reduce human error, and shortening the time it takes to assess the impact of paddy leaf disease. It also increases the precision. This paper investigates paddy leaf detection and identification using k-means techniques.

In (Islam, Rashedul, and Md Rafiqul Islam., 2015), the image was segmented into three images based on colour using the K-means clustering process. The percentage of affected pixels is determined using unaffected leaf regions and disease-affected regions from these images. By measuring the number of pixels that are affected, the magnitude of the disease can be determined, allowing effective care measures to be taken.

In (Lu, Cunlou, Shangbing Gao, and Zecheng Zhou., 2013), the fuzzy least-square vector machine (FLSVM) algorithm has been used to develop a new method for detecting maize disease. It utilizes YCbCr colour space methodology to segment infection spots, the co-occurrence matrix spatial grey level layer to extract infections spot texture features, and FLSVM to classify maize diseases based on their texture characteristics. The sample mean and the center of each class are determined in this process; the distance between the sample and the center is calculated using the distance sample's initial membership, and the sample membership degree is calculated using the fuzzy K nearest neighbor method by finding K neighbors for each sample point.

Extensive tests on public datasets show that the algorithm can accurately recognize disease images, with a 98 percent or higher accuracy rate.

In (Abdullakasim, Wanrat, et al., 2011), developed a meth of for the identification of brown leaf spot infection in cassava, an image analysis method relied on several colour indices and neural network classification has been suggested. The findings revealed that getting visible symptoms of BLS-infected cassava plants is possible in general.

In (Bhoi, Sourav Kumar, et al., 2021), to identify pests in rice during field processing, and IoT supported UAV (Unmanned Aerial Vehicle) relied paddy pest prediction prototype using Imagga cloud is recommended. For giving paddy pest pictures to the Imagga cloud and giving pest data, the IoT-supported UAV aims on the artificial intelligence (AI) methodologies and Python programming concepts. The Imagga cloud detects the pest by combining the tags' trust values. The image's object is represented by the name. To define the insect, the tag with the highest confidence value and above the threshold is chosen as the destination tag. If a pest is discovered, the data is forwarded to the property owner for the action further. The suggested procedure is capable of identifying any type of pest that affects production.

In (Sethy, Prabira Kumar, et al., 2020), it examines almost every paper published between 2007 and 2018. This survey assists researchers in gathering information about various paddy dysfunctions based on their growth level. Image pre-processing, segmentation of pictures, feature extraction from the images, feature selection, and classification techniques are all briefly discussed in this article. Various issues concerning paddy disease diagnosis were addressed, and a common structure was proposed. Other imaging methods, such as hyperspectral and thermal imaging, are also discussed. The quality of the image, the number of training images, and the characteristics of the sample all play a role in these imaging systems.

In (Badage, Anuradha., 2018), the proposed system offers a solution for tracking the cultivated area regularly as well as automatic disease detection using remote sensing images. The proposed system alerts agriculturists to crop diseases, allowing them to take appropriate action. The proposed system's aim is to identify diseases early, as soon as they started to spread on the outer layer of the leaves. The suggested framework has 2 phases: the 1st phase deals with training image data sets, and the 2nd phase deals with real-world image data sets. This involves both stable and infected data sets for training. The 2nd step entails tracking the crop and using Canny's edge detection algorithm to classify the disease.

In (Pinki, Farhana Tazmim, Nipa Khatun, and SM Mohidul Islam., 2017), suggested that the 3 general paddy leaf infections (Brown spot, Leaf blast, and Bacterial blight) are diagnosed by utilizing an automated method, and pesticides and/or fertilizers are recommended based on the severity of the infections. The affected part of the paddy leaf picture is separated using K-means clustering. These infections are categorized based on their visual contents such as color, texture, and finally the shape. The Support Vector Machine (SVM) classifier recognizes the form of paddy leaf disease. Following awareness, a predictive remedy is proposed, which can assist agriculture-related individuals and organizations in taking effective measures to combat these diseases. In this paper, the proposed method utilizes a deep neural network model, i.e., Convolutional Neural Network (CNN) for paddy leaf disease classification. Loading data, augmentation of images, building the CNN model, measuring the Loss function, training the neural network, and finally classifying paddy leaf diseases are all part of the system.

In (Deng, Ruoling, Ming Tao, Hang Xing, Xiuli Yang, et. al., 2021) proposed that the paddy infection has bad effects on production of crop, and also the right detection of rice infection is the main key to keep away from the effects. However, the keep going infection detection procedures for paddy are not correct and also not much more efficient, as well as also often special instruments are needed. In this paper, an automatic detection procedure was designed and also implemented in a app. The procedure was designed using deep learning relied on a large image dataset that contained 32,026 pictures of 6 categories of rice infections: leaf blast, false smut, neck blast, sheath blight, bacterial stripe disease, and finally the brown spot. The heart of the procedure was the Ensemble Model in which sub-models were integrated. Lastly, the Ensemble prototype was validated by using a different set of pictures. Outcome showed that the 3 best sub-models were DenseNet-121, SE-ResNet-50, and ResNeSt-50, in terms of different attributes, such as, learning rate, precision, recall, and infection recognition accuracy. Therefore, these 3 sub-models were selected and integrated in the Ensemble prototype. The Ensemble prototype minimized confusion among the several types of infection, reducing misdiagnosis of the infection. By utilizing the Ensemble prototype to detect the 6 categories of rice infections, an overall accuracy of 90.7% was obtained, which is suppose to be best, considering the aspects likeliness in some types of rice infections. The app recommended the farmers to utilize the prototype on the internet server through a n/w, that was convenient and more efficient for the area of detection of rice leaf blast, false smut, stripe disease, and the brown colour spot.

In (Daniya, T., and S. Vigneshwari, 2021) suggested that the infections in plants gives a major impact on initiating safety in the yeild of food and they can lead to a decrease in the amount and quality of the crop product. In most of the cases, plant infections lead to no harvesting of grains. Therefore, an automatic detection of plant infection is recommended for detecting agronomical data. Different methods as well as procedures are obtained by several researchers for plant infection detection where deep learning is the most common method due to its good performance. Deep learning is used to detect infection from pictures of the rice crops. In this case, the paddy plant pictures undergo image pre-processing to eliminate noise and artifacts contained in the pictures. After that, the image segmentation is implemented with SegNet to obtain different segments. These several segments are again taken for bring out the characters, convolution neural network (CNN) characteristics and texture characteristics. These characteristics are waged for plant infection prediction where in the deep recurrent NN (Deep RNN) is used. The DRNN is trained with the suggested RideSpider Water Wave (RSW) procedure. The suggested RSW is come up with by combining RWW in Spider monkey optimization. The suggested RWS-relied Deep RNN gives better performance with the largest accuracy of 91.50% and the sensitivity of 83.89% and specificity of 94.02%.

In (Vasantha, Sandhya Venu, 2021) shown that paddy is a category of cereal grain which is used as staple diet by generally half of the world's people, specifically in Africa and In Asia. The paddy of Oryza Sativa is manifests to both abiotic environment such as frosty, drought, salinity, etc., and biotic stresses such as insects, pests, bacterial, viral and fungal infections. Again also it has become one of the most demanding jobs for the agronomists to detect the category of infections, the paddy has affected with and what is the fact that affects the production of the paddy if not predicted or detected in time. This article gives the possible suggestions for the solutions by using different machine learning procedures and the comparative analysis of procedures for detecting the categories of infections which has affected the paddy relied on the paddys picture information and mostly it represents recently used procedures with their performance measures.

In (Kathiresan, Gugan, M. Anirudh, M. Nagharjun, and R. Karthik., 2021) suggested that the latest evolution in deep learning methods have had a massive impact in the area of agricultural infection prediction. The bad impact of pests and bacterial infections to paddy plants are known, and for areas where paddy is staple, this is problem that carries a lot of weights. This methodology suggests a higher accuracy, transfer learned prototype that can give an active solution for agronomists and agricultural institutions to predict rice leaf infections. This article also uses a general adversarial n/w to balance the no. of infections samples. Here the authors also compare their prototype to other transfer learning architectures. The proposed prototype was implemented on GAN information set, and it reaches an average accuracy of 98.78% outperforming paradigm differentiation architectures. The prototype is also compared on 4 several datasets, without the GAN augmentation, establishing benchmark performance of 98.37% average accuracy.

In (Kumar, V. Vinoth, KM Karthick Raghunath, et. al, 2021) proposed that a particular no. of the world's residents is relied on paddy crops for living. In inclusion to sugar cane and corn, paddy is said to be the 3rd most growing crop food in the era. As a consequence of intensive usage of fertilizers, paddy plant infections have also increases faster in current era. Exploring the possible infection spread and categorizing to predict the consequent impact at an early stage will prevent the loss and also increases the rice yeild. The main job of these work is to recognize several categories of infections that affect the paddy crop plants, such as spots of brown color, bacterial blight, and leaf blasts. Both prediction and recognition are performed relied on the risk analysis of paddy crop leaf pictures. The procedure has the benefits of both fuzzy logic and DCNNs when handling with unorganized information, extracting important attributes from imprecise and ambiguous information. From the crop area, continuous picture information are captured through sensors of an image and given as a primary i/p to the suggested prototype to analyze the risk and then to distinguish them for recognition/prediction of the infection. The prediction or identification rate of the DCNFM is set up to be 98.18% which is evenly got to be effective in contrast with the Covolutiona lNN.

In (Anandhan, K., and Ajay Shanker Singh, 2021) proposed that now-a-days the paddy is on demand due to increase in population. Paddy is the one of the important grains for half of the world's people. In India many of the agronomists faces the problem to detect paddy leaf infection at a premature stage. The deep learning models were used to detect a rice blast disease such as Brown spot, Sheath blight, Blast and Leaf streak disease at early stage, so we can control disease spread over all plant to increase the rice production. The suggested process is implemented to detect a different categories rice plant leaf pictures disease prediction using mask R-CNN, Faster R-CNN procedures. We have studied five different paddy diseases and applied in southern region (India) to improve the paddy quantity. We collected a paddy leaf images through a Sony Camera & Mobile device (Vivo V9) and conducted experiments to train the model using 1500 images. Our experimental results analysis mask R-CNN is best suitable for to detect and identify the various rice blast disease such as Blast-96%, Brown spot-95%, and Sheath blight-94.5%. The results imply capable of disease identification from real-time capture leaf images and improve the accuracy applying different areas to control the spread rice blast disease.

In (Patil, Nilam Sachin, 2021) they suggested that in the area of horticulture, especially in rice plants, there is a huge demand for researchers to differentiate rice infections at the beginning phases. This is possible if there are automatic procedures that can guide the agronomists to detect the rice infections from the rice leaf pictures of the plants. The detection of agricultural plant infections by using the image preprocesses and machine learning methods can reduce the reliance on the agronomists to save the production of rice grains. In this article, authors put an attempt to process the pictures to make the character set for Classifiers and then feature extraction methods are utilized to extract the required attributes from the pictures that is already processed. The character-set is then given to the classifiers for detection of rice leaf infection. The cascaded classifier has been used to predict the infection of rice leaf. An effort has also been assembled to use genetic procedure with NN procedure to detect the infection of rice leaf. The suggested automated system can be utilized on Android phones, Windows and Apple platform for quickly detecting the rice leaf infections and the whole methods has been implementation by utilizing MATLAB software. The suggested system that is automated can also assists the agronomists to differentiate the infection in rice leaves at beginning to save the rice from destruction.

In this paper authors used the machine learning procedures such as AdaBoost and Bagging Classifier with GA for the detection of the rice leaf infection. The pictures of rice leaves have been pre-processed to lessen the unwanted attributes, to increase the picture characteristics for further processing. This includes resizing of pictures, brightness increase, filtering, corrections, noise reduction, geometric transformations, and grey scale transformations. After the processing of pictures, feature extraction procedures are implemented, so that the relevant character dataset can be attained from the pre-processed pictures. After that, the cascaded classifiers with modifications with respect to the problem statement are used to get the better accuracy. GA has also tried with NNA for detecting the pictures of rice leaf infections. The suggested procedure can also be utilized on Android phones and Windows platform for fast detecting the rice leaf infections. It can also assist the agronomists to differentiate the infected rice leaves at beginner stage to save the rice grain from damage.

In (Moparthi, Nageswara Rao, Debnath Bhattacharyya, 2021) suggested that the infection in plants corollary as a threat to the food industries. The suggested work assists in prediction of rice infection and gives suggestions which might protect in contradiction of the rice desease. The data from the internet is classified and also the totally several plant categories are known and are relabeled so that the author can create accurate data then get a specimen dataset that comprises of different crop infections which will assists in predicting the accuracy levels of the app. So by a training dataset they will train their classifier so the production will be predicted with best truthfulness. They tend to practice the CNN that includes several layers that are utilized for detection of infections in rice leaf.

In (Dhingra, Gittaly, Vinay Kumar, and Hem Dutt Joshi, 2019) suggested that the herbal items are less price, good to health, and have no or less complicacy. Thus, the demand of herbs relied medical things, health product; nutritional supplement, cosmetic etc. are increasing. The degree of excellence of a leaf is described by the quality of the leaf or the possibility of being free from the defects and substantial variations and also the infections in leafs causes the threats in economy of the country as well as the production status of agricultural industries. The catching of infections in leaves by utilizing digital image processing reduces the dependency on the agronomists for the safe of agricultural productions. Therefore, the leaf infection prediction and categorizations is the main aim of their suggested work. Her, in this article, the authors use a fuzzy fuzzy set extended form neutrosophic logic relied segmentation methods to measure the interest regions. The divided neutrosophic picture is distinguished by 3 membership dats: true, false and intermediate area. Relied on segmented areas, new attributes subset using texture, colour, histogram and infection sequence area are calculated to predict leaf as infected or healthy. Also, 9 several classifiers are utilized to check and demonstrate the discrimination power of merged feature effectiveness, where random forest dominates the other procedures. The suggested work is validated with 403 cases (201 healthy, 202 diseased). The suggested method could be utilized as an effective tool for infection prediction in leafs. A brand-new character set is promising and 97.40% classification accuracy is obtained.

In (Azimi, Shiva, Taranjit Kaur, and Tapan K. Gandhi, 2021) suggested that the agricultural yield reduces significantly because of the nutrient's reductions in plants. Nitrogen is an affective nutritive, is a key factor of growth restricting and also the major part of amino acids, nucleic acids, proteins and chlorophyll. Nitrogen inadequency also affects particular visible plant traits such as area, colour, the no. of leaves and tallness of the plant, etc. In imaging technology, with the evolution of computer vision relied plant phenomics has become a main area of plant research and management. Such imaging-relied procedures are non-destructive and much faster with higher levels of automation. In this work, authors have given an image-based plant phenotyping method for stress classification in plant shoot pictures. Results shows that a simple 22-layered deep learning model is contrasted to the established state of art deep learning models like ResNet18 and NasNetLargein yielding ceiling level stress classification from plant shoot images. The suggested prototype also out-performs traditional Machine Learning methods by attaining an average of 8.15% accuracy.

Proposed Model

For paddy leaf disease classification, the proposed approach uses a deep neural network model called Convolutional Neural Network (CNN). The framework includes data loading, image augmentation, constructing the CNN model, calculating the Loss function, training the neural network, and finally classifying paddy leaf diseases. Figure 6 depicts the Proposed Model.

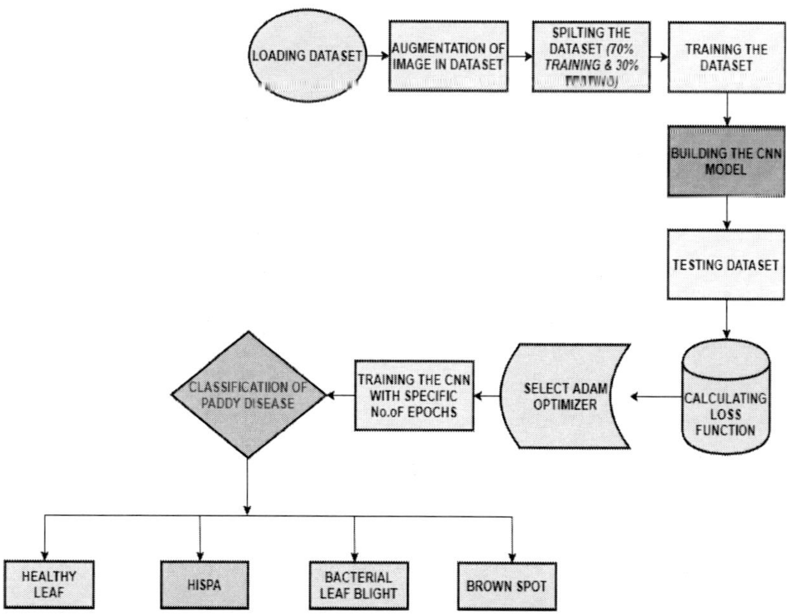

Figure 6. Flow diagram of proposed model.

Since CNN does not require a Feature Extraction step, Deep Learning technique is favoured in this paper over Traditional Machine Learning algorithms. The layers in Deep Learning can learn an implicit representation of the raw data on their own. In other words, the feature extraction step is already included in the Deep Learning process. When the inputs are in the form of images, it is preferable to use CNN as a deep learning model for the differentiation of paddy leaf diseases because CNN's function is to decrease the pictures into a form that is easy to operate without sacrificing characteristics that are crucial for getting a better result of prediction.

Collection of Image Data Sets

An information set in computer vision is a set of digital images that the researchers utilize to test, train and evaluate the performance of their work. There is no concept like database are the best or file system is the best - place to store pictures. It just relied on how we are designing the structure of our app. Image Datasets of various paddy crop diseases were gathered from both online and offline sources, including the UCI Machine Learning Repository, APS Image Dataset, and Flickr, among others.

Image Augmentation

Image augmentation methods are used effectively to increase the size of the training informations, which aids the output of the neural network classifier. It generates training images artificially by combining multiple processing methods such as random rotations, zooming the images, shifts (left and right shifts), shears and flips, and so on. In a simple example, if the number of datasets is limited, data augmentation can be used to produce more images. It aids in the collection of further data so that the CNN model can be properly trained. In the convolutional layer of CNN, there are filters. Those layers would be able to extract more and more functionality from the photos, making the system more robust and accurate.

Break the Datasets into Training Set and Testing Sets

Breaking the information into training and testing sets is a major part of evaluating the data mining prototypes. Basically, when we divide an image data set into training and testing set, many of the information is utilized for training, and a very little part of the information is utilized for testing. We must to have a mechanism to assess how well our prototype is generalizing. Therefore, we must require separating the input information into training, validation, and testing sub-sets to prevent our prototype from overfitting and to evaluate our model efficiently.

The training and testing of the augmented image dataset are covered in this process. With a division ratio of 70% or 30%, the augmented picture data of Paddy leaf disease of four different groups are randomly divided into two sets, such as training set data and testing set data (0.7 or 0.3).

CNN Architecture

CNN design is created by a stack of recognizable layers that transform the input data into an output data through a different function.

These designs are also referred as NN. It tells the computer to do what normally comes to human beings. CNN is a category of DNN that can identify and differentiate specific characters from pictures and are broadly utilized for inspecting visual pictures. Their requisition scope from picture and video identification, image differentiation, image analysis of medical images, computer vision and natural language processing.

The word 'Convolution" in CNN describes the mathematical function of convolution which is a category of linear performance wherein 2 functions are multiplied to obtain a 3rd function which describes how the shape of one function is altered by the other. In simple words, 2 pictures which can be presented as matrices are multiplied to give an outcome that is utilized to extract characters from the picture.

Figure 7 depicts the CNN Architecture for the Proposed Model. CNN is a multi-layered neural network with a unique architecture for detecting complex features in data. It is primarily used for image recognition, classification, and other tasks. An i/p layer, several hidden layers, and an o/p layer make up a CNN. The convolution layer, Rectified Linear Unit, pooling layer, and completely linked layer make up the hidden layer. The image is passed into the CNN model through the Input Layer, and the o/p Layer generates the paddy leaf disease prediction. The following are the layers of a Convolutional Neural Network.

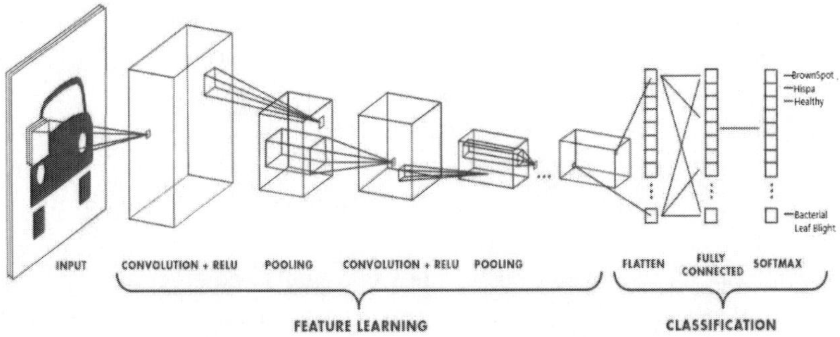

Figure 7. CNN architecture.

Convolution Layer

The image is sent to the Convolutional Layer after being loaded into the CNN model through the input layer. The convolution operation is performed on the input data through the use of filters to generate the output of that layer, that is, several filters slide over the previous layer's feature map to produce output characteristics maps.

Rectified Linear Unit (ReLU) Layer

The ReLU layer comes after the Convolutional Layer. It is, in essence, an activation function. Sigmoid, Tanh, ReLU, Softmax, and other forms of activation functions are among the options. The ReLU activation function is used in hidden layers in our model. It's the most common activation method. It's the most common activation method. The picture with negative pixel values is replaced with pixel value 0 in the ReLU layer, while the remaining pixels remain unchanged. This ReLU layer aims to eliminate negative values.

Pooling Layer

The Pooling Layer comes after the ReLU Layer. The pooling layer minimizes the picture's dimension by reduction operations along the image's dimensions, such as weight and height. The primary aim of the pooling process is to squize the picture size as much as possible. This layer utilizes executing window to search through the image and compresses the picture extraction functionality. The most popular methods for pooling layers are average pooling and maximum pooling. Maximum Pooling layer is used in our proposed model because it selects the brighter pixels from the image, which helps us identify the features.

Fully Connected Layer
The final and last layer, the Fully Connected layer, is where the actual classification takes place. These layers are stacked several times after the convolution + RELU+ Pooling layers before the picture is reduced to a vector. This layer is where the real classification takes place. Every neuron in this layer is interconnected, and this layer produces an N-dimensional vector, with the vectors of the features extracted from the picture contained in each neuron.

Working of CNN

Input images are used in the proposed model. On that Input image, a filter and convolution operation is applied. On the output received, the ReLU Activation Function is used. Then, to minimise the size of the dimensionality, Max Pooling is used. To obtain a filtered and shrunk image, several convolutional layers are added. The output is then flattened and supplied into the fully connected layer. The image was then classified using the Softmax classification technique.

Experimental Results

The concluding result is that the proposed system obtains an overall accuracy of 98.81 percent and a minimum loss of just 0.01 after CNN training. The individual accuracy achieved by each class i.e., Healthy Leaf, Hispa, Bacterial Leaf Blight, and Brown Spot is 86%, 95%, 100%, and 100% respectively. The Healthy Leaf class has an accuracy of 86%, with the remaining 13% of the Healthy Leaf class predicting healthy leaf as Hispa contaminated leaf. It's because white patches on paddy leaves are the most common symptom of Hispa Infected disease. For classifying Hispa infected Disease, the model uses a white patch on a paddy leaf as a feature. The model classifies paddy leaf as Hispa Infected Disease Leaf even though there are just a few patches on it, such as one or two small white patches. However, a few tiny patches will not turn a healthy leaf into a Hispa infected leaf. This is because a Hispa-infected leaf has a large number of white patches. The paddy leaf becomes diseased as a result of the large number of white patches that infect it. This problem can be overcome by training the system with more no. of images.

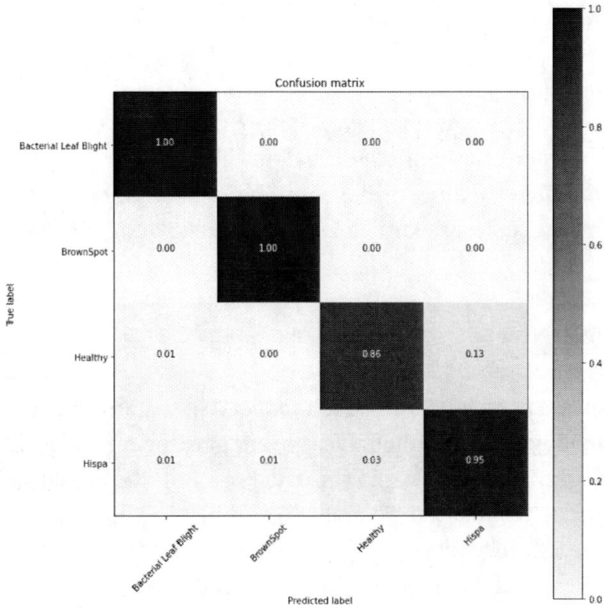

Figure 8. Confusion matrix.

The confusion matrix for the suggested prototype is described in Figure 8 (Das, Debasish, Mahinderpal Singh, Sarthak Swaroop Mohanty, and S. Chakravarty, 2020; Kishore, KantiMahanti Sai, Manoj Kumar Behera, S. Chakravarty, and Satyabrata Dash, 2020). The confusion matrix's main goal is to produce a resultant matrix as an output that defines the model's results. The number of photographs, as well as the input and output results, are all registered. As shown in the equation, the accuracy of disease prediction was determined by taking into account the following parameters: Specificity, Sensitivity or Recall, and Precision (i-iii).True Positive (TP), True Negative (TN), False Positive (FP), and False Negative (FN) are the 4 most common conditions used to predict sensitivity, specificity, precision, and accuracy. The accuracy is measured using the equation after the specificity and sensitivity have been determined (v) and F1_score is determined after Precision is determined (vi).

$$\text{Sensitivity (or) Recall} = TP/(TP + FN) \quad \text{(i)}$$

$$\text{Specificity} = TN/ (FP + TN) \quad \text{(ii)}$$

$$\text{Precision} = TP/(TP + FP) \quad \text{(iii)}$$

$$\text{Accuracy} = TP + TN/(TP + FP + TN + FN) \quad \text{(iv)}$$

$$\text{F1\_Score} = 2\ TP/(2TP + FP + FN) \quad \text{(v)}$$

In Table 1 the model performance is tabulated. The data present in Table 1 is represented in graphical manner in Figure 9.

Figure 10 depicts the proposed model's Training and Validation accuracy (Accuracy vs Epochs). Figure 11 depicts the training and validation failure of our proposed model (Loss vs Epochs). From the graphs, it concluded that the overall accuracy is 98.8% with a loss of 0.001.

This model will predict what kind of disease is afflicting the paddy crop instantly. Figure 12 depicts some of the random predictions of paddy leaf disease photographs.

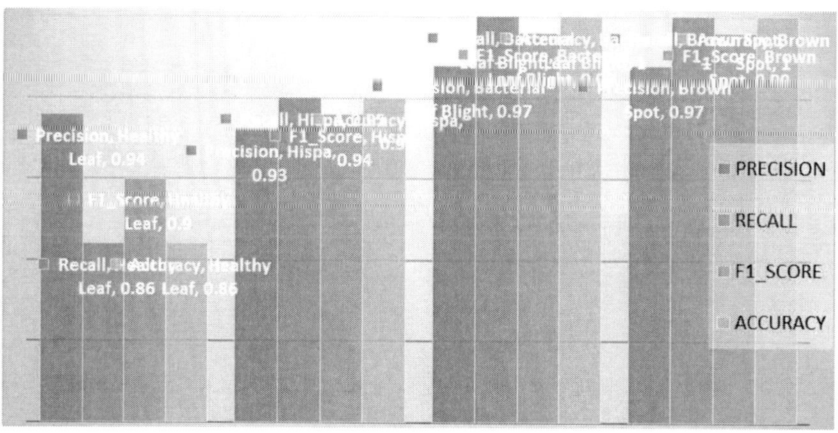

Figure 9. Performance Measure of all Class.

Table 1. The model performance

Class	Accuracy	Precision	Recall	F1_Score
Healthy Leaf	0.86	0.94	0.86	0.90
Bacterial Leaf Blight	1.00	0.97	1.00	0.99
Brown Spot	1.00	0.97	1.00	0.99
Hispa	0.95	0.93	0.95	0.94

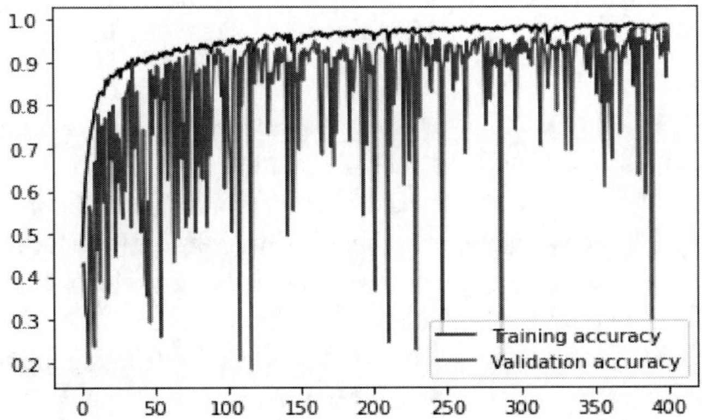

Figure 10. Accuracy vs Epochs.

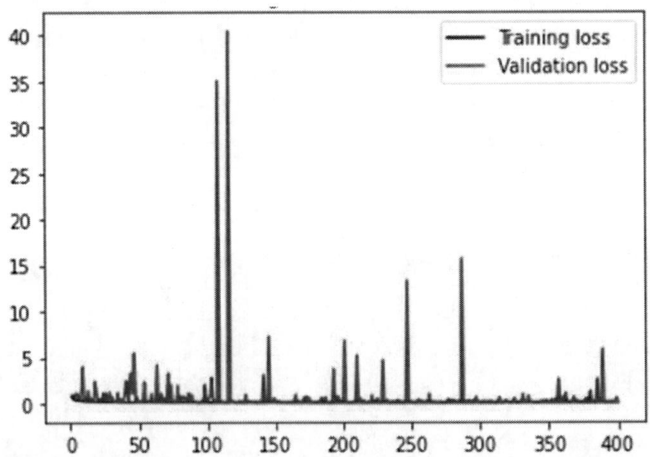

Figure 11. Loss vs Epochs.

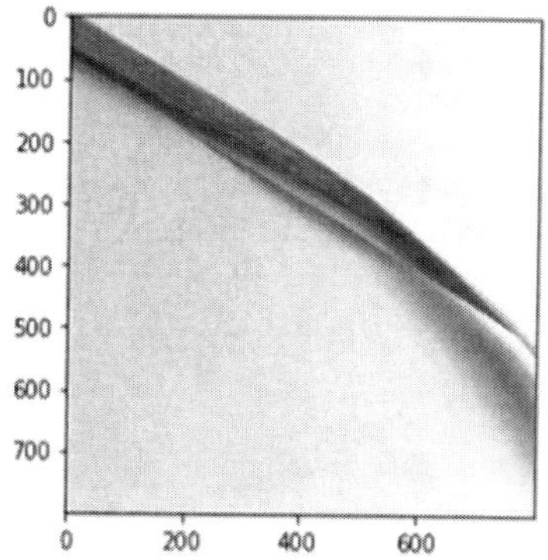

Actual: Bacterial Leaf Blight
Detected: Bacterial Leaf Blight

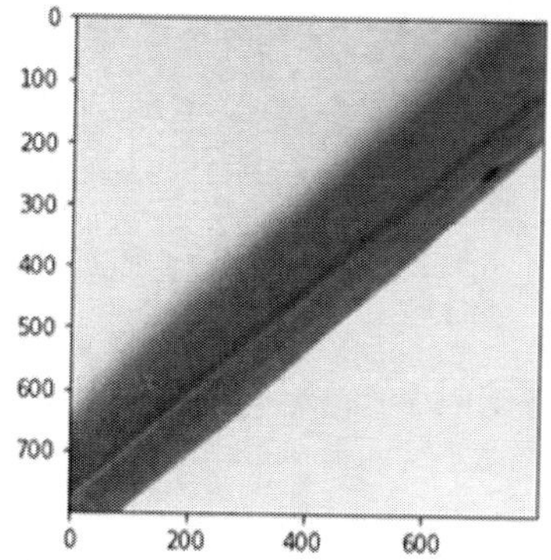

Actual: BrownSpot
Detected: BrownSpot

Figure 12. Prediction of diseases.

Conclusion: Discussion

Agriculture is vital to the survival of human beings and animals; therefore, it must be secured and conserved. In the cultivational domain, a lack of technical knowledge and leaders can lead to major difficulties with plant health. In the last few decades, there has been a significant increase in the use of technology in agriculture, which has improved the method of detecting and diagnosing various fatal diseases. The problem is solved by creating a deep learning model that uses the Convolutional Neural Network (CNN) procedure to predict paddy leaf infections such as Brown coloured Spot, Bacterial Leaf Blight, Hispa Infected, and Healthy paddy leaf from images with a 98.8% overall accuracy. This prototype's potential and future are extremely promising. It can be integrated into a website and turned into an app. The agronomist or person will upload a well-photographed paddy leaf to the website or established app, which will instantly detect the infection that the rice crop is messed with. It will tell us how likely the plant is to be infected with diseases.

Disclaimer

None

References

Abdullakasim, Wanrat, Kittipong Powbunthorn, Jintana Unartngam, and Tomohiro Takigawa. "An images analysis technique for recognition of brown leaf spot disease in cassava." *TarımMakinalarıBilimiDergisi* 7, no. 2 (2011): 165-169.

Anandhan, K., and Ajay Shanker Singh. "Detection of paddy crops diseases and early diagnosis using faster regional convolutional neural networks." In *2021 International Conference on Advance Computing and Innovative Technologies in Engineering* (ICACITE), pp. 898-902. IEEE, 2021.

Azimi, Shiva, Taranjit Kaur, and Tapan K. Gandhi. "A deep learning approach to measure stress level in plants due to Nitrogen deficiency." *Measurement* 173 (2021): 108650.

Badage, Anuradha. "Crop disease detection using machine learning: Indian agriculture." *Int. Res. J. Eng. Technol* 5, no. 9 (2018): 866-869.

Bhoi, Sourav Kumar, Kalyan Kumar Jena, Sanjaya Kumar Panda, Hoang Viet Long, Raghvendra Kumar, P. Subbulakshmi, and Haifa Bin Jebreen. "An Internet of Things assisted Unmanned Aerial Vehicle based artificial intelligence model for rice pest detection." *Microprocessors and Microsystems* 80 (2021): 103607.

Daniya, T., and S. Vigneshwari. "Deep Neural Network for Disease Detection in Rice Plant Using the Texture and Deep Features." *The Computer Journal* (2021).

Das, Debasish, Mahinderpal Singh, Sarthak Swaroop Mohanty, and S. Chakravarty. "Leaf Disease Detection using Support Vector Machine." In *2020 International Conference on Communication and Signal Processing (ICCSP)*, pp. 1036-1040. *IEEE*, 2020.

Deng, Ruoling, Ming Tao, Hang Xing, Xiuli Yang, Chuang Liu, Kaifeng Liao, and Long Qi. "Automatic diagnosis of rice diseases using deep learning." *Frontiers in Plant Science* (2021): 1691.

Deshmukh, Radhika, and Manjusha Deshmukh. "Detection of paddy leaf diseases." *International Journal of Computer Applications* 975 (2015): 8887.

Devaraj, Abirami, Karunya Rathan, SarvepalliJaahnavi, and K. Indira. "Identification of plant disease using image processing technique." In *2019 International Conference on Communication and Signal Processing* (ICCSP), pp. 0749-0753. IEEE, 2019.

Dhingra, Gittaly, Vinay Kumar, and Hem Dutt Joshi. "A novel computer vision based neutrosophic approach for leaf disease identification and classification." *Measurement* 135 (2019): 782-794.

Hasan, Md Jahid, Shamim Mahbub, Md Shahin Alom, and Md Abu Nasim. "Rice disease identification and classification by integrating support vector machine with deep convolutional neural network." In *2019 1st international conference on advances in science, engineering and robotics technology* (ICASERT), pp. 1-6. IEEE, 2019.

Islam, Rashedul, and Md Rafiqul Islam. "An image processing technique to calculate percentage of disease affected pixels of paddy leaf." *International Journal of Computer Applications* 123, no. 12 (2015).

Kathiresan, Gugan, M. Anirudh, M. Nagharjun, and R. Karthik. "Disease detection in rice leaves using transfer learning techniques." In *Journal of Physics: Conference Series*, vol. 1911, no. 1, p. 012004. IOP Publishing, 2021.

Kishore, KantiMahanti Sai, Manoj Kumar Behera, S. Chakravarty, and Satyabrata Dash. "Hyperspectral Image Classification using Minimum Noise Fraction and Random

Forest." In *2020 IEEE International Women in Engineering (WIE) Conference on Electrical and Computer Engineering (WIECON-ECE)*, pp. 296-299. IEEE, 2020.

Kumar, V. Vinoth, KM Karthick Raghunath, N. Rajesh, Muthukumaran Venkatesan, Rose Bindu Joseph, and N. Thillaiarasu. "Paddy Plant Disease Recognition, Risk Analysis, and Classification Using Deep Convolution Neuro-Fuzzy Network." *Journal of Mobile Multimedia* (2021): 325-348.

Kurniawati, NunikNoviana, Siti Norul Huda Sheikh Abdullah, Salwani Abdullah, and Saad Abdullah. "Investigation on image processing techniques for diagnosing paddy diseases." In *2009 international conference of soft computing and pattern recognition,* pp. 272-277. IEEE, 2009.

Lu, Cunlou, Shangbing Gao, and Zecheng Zhou. "Maize disease recognition via fuzzy least square support vector machine." *Journal of Information and Computing Science* 8, no. 4 (2013): 316-320.

Moparthi, Nageswara Rao, Debnath Bhattacharyya, G. Balakrishna, and J. Siva Prashanth. "Paddy Leaf Disease Detection using CNN." *NVEO-NATURAL VOLATILES & ESSENTIAL OILS Journal| NVEO* (2021): 800-806.

Narmadha, R. P., and G. Arulvadivu. "Detection and measurement of paddy leaf disease symptoms using image processing." In *2017 International Conference on Computer Communication and Informatics* (ICCCI), pp. 1-4. *IEEE,* 2017.

Narmadha, R. P., and G. Arulvadivu. "Disease symptoms identification in paddy leaf using image processing." *Advances in Computational Sciences and Technology* 10, no. 8 (2017): 2213-2223.

Patil, NilamSachin. "Identification of Paddy Leaf Diseases using Evolutionary and Machine Learning Methods." *Turkish Journal of Computer and Mathematics Education* (TURCOMAT) 12, no. 2 (2021): 1672-1686.

Pinki, Farhana Tazmim, Nipa Khatun, and SM Mohidul Islam. "Content based paddy leaf disease recognition and remedy prediction using support vector machine." In *2017 20th International Conference of Computer and Information Technology* (ICCIT), pp. 1-5. IEEE, 2017.

Ramesh, Shima, Ramachandra Hebbar, M. Niveditha, R. Pooja, N. Shashank, and P. V. Vinod. "Plant disease detection using machine learning." In *2018 International conference on design innovations for 3Cs compute communicate control* (ICDI3C), pp. 41-45. IEEE, 2018.

Ramesh, S., and D. Vydeki. "Recognition and classification of paddy leaf diseases using Optimized Deep Neural network with Jaya algorithm." *Information processing in agriculture* 7, no. 2 (2020): 249-260.

Sethy, Prabira Kumar, Nalini KantaBarpanda, Amiya Kumar Rath, and Santi Kumari Behera. "Image processing techniques for diagnosing rice plant disease: a survey." *Procedia Computer Science* 167 (2020): 516-530.

Vasantha, Sandhya Venu, BejjamKiranmai, and S. Rama Krishna. "Techniques for Rice Leaf Disease Detection using Machine Learning Algorithms." *Int. J. Eng. Res. Technol* 9, no. 8 (2021): 162-166.

Chapter 14

Automatic Face Detection Attendance System

Ritu Maity[1,*], MD and Shamaun Alam[2]
[1]School of Mechanical Engineering, KIIT University, Bhubaneswar, India
[2]IT Department, Utkal University, Bhubaneswar, India

Abstract

Automatic Face detection attendance systems have seen drastic improvements in recent years. Attendance systems for employees and students in various organizations are done through RFID cards or by manual methods which results in fake attendance and the manual process consumes time and manpower involvement. The setbacks in the systems can be made better by introducing a smart attendance system that reduces human effort and time and also provides accurate results. In recent years with help of various deep learning models, we can get quite accurate results as compared to traditional systems. This project consists of four stages image acquisition, face detection, recognition, and attendance updating.

Face detection is done through the VGG face model which creates embedding's for faces and accordingly labels the same and updates the attendance.

We have used the raspberry pi camera to capture the images and using the deep learning model we can detect faces and predict their roll numbers which are displayed on the LCD screen and we have developed GUI pages to maintain an attendance database for the same.

Keywords: face detection, VGG Face model, image acquisition, deep learning models, database

---

* Corresponding Author's Email: maity.ritu07@gmail.com.

In: Computational Intelligence for Sustainable Development
Editor: Brojo Kishore Mishra
ISBN: 979-8-88697-198-9

Introduction

Currently, in this digital era, the requirement of automatic face detection systems has become vital. Starting from colleges, universities, and offices everywhere tracking attendance of students and employees and maintaining the record becomes a time-consuming task [1].

Attendance is one of the vital parts of any organization, so it is important to make this system digitalized and intelligent so that no manipulation can take place in the attendance tracking. Different kinds of already available attendance systems are given below:

- Manual Attendance System
- RFID system
- Biometric system
- Bluetooth system
- Iris-Based Recognition System
- Automated Attendance System (AAS)

1.1. Manual Attendance System (MAS)

It is a method in which the teacher takes attendance manually in registering for students. It is a traditional method practiced by various colleges and universities which is a time-consuming process [2].

1.2. RFID System

It is a system in which an employee or student has to carry an RFID card which he/she needs to punch near a machine that detects the RFID rays and marks the attendance [2].

1.3. Biometric System

It is a system that will scan our fingerprint for attendance which is considered to be a unique part of an individual so that it will not match from one person to another [3].

1.4. Bluetooth System

It is a system that uses a smart Bluetooth device to mark attendance. It can at a time take eight connections [4].

1.5. Iris Based Recognition System

In this system iris of the human eye is used to detect the attendance. Initially, all person's iris image is captured and the computer is trained with the person's details and iris image. It is also an effective method of smart attendance tracking system. All kinds of false attendances or proxies can be stopped with this system [5].

1.6. Automated Attendance System (AAS)

The Automated Attendance System (ASS) can be used in tracking the attendance with help of a high-resolution digital camera which can discover and acknowledge the faces of the students and the machine would relate the acknowledged face with students' face images reserved in the database. Once the face of the student is captured then corresponding details of the student are entered concerning the image. This method has the possibility of machine error where the camera may miss capturing some of the images [6].

2. Literature Survey

Various research works have been conducted in the area of automated face recognition systems. Authors in [2] have proposed an automated attendance system based on radio frequency identification. This RFID system detects each student entering the classroom and getting out of the classroom and maintains all the records.

Authors in paper [7] have proposed an automated face recognition system using the OpenCV package and the local binary pattern histogram method was used for face recognition. In this method histogram for each face is calculated and matched with the existing histogram for each face.

The paper [8] proposes an attendance system based on the iris system. The image of the eyes of each student was recorded and based on unique iris template attendance was recorded. It was a web-based system developed. Authors in paper [9] developed an attendance system based on a Histogram of oriented gradients features along with a support vector machine classifier that was used for building the face detection algorithm. MATLAB environment was used for capturing various real-time scenarios, scalings, etc. In paper [10] researchers used Eigen face and fisher face provided by OpenCV package and compared with ROC curve and then implemented it in attendance system for developing an automated face recognition attendance system.

Researchers in paper [11] proposed a face recognition system by combining discrete wavelet transform and discrete cosine transform methods to extract face features.

Various models were proposed for face recognition systems like One-shot learning [12], CNN, Viola-Jones algorithm, face net classifier, kernel method, Haar classifier, etc. First, these models were studied then we decided to calculate the face embedding and then use the CNN technique for face recognition. For the Hardware system, we had also gone through various existing systems and tried to develop a unique and effective hardware system for our automated face detection attendance system project.

Many research papers have proposed only the web-based model but in this work, we have developed a full-fledged hardware setup for our face recognition attendance system.

3. System Architecture

In this project first user registration was done by taking three face images with different emotions thereafter calculating the embedding for those images and storing them in the database against their employee id. We had developed the hardware setup in which we had used a Raspberry Pi camera, Raspberry pi, LCD, LED, Ultrasonic sensors, and buzzer. First, the user registration was done by capturing three images of each individual and they were registered in the database as per their employee id.

For taking the attendance once the face image was captured in a Raspberry pi camera then the images were sent to the server, then the face embeddings were calculated and the face embedding was matched with the existing database of employees.

If the face embedding were matched with existing embedding then attendance will be marked and for matching the face embedding a threshold of 0.8 was used.

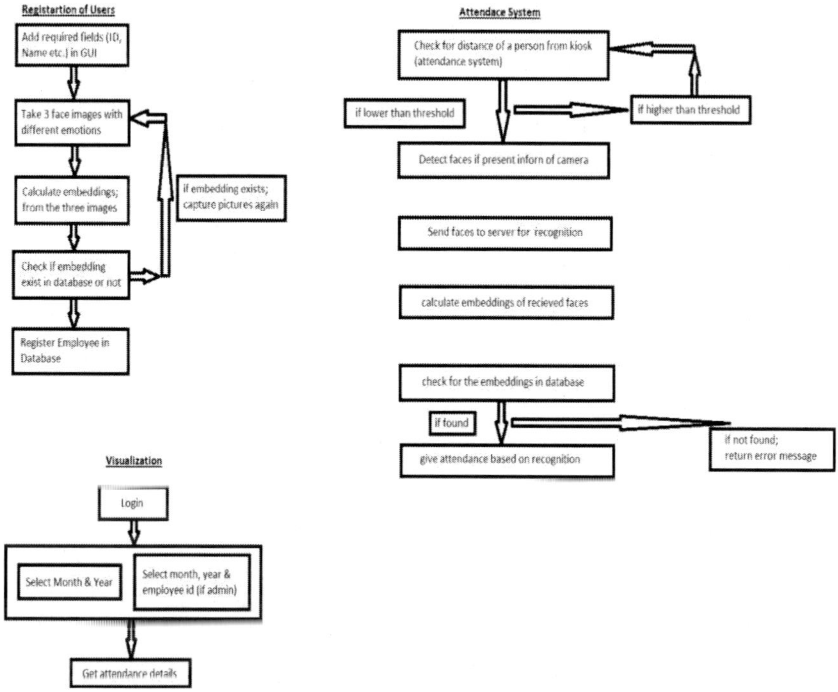

Figure 1. Flowchart showing system architecture.

If the face embedding matched with existing face embedding with an accuracy of more than 80 percent then we will mark the attendance otherwise will get an error message. For this whole setup first, the setup box was designed in SOLIDWORKS software and then 3d printing of the box was done, polypropylene material was used for making the box. Inside the box the total circuit design was done which consists of Raspberry pi, raspberry pi camera, LCD, ultrasonic sensor, and buzzer. An ultrasonic sensor was used to detect the person standing in front of the camera at a particular distance for taking attendance.

Buzzer was used to indicate whether attendance is marked or not. In our setup we have also used LED lights for indication which will help to know if green led glows it means attendance is tracked and if red led glows it means attendance is not marked.

Figure 2 shows the cad model developed in CAD software for making the setup. Figure 3 shows the setup box for the attendance system was 3d printed and assembled with electronics parts and made ready.

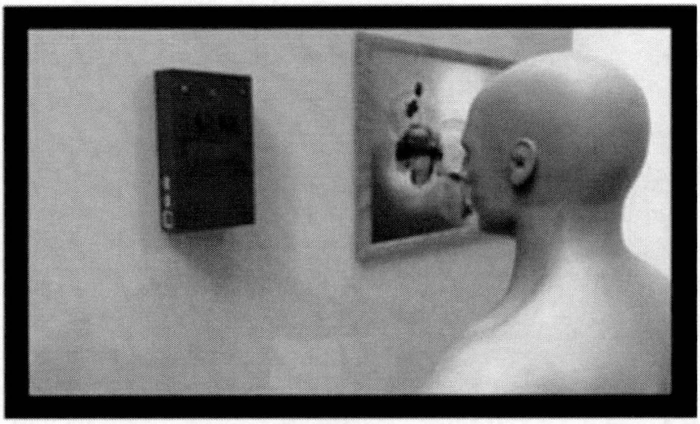

Figure 2. CAD model developed for the setup.

Figure 3. Hardware setup develop.

3.1. A Deep Learning Model Developed for Face Attendance

Deep learning is one of the most used technologies of the current era. In this project, we have used the concept of CNN architecture. CNN architecture consists of a convolutional layer, pooling layer, and hidden layers. The image is first convolved to extract features of the image after that it is passed through the pooling layer where negative pixels are removed and then it is passed through fully connected layers to generate the final output. The fig4 shows the CNN architecture layers [13].

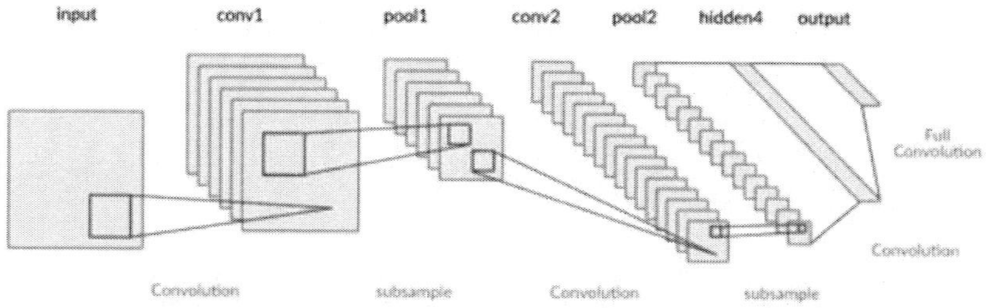

Figure 4. CNN Architecture [14].

In this project, we have used the VGG-16 face network for our model. VGG [14]is "Very Deep Convolutional Networks for Large-Scale Image Recognition". The model achieves 92.7% top-5 test accuracy in ImageNet, which is a dataset of over 14 million images belonging to 1000 classes. It improves AlexNet by replacing large kernel-sized filters (11 and 5 in the first and second convolutional layer, respectively) with multiple 3x3 kernel-sized filters one after another.Fig6 shows the VGG network. In this project initially, the face embedding was calculated using the face embedding network model and then for each face facial embedding was calculated which was further used to develop the deep learning model ie VGG-16 network.

In this model initially, a 224X224 RGB image is passed as the input layer, image is passed through a stack of the convolutional layer. In every layer, it utilizes a 1X1 convolutional layer which can be seen as a linear transformation. After the image is passed through the convolutional and ReLu layer, it is passed through the max-pooling layer where a pooling window of 2X2 is passed to off all negative values with a stride of 2. After which three fully connected layers form a stack of Convolutional layer. All hidden layers have ReLu as an activation function. Though VGG network weight is more so it takes larger time for training but many times it outperforms various well-defined models. Using this VGG network we got an accuracy of more than 95%.

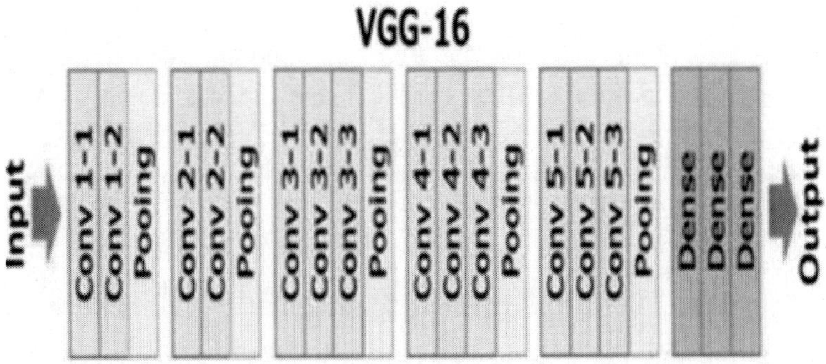

Figure 5. Shows VGG-16 Architecture [14].

3.2. Facial Embedding Calculation

Every facial embedding dataset consists of three triplets images. The set consists of one-class triplets, two-class triplets, and three-class triplets. In one class of triplets, all images are of the same expression let's say happiness, in two class triplets two face images will be of happiness and one will be of anger it will be a combination of emotions but at least two faces will be of the same type, in three-class triplets all three images will be of different type of emotions. Here a facial embedding network was used which was built using a face net model. NN2 version of pre-trained face net model was used in which we received output of 7X7 feature map with 1024 features. Then the features were sent to Densenet which consists of a 1X1 convolutional layer and a dense block with five layers. Output from Densenet is sent to a 7X7 average pooling layer followed by a fully connected layer. ReLu activation function was used [15].

The face embedding network uses NN2 face net architecture and dense net model to generate an embedding matrix for faces. The features extracted from face images are passed through face net architecture from which features extracted were passed through dense net and pooling layers and then fully connected layer to finally generate embedding out of the image. For training the embedding network triplet loss function was used which proposes to find the distance between two images that form the most similar pair to be as small as possible in comparison to the distance between two images from the third image. Once the facial embeddings were found then a dataset was formed having images and their facial embedding values which were passed into the VGG model formed to train the network and get the desired results.

3.3. Hardware Setup

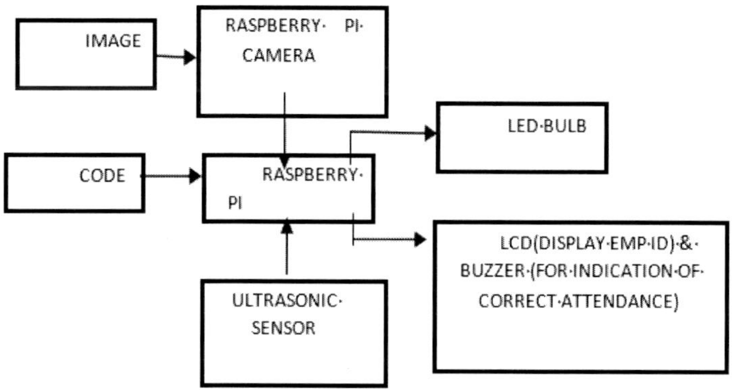

Figure 6. Flow chart showing the workflow of the hardware setup.

Figure 6 shows the flow chart of the hardware setup developed in the project and details of hardware parts are mentioned below.

Raspberry Pi

Raspberry pi is a microcontroller which is a series of board computer which is used for communication between hardware and software programs. In this project, we have used Raspberry pi 3 Model B+ with a 1.4GHz processor, and memory of 2GB.

Raspberry Pi Camera

A Raspberry pi camera was used to take the images and send them to the raspberry pi platform for further processing. The camera was 8 Megapixels. It was compatible with our Raspberry pi.

Ultrasonic Sensor

An ultrasonic sensor was used to detect distance from a physical object using ultrasonic waves. Here the ultrasonic sensor that we have used can detect distance in the range of 2cm-400cm, the voltage in the range of 5V DC, and Precision up to 0.3cm. Here the use of an ultrasonic sensor was important as it was required to use the sensor to detect humans or any object then only the image of the person will be captured using raspberry pi. It was also important to limit the distance otherwise each time the person passes near the device the attendance will be tracked but we wanted only when the person is within 100cm range then only the image should be captured for attendance for which this sensor was used.

Led Bulb

Digital control bulbs for indication of whether attendance is marked or not.

LCD

LCD unit was used for displaying the employee id if attendance was marked and showing an error message if attendance was not marked. A 16X2 LCD screen was used for display.

Buzzer

The buzzer is an audio signaling device that is used to send an indication if attendance is marked it will give a single beep and if attendance is not marked it will give a double beep sound. In this setup, we have used an ultrasonic sensor and raspberry pi camera to detect humans standing in front of the box for taking attendance. Then information from the ultrasonic sensor and raspberry pi camera is sent to Raspberry pi. The deep learning model developed is deployed into the raspberry pi module and once detection is done successfully by the model raspberry pi sends information in form of employee id which is displayed on the LCD screen and as well an LED bulb blinks green if attendance is marked and LED blinks red if face not recognized. As well as the buzzer was also given in output so that for correct face detection buzzer buzzes once and if not detected correctly it buzzes more than two times.

Once t face is detected successfully attendance is stored in the database developed in this project so that concerned person can check the attendance of their employees/students.

4. Results and Discussion

In this project we had collected images of 25 employees, for each person we had recorded three images, then we have calculating the embeddings of the face using the VGG network model and compare them with already stored image embeddings if it matches we have predicted to be given person image and his/her attendance was marked and stored in the database. Table 1 below shows the summary of model training and the convolutional layers used.

Table 1. Summary of model training and the convolutional layers used

conv5_2_3x3 (Conv2D)	(None, 7, 7, 512)	2359296	activation_44[0][0]
conv5_2_3x3/bn (BatchNormalizat	(None, 7, 7, 512)	2048	conv5_2_3x3[0][0]
activation_45 (Activation)	(None, 7, 7, 512)	0	conv5_2_3x3/bn[0][0]
conv5_2_1x1_increase (Conv2D)	(None, 7, 7, 2048)	1048576	activation_45[0][0]
conv5_2_1x1_increase/bn (BatchN	(None, 7, 7, 2048)	8192	conv5_2_1x1_increase[0][0]
add_15 (Add)	(None, 7, 7, 2048)	0	conv5_2_1x1_increase/bn[0][0] activation_43[0][0]
activation_46 (Activation)	(None, 7, 7, 2048)	0	add_15[0][0]
conv5_3_1x1_reduce (Conv2D)	(None, 7, 7, 512)	1048576	activation_46[0][0]
conv5_3_1x1_reduce/bn (BatchNor	(None, 7, 7, 512)	2048	conv5_3_1x1_reduce[0][0]
activation_47 (Activation)	(None, 7, 7, 512)	0	conv5_3_1x1_reduce/bn[0][0]
conv5_3_3x3 (Conv2D)	(None, 7, 7, 512)	2359296	activation_47[0][0]
conv5_3_3x3/bn (BatchNormalizat	(None, 7, 7, 512)	2048	conv5_3_3x3[0][0]
activation_48 (Activation)	(None, 7, 7, 512)	0	conv5_3_3x3/bn[0][0]
conv5_3_1x1_increase (Conv2D)	(None, 7, 7, 2048)	1048576	activation_48[0][0]
conv5_3_1x1_increase/bn (BatchN	(None, 7, 7, 2048)	8192	conv5_3_1x1_increase[0][0]
add_16 (Add)	(None, 7, 7, 2048)	0	conv5_3_1x1_increase/bn[0][0] activation_46[0][0]
activation_49 (Activation)	(None, 7, 7, 2048)	0	add_16[0][0]
avg_pool (AveragePooling2D)	(None, 1, 1, 2048)	0	activation_49[0][0]
global_average_pooling2d_1 (Glo	(None, 2048)	0	avg_pool[0][0]

Total params: 23,561,152
Trainable params: 23,508,032
Non-trainable params: 53,120

Once the model has trained on the VGG network then accuracy checking of the same was done. For accuracy checking our data was split into training data, testing data, and validation data during the training of the model, out of which validation data and training data were compared to check how accurate the model is. The graph above shows the relationship between no of epochs and loss which is nothing but error. We can see in the above curve shown in figure 7 that as no of epochs increases loss decreases which shows model accuracy increases. The overall accuracy of more than 95% was achieved for our face recognition model.

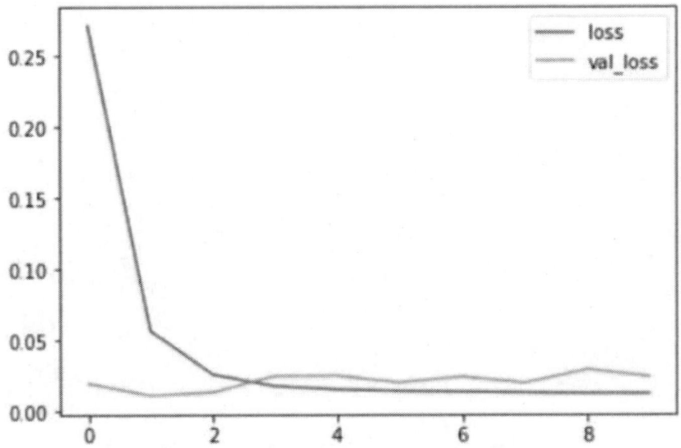

Figure 7. Result of a network trained showing the relationship between no of epochs and loss for training and validation data.

4.1. Database Handling

Here in this Project for the face detection attendance system, we are using PyQt5 for GUI design and SQLite for database creation. In the GUI part, we have designed a few UI pages for the smooth flow of our attendance system and we have created three tables named student/employee attendance, student/employee details, and student/employee_face_data respectively. Once the face detection was done attendance was marked and stored in the database.

4.1.1. PyQt5

It is one of the powerful Graphic user interface toolkits for developing GUI platforms. It is an integration of python programming Qt library to create graphical applications. It is a cross-platform tool as we have used python for developing our face recognition model so integrating with the GUI platform which supports python made our project more smooth for transition and database handling and execution. In this project, we have used the Qt designer platform shown above in figure 8 to design various GUI pages starting from the user registration page, user login page, user account, and admin account.

All the GUI pages were built using the platform shown above. Figure 9 below shows the GUI page developed for the login of the user. So, that each employee can view their profile and can also check their monthly attendance, and punch in and out time.

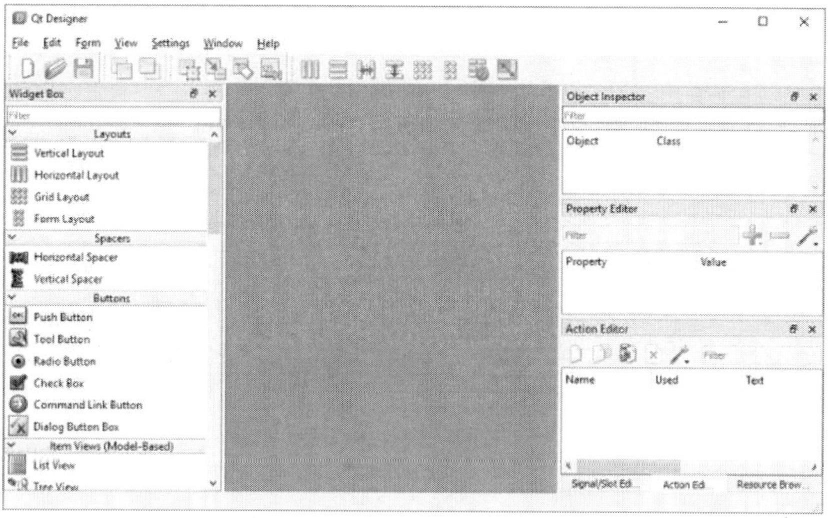

Figure 8. The Qt designer platform.

We had developed a home page tab where you will get the option to select month and year and accordingly a particular user can view their attendance for a particular month, then in this environment, we have a change password tab where you can click and change your password. Similarly, an admin side page was also developed where the admin can view the attendance of all employees and can see all the details of punch in and out the time of different employees. We had also provided options to download the attendance report for a particular month of one particular employee as well as reports for different employees' attendance can also be fetched. We had also created pages for employees' initial registration process where all details of employees will be fed into the system and as well corresponding to each employee id there will be three images of each employee all these things will be done through registration page developed.

A full fledge system was developed starting from user registration pages, to the user login page, and admin login pages which will have both user/employees as well as admin members to track attendance every month.

This system can also be deployed in schools and colleges for tracking the attendance of students by using the same procedure. The whole system was developed keeping in view the scalability of the users, as well as security features, were incorporated to ensure anytime type of unethical practices. Users can not edit their attendance once recorded they can only view their attendance report.

Once this whole system was developed we tested the same by making a few user registrations and then recording the attendance of users to check if the GUI pages developed and the database management system which was developed can it record all the data as per requirement.

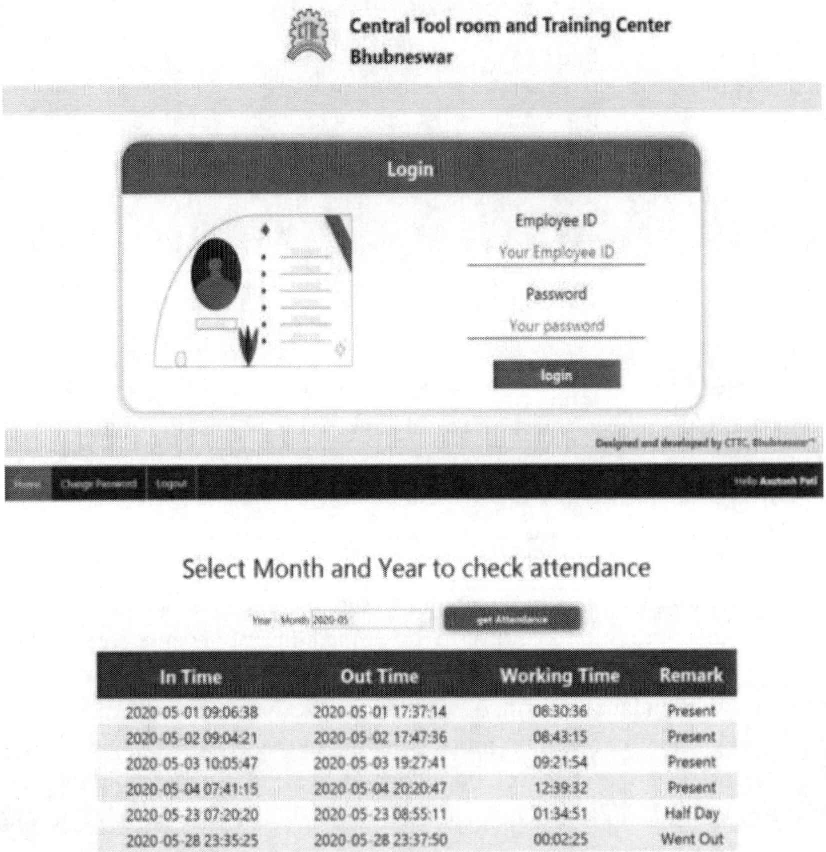

Figure 9. GUI pages designed for login and maintaining database.

4.1.2. SQLite

SQLite is a relational database management system that is present in the C library and is mostly used for database creation and handling. It is one of the stable reliable database management systems.

It is also cross-platform. In our project, we have used SQLite for database creation. Figure 10 below shows the images of the database created using SQLite.

In this project, we have used SQLite to create all databases. The above image shows the creation of a table in SQLite and how to enter details and create a database in the SQLite platform.

Figure 11 shows the table created using the SQLite tool to maintain the attendance database.

The complete database development was done in this project using the SQLite tool and as well as all the GUI page designing was done using the PyQt5 model.

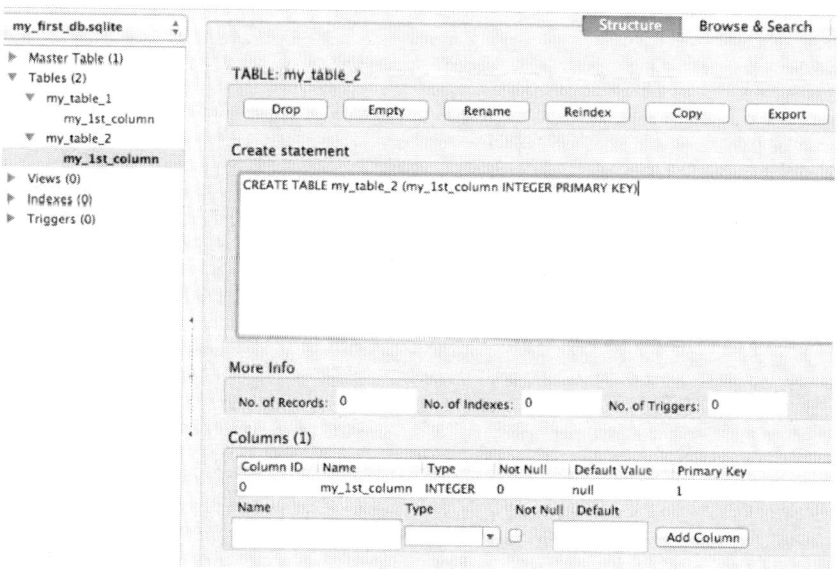

Figure 10. The SQLite platform.

File Edit View Tools Help

New Database | Open Database | Write Changes | Revert Changes | Open Project | Save Project

Database Structure | Browse Data | Edit Pragmas | Execute SQL

Table: student_details

	Sid	Sname	Sclass_id	Sphone_no	Spassword
	Filter	Filter	Filter	Filter	Filter
1	001	Bibhu	MCCAI	9861943043	123456
2	002	Rajat	MCCAI	9348531108	123457
3	003	Priyanka	MCCAI	9437799627	123458
4	MCCAI-14	BIBHUPRASAD...	MCCAI	9348294071	jitu1234

```
        cur = sql.connect('attendence_sys.db',detect_types=sql.PARSE_DECLTYPES)
        query = """INSERT INTO student_details(Sid,
                                               Sname,
                                               Sclass_id,
                                               Sphone_no,
                                               Spassword) values(?,?,?,?,?)"""
        cur.execute(query,(Sid,Sname,Sclass_id,Sphone_no,Spassword))

        query = """INSERT INTO student_face_data(Sid,arr) VALUES(?,?)"""

        cur.execute(query,(Sid,emb))

        cur.commit()
        cur.close()
```

Figure 11. Figure shows the table formed for maintaining the attendance database.

Conclusion

In this project, we have developed a model for an automatic face detection attendance system for which we had developed complete hardware and we had used VGG face deep learning model to predict faces accurately and deployed the same in hardware.

Then the setup was tested and it was found the setup was able to give accurate results, faces were detected with more than 95% accuracy and attendance was marked and stored in the database effectively. Figure 12 below shows the final setup developed and installed as well as tested. The following were the outcomes of the project:

a) We had developed a complete hardware setup for tracking the attendance of employees/students in any organization. The setup box consists of a Raspberry pi, raspberry pi camera, LCD screen, buzzer, and LED complete hardware setup was enclosed inside a 3d printed box.
b) The face recognition algorithm was developed using the embedding network concept and VGG model. The facial embedding was calculated so that we can get accurate results for the face detection system.
c) All the user registration pages, user login pages, and admin pages were developed using PyQt5, and the database management system was built using SQLite.
d) The complete integration of hardware and software was done to develop the whole system.
e) A complete full-fledged system was developed which was first tested in our local server by registering 10 users and tracking the attendance it was found the system was able to give accurate results and also the database was created as per the admin requirement.
f) The above system was then implemented in our production server for tracking attendance of our all employees and it was found that it was able to detect the persons and track attendance as expected

Figure 12. Figure shows final testing of Automatic face detection project.

Future Scope

This project was built and tested for 100 users further it can be tested for more users to check the scalability and performance of the model. We had developed one setup box further we had a plan to implement multiple setup boxes in our workspace and check the interconnectivity of those setup boxes with each other. Further, we will also work on the model developed to reduce the lag period for face detection which was near about 20secs in our existing system. Different other face recognition models can be explored to get better results.

Disclaimer

None

References

[1] Mayur Surve, Priya Joshi, Sujata Jamadar, Minakshi Vharkate. "Automatic Attendance System Using Face Recognition Technique". *International Journal of Recent Technology and Engineering (IJRTE),* Volume 9, Issue-1, May 2020, pp. 2134-2138.

[2] Dhanush Gowda HL, Vishal K, Kumari Dubey Neha, Pooja MR. "Face Recognition based Attendance System". *International Journal of Engineering Research and Technology (IJERT),* Volume 9, Issue-06, June 2020, pp. 761-767.

[3] Lodhaa Riya, Gupta Suruchi, Jaina Harshil, Narula Harish. "Bluetooth Smart based Attendance Management System". *International Conference on Advanced Computing Technologies and Applications (ICACTA2015),* 2015, pp. 524-527.

[4] Dwi Sunaryonoa, Joko Siswantorob, Radityo Anggoro. "An android-based course attendance system using face recognition". *Journal of King Saud University,* 2021, pp. 304-312.

[5] Seifedine Kadry and Mohamad Smaili. "Wireless attendance management system based on iris recognition". ISSN 1992-2248, *2010 Academic Journals,* pp. 1428-1430.

[6] Smitha, Pavithra S Hegde, Afshin, "Face Recognition based Attendance Management System". *International Journal of Engineering Research and Technology (IJERT),* Volume 9, Issue-05, May 2020, pp. 1190-1192.

[7] Manikandan J, Lakshmi Prathyusha S, Sai Kumar P, Jaya Chandra Y, Umaditya Hanuman M. "Face Detection and Recognition using Open CV Based on Fisher

Faces Algorithm". *International Journal of Recent Technology and Engineering (IJRTE)*, ISSN: 2277-3878, Volume-8 Issue-5, January 2020, pp. 1204-1208.

[8] Okokpujie, Kennedy O. "Design and implementation of a student attendance system using iris biometric recognition." *International Conference on Computational Science and Computational Intelligence (CSCI), IEEE,* 2017.

[9] Rathod, Kumar Hemant. "Automated attendance system using machine learning approach". *International Conference on Nascent Technologies in Engineering (ICNTE), IEEE,* 2017, pp.1-4.

[10] Siswanto, Adrian Rhesa Septian, Anto Satriyo Nugroho, and Maulahikmah Galinium. "Implementation of face recognition algorithm for biometrics-based time attendance system." *International Conference on ICT For Smart Society (ICISS), IEEE,* 2014.

[11] Lukas Samuel. "Student attendance system in the classroom using face recognition technique.", *International Conference on Information and Communication Technology Convergence (ICTC), IEEE,* 2016.

[12] Tanmoy Paul, Ummul Afia Shammi, Mosabber Uddin Ahmed. "A Study on Face Detection Using Viola-Jones Algorithm for Various Backgrounds, Angels and Distances", *Biomedical Soft Computing and Human Sciences,* Vol. 23, No.1, May 2018, pp. 27-36.

[13] https://towardsdatascience.com/applied-deep-learning-part-4-convolutional-neural-networks-584bc134c1e2.

[14] https://neurohive.io/en/popular-networks/vgg16/.

[15] Sourabh Sarkar, Geeta Sikka, "A comparative study of classifiers used in facial embedding classification", *IEEE,* 2018.

[16] https://www.jameco.com/Jameco/workshop/circuitnotes/raspberry-pi-circuit-note.html.

[17] https://thepihut.com/products/raspberry-pi-camera-module.

[18] https://robu.in/product/hc-sr04-ultrasonic-range-finder/.

[19] https://www.indiamart.com/proddetail/16x2-lcd-display 11673379433.html

[20] https://www.electronicscomp.com/active-buzzer-module-5v-india?gclid=CjwKCAiA-9uNBhBTEiwAN3IlNNYhFFfdTI_atF4I9GYN7yYWTllGhA2jvEmlF2Noi7hE5wioqn0NxRoC0YgQAvD_BwE.

Chapter 15

Role of Fuzzy Systems in the Field of Image Processing and VANET

Aradhana Behura[1,*] and Shreya[2,†]
[1]Silicon Institute of Technology, BBSR, India
[2]GITA Engineering College, Bhubaneswar, India

Abstract

Nowadays, fuzzy logic is very important for smart technology. This fuzzy-based chapter represents fundamental concepts of fuzzy theory, such as what you mean by fuzzy, how we will represent fuzzy logic or concept (mathematical representation), the composition of relations, measurement of fuzziness, and inaccuracy of fuzzy set and fuzzy proposition. This chapter focused on how fuzzy concepts help improve service parameters' quality in vehicular ad hoc networks and image processing. The fuzzy inference system discussed components of fuzzy logic controller and structural optimization based FLC (Fuzzy Logic Controller) – GA (Genetic Algorithm). At the end of the chapter, we will discuss the fuzzy C- means Clustering algorithm, which provides a better result than other clustering algorithms in real-time systems such as disease prediction in leaf and efficient data transmission in VANET.

Keywords: fuzzy controller, C-means, fuzzy Inference, leaf disease prediction, vehicular ad hoc network, optimization algorithm

---

* Corresponding Author's Email: aradhanabehura@gmail.com.
† Corresponding Author's Email: shreyasinha295@gmail.com.

In: Computational Intelligence for Sustainable Development
Editor: Brojo Kishore Mishra
ISBN: 979-8-88697-198-9

1. Introduction

The model of Fuzzy theory was presented in the year 1965 by scientist Lofti A. Zadeh in his investigation broadsheet "Fuzzy Sets," and he is deliberated as the inventor of the Fuzzy concept or logic [1-3]. The fuzzy concept refers to the decision-making human-based approach and contracts with imprecise and vague data. Fuzzy lies between 0 to 1. This is an unrefined simplification of real-life situations and is based on the degrees of membership, usually 1/0 Boolean logic or false/true [4-8]. Consider the question, "Is the water colorless? It is definite Yes/True, or No/False as per the situation or condition. If "yes"/" true" is an assessment of 1 and "no"/" false" is accorded an assessment of 0, these types of statements are outcomes in 0 or 1 type of condition. Such types of fuzzy logic that load a binary means (0/1) type is labeled as a crisp set in the field of fuzzy theory. Suppose there is a question find out all power sets & cardinality of the presented set X = {2,4,6} and find the cardinality of the power set.

Answer:
Nx = 3
P(x) = {∅,{2},{4},{6},{2,4},{4,6},{2,6}, {2,4,6}}

First, we will discuss the difference of Crisp set or Classical set and fuzzy set. A **set** is a group of different or unordered elements. The example of set is described below:

Example:

- A set of Cadbury chocolate from the basket.
- A set of districts in the state odisha.
- A set of temples in India.
- Set of lowercase in alphabet.

We will discuss about fuzzy Sets and can be denoted in the formatted ways:

- *Tabular Form:*
  Following are the illustrations of set in the tabular form –
  - Set of vowels in the alphabet, k = {A,E,I,O,U}
  - A Set of even numbers greater than 10 and less than 15, B = {12,14}

- *Set builder notation:*
  The definition of set is defined as AA = {xx:p(xx)}
  1. The set {a,e,i,o,u} is transcribed as
     AA = {xx:xx is a vowel in alphabet}
  2. The set {1,3,5,7,9} is engraved as
     BB = {xx:1 $\leq$ xx< 10 and (xx%2) $\neq$ 0}

If x is an element of the set SS, and it is represented by xx∈SS and if yy element is not a member of set SS, this is symbolized by yy∉SS.

Example 3 – If SS = {1,1.2,1.7,2},1 ∈SS but 1.5 ∉ SS

- **Description of the Overlapping Set:**

  If twodifferent sets have at least only one mutual element is known as overlapping sets. This types of sets are defined as –
  nn(AA∪BB) = nn(AA) + nn(BB)−nn(AA∩BB)nn(AA∪BB) = nn(AA) + nn(BB)−nn(AA∩BB)
  nn(AA∪BB) − nn(AA−BB) + nn(BB−AA) + nn(AA∩BB)nn(AA∪BB) = nn(AA−BB) + nn(BB−AA) + nn(AA∩BB)
  nn(AA) = nn(AA−BB) + nn(AA∩BB)nn(AA) = nn(AA−BB) + nn(AA∩BB)
  nn(BB) = nn(BB−AA) + nn(AA∩BB)nn(BB) = nn(BB−AA) + nn(AA∩BB)

Example 4 – Let, dA = {11,22,66} and dB = {66,12,44}. Here '66' is a common element among dA and dB set, so these types of sets are known as overlapping sets.

- **Disjoint Set:**

  Two sets AC and BC are known as disjoint sets if they do not have one element in common. So, disjoint sets are defined as –
  nn(AC∩BC) = ϕ
  nn(AC∪BC) = nn(AC) + nn(BC)
- **Fuzzy Logic:**

  Classical set covers components that satiate precise things of membership function while fuzzy set covers features that fulfil imprecise things of the membership function. From Figure 1, we can know the fundamental concept of the fuzzy logic.

Figure 1. Fundamental of fuzzy logic.

- **Mathematical Concept**

  A fuzzy set CA˜CA~ in the universe of information U can be defined as a set of ordered pairs and it can be represented mathematically as –

  CA˜={(y,μCA˜(y))|y∈U}

  Here μCA~(y) = degree of membership of yy in \wide tilde{CA}, assumes values lies from 0 to 1, i.e., μCA~(y)∈ [0,1].

- **Representation of Fuzzy Set**

  Now we will consider two types of case which is described below in case 1 and case 2.

  Case 1:

  When universe of information UU is discrete and finite –

  CA˜={μCA˜(y1)/y1 + μCA˜(y2)/y2 + μCA˜(y3)/y3 + ...}

  $= \sum_{k=1}^{n} \mu\tilde{CA}(Yi)\ /Yi$

  Case 2:

  When universe of information UU is continuous and infinite –

  CA˜={∫μCA˜(y)/y}

  In this type of demonstration, the summation symbol characterizes the gathering of each and every element.

1.1. Composition of Relations

Suppose R is a fuzzy relation defined on X ×Y and S is a fuzzy relation defined on Y × Z then R. S is a fuzzy relation on X × Z. The fuzzy max-min composition id defined as:

$U_{R.S}$ (X,Z) = max (min(u_R(X, Y), U_S (Y, Z))

Example:

For X = {X1, X2, X3}, Y = {Y1, Y2}, Z = {Z1, Z2, Z3}

R:

	Y1	Y2
X1	0.6	0.2
X2	0.3	0.10
X3	0.9	0.7

S:

	Z1	Z2	Z3
Y1	0.7	0.5	0.8
Y2	0.6	0.9	0.10

R. S =

	Z1	Z2	Z3
X1	0.6	0.5	0.6
X2	0.6	0.9	0.10
X3	0.7	0.7	0.8

$U_{R.S}$ (X1, Z1) = max (min(0.6,0.7), min(0.2,0.6)) = max (0.6, 0.2) = 0.6
$U_{R.S}$ (X1, Z2) = max (min(0.6,0.5), min(0.2,0.9)) = max (0.5, 0.2) = 0.5
$U_{R.S}$ (X1, Z3) = max (0.6, 0.2) = 0.6
$U_{R.S}$ (X2, Z1) = max (0.3, 0.6) = 0.6
$U_{R.S}$ (X2, Z2) = max (0.2, 0.9) = 0.9
$U_{R.S}$ (X2, Z3) = max (0.3, 0.10) = 0.10
$U_{R.S}$ (X3, Z1) = max (0.7, 0.6) = 0.7
$U_{R.S}$ (X3, Z2) = max (0.5, 0.7) = 0.7
$U_{R.S}$ (X3, Z3) = max (0.8, 0.7) = 0.8

Question: It is required to represent a variable x by a binary bit string in the range [-1, 2] with precision up to 6 places after decimal points. Determine the minimum number of bit required.

Solution: x [-1, 2]

p=6

q = string length?

x_i^L=-1, x_i^U=2

$$(x_i^U - x_i^L)10^P \leq (2^q - 1)$$
$$= 2 - (-1)10^6 \leq (2^q - 1)$$
$$= 3 \times 10^6 \leq 2^q - 1$$

= q=22 bit (max. string length)

Question: If $\mu_{young}(x) = \frac{1}{1+\left(\frac{x}{20}\right)^4}$, determine the membership function for the set "young but not so young".

Solution: Not so young = Not young

$\mu_{\overline{young}}(x) = 1 - \mu_{young}(x)$

$$= 1 - \frac{1}{1+\left(\frac{x}{20}\right)^4}$$

$$= \frac{1+\left(\frac{x}{20}\right)^4 - 1}{1+\left(\frac{x}{20}\right)^4}$$

$$= \frac{\left(\frac{x}{20}\right)^4}{1+\left(\frac{x}{20}\right)^4}$$

1.2. Measure of Fuzziness and Inaccuracy of Fuzzy Set

Let X = {X1,X2…….XN} be a discrete universe of discourse. The entropy of a fuzzy set A(X) could be determined as follows:

$H(A) = -\frac{1}{n}\sum_{i=1}^{n} uu_A$ (xi) log { uu_A (xi)} + { 1 -uu_A(xi) log {1-uu_A (xi)}]

Example 1:

A(x) = {(x1, 0.1), (x2, 0.3), (x3,0.4), (x4,0.5)} then calculate the entropy value ?

$H(A) = -\frac{1}{4}$ [{ 0.1 × log (0.1) + 0.9 log (0.9) + (0.3 log 0.3 + 0.7 log 0.7)+ 0.4 log 0.4 + 0.6 log 0.6 + 0.5 log 0.5 + 0.5 log 0.5 }] = 0.24

Inacurracyof two fuzzy sets:

$I(A;B) = -\frac{1}{n}\sum_{i=1}^{n} uu_A$ (xi) log { uu_B (xi)) + { 1 - uu_A(xi) log {1- uu_B (xi)}]

Inacuraccy of the set B(x) can be measured w.r.t set A(x).

Example 2:

A(x) = { (x1, 0.1), (x2, 0.2), (x3, 0.3), (x4, 0.4)}

B(x) = { (x1, 0.5), (x2, 0.7), (x3, 0.8), (x4, 0.9)}

Inaccuracy of B(x) w.r.t A(x),

$I(A;) = -\frac{1}{4}$ [(0.1 log 0.5 + 0.9 log 0.5) + (0.2 log 0.7 + 0.8 log 0.3) + (0.3 log 0.8 + 0.7 log 0.2) + 0.4 log 0.9 + 0.6 log 0.1)] = 0.47

1.3. Fuzzy Proposition

Fuzzy proposition means the truth values varies between 0 to 1(membership value).

Example 1:

Munmun is sincere girl in the class.

T ($\tilde{P}$) = 0.75: It means partly true

T ($\tilde{P}$) = 0.25: It means partly false

T ($\tilde{P}$) = 1: It means absolute true

T ($\tilde{P}$) = 0: It means absolutely false

Example 2:

$\tilde{P}$: Aradhana is efficient, T($\tilde{P}$) = 0.9

$\bar{Q}$: Anurag is efficient, T($\bar{Q}$) = 0.66

i. $\overline{\overline{PP}}$: Aradhana is not efficient.
   T ($\overline{\overline{PP}}$) = 1- T($\tilde{P}$) = 1- 0.9 = 0.1
ii. (II) $\tilde{P} \cap \bar{Q}$: Aradhana is efficient & so is Anurag.
   T ($\tilde{P} \cap \bar{Q}$) = min (T($\tilde{P}$), T ($\bar{Q}$)) = min (0.9, 0.66) = 0.66
iii. (III) T ($\tilde{P} \cup \bar{Q}$): Either Aradhana or Anurag is efficient.
   T ($\tilde{P} \cup \bar{Q}$) = max (T($\tilde{P}$), T ($\bar{Q}$)) = max (0.9, 0.66) = 0.9
iv. (IV) $\tilde{P} \Rightarrow \bar{Q}$: If Aradhana is efficient then so is Anurag.
   T ($\tilde{P} \Rightarrow \bar{Q}$) = max (1- T($\tilde{P}$), T ($\bar{Q}$)) = max (0.1, 0.66) = 0.66

1.4. Fuzzy Inference Model

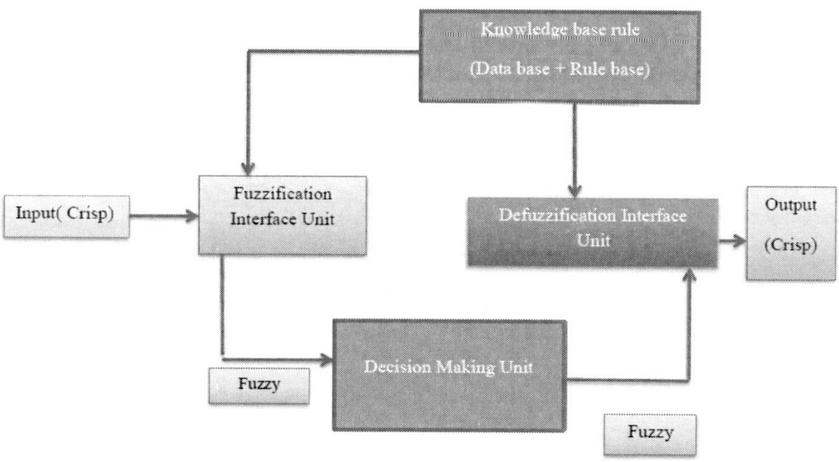

Figure 2. Pictorial representation of Fuzzy Inference System.

1.4.1. Components of Fuzzy Logic Controller (FLC)

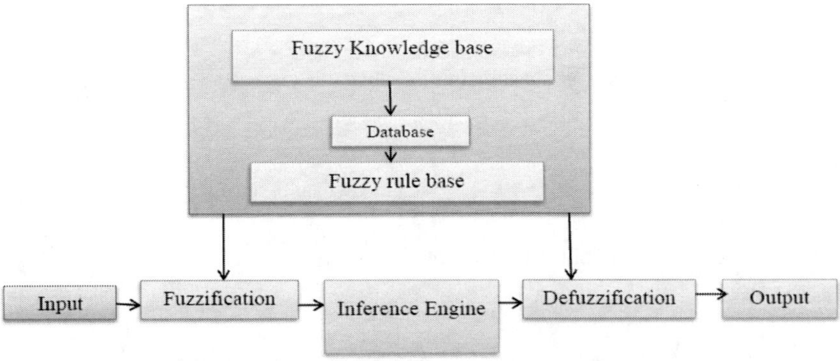

Figure 3. Component of Fuzzy logic controller.

Figure 2 and Figure 3 describe the fundamentals of a fuzzy-based controller, just as the technique for producing a fuzzy membership matrix for fuzzy guidelines or rule base [9, 10]. In this fuzzy control, the assignment of producing control choices are PC robotized, along these lines mitigating the issues of control speed and working with synchronous observing and control of numerous factors. The Fuzzification technique is used to evaluate fuzzy input value, fuzzy inference rule base systems are used to measure the activation strength, and the defuzzification technique is used to evaluate the actual output [11, 12]. Since low-level control activities are normally crips, the control induction should be defuzzified for invitation purposes. A few strategies are accessible for the defuzzification of a fuzzy control derivation. The centroid strategy utilizes the membership capacity of a control deduction. Consequently, the defuzzified esteem relies upon both the size and state of the membership function. The mean of maxima strategy relies just upon the pinnacle upsides of the control inference membership function.

1.5. Structural Optimization Based on Fuzzy Logic Controller (FLC) – GA (Genetic Algorithm)

This design methodology (means use of FLC) of Figure 4 used to diminish the possibility of premature problem clarification affected by the improper fuzzy rule. Genetic Algorithm provides good objective function as well as fitness function. Figure 6 describes about how FLC & GA helps for structural optimization.

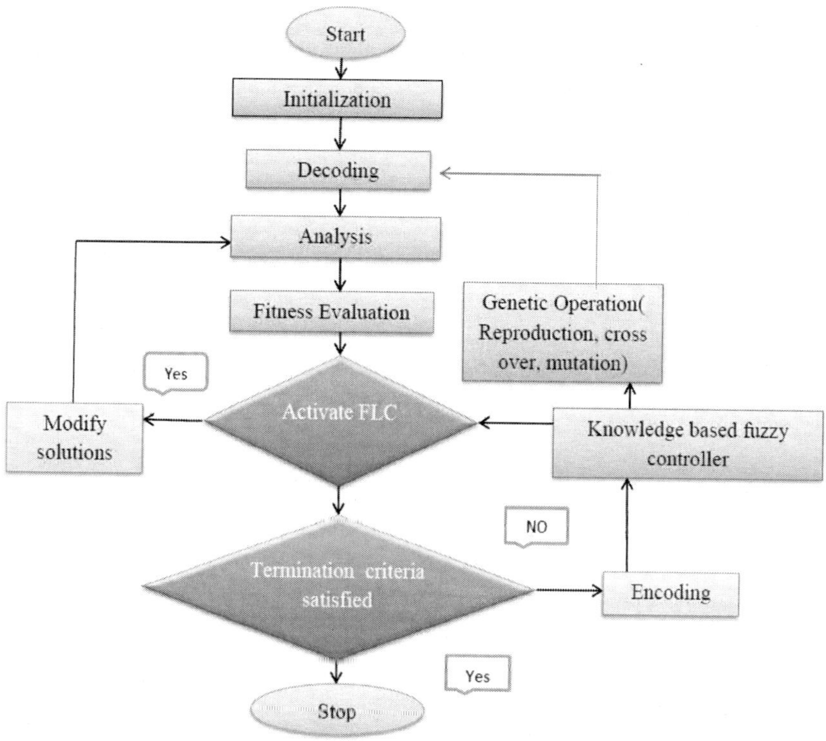

Figure 4. Structural Optimization based on FLC – GA (Genetic Algorithm).

2. Role of Fuzzy in the Field of Clustering

2.1. A Numerical Example of Fuzzy C-Means Clustering Algorithm

Clustering technique used to group similar types of data points. Table 1 contains coordinates of ten points lying on a free-form surface shown in the figure carry out clustering using the fizzy c-means algorithm [13, 14, 15]. Assume g=1.25 and termination criterion E = 0.01.

Table. 1. 10 points lying on a free-form surface

Points	X-coordinate	Y-coordinate	Z-coordinate
1	0.2	0.4	0.6
2	0.4	0.3	0.8
3	0.8	0.2	0.5
4	0.9	0.5	0.4
5	0.6	0.6	0.6
6	0.3	0.4	0.5
7	0.7	0.6	0.5
8	0.2	0.5	0.3
9	0.3	0.6	0.8
10	0.8	0.3	0.1

Solution:

Number of data points to be clusteredμ = 10

Dimension of the data L = 3

Level of cluster fuzziness y = 1.25

Let us assume that the number of clusters to be made C = 2. Let us also suppose a membership matrix [μ] of size 10 × 2 is generated at random, as given below

$$\mu = \begin{bmatrix} 0.680551 & 0.319449 \\ 0.495150 & 0.504850 \\ 0.821897 & 0.178103 \\ 0.303795 & 0.696203 \\ 0.333966 & 0.666034 \\ 0.431538 & 0.568462 \\ 0.415384 & 0.584616 \\ 0.509643 & 0.490357 \\ 0.469850 & 0.530150 \\ 0.189164 & 0.810836 \end{bmatrix}$$

Now, the first dimension of first cluster center $CC_{11}$is determined like the following:

$$CC_{11} = \frac{\sum_{i=1}^{N} \mu_{il}^{g} x_{il}}{\sum_{i=1}^{N} \mu_{i1}^{g}} = \frac{A}{B}$$

A= $0.680551^{1.25} \times 0.2 + 0.495150^{1.25} \times 0.4 + 0.821897^{1.25} \times 0.8 + 0.303795^{1.25} \times 0.9 + \quad 0.333966^{1.25} \times 0.6 + 0.431538^{1.25} \times 0.3$

$+0.415384^{1.25} \times 0.7 + 0.509643^{1.25} \times 0.2 + \quad 0.469850^{1.25} \times 0.3 + 0.189164^{1.25} \times 0.8 = 1.912120$

$B=0.680551^{1.25} + 0.495150^{1.25} + 0.821897^{1.25} + 0.303795^{1.25} + 0.333966^{1.25} + \quad 0.431538^{1.25} + 0.415384^{1.25} + 0.509643^{1.25} + 0.469850^{1.25} + 0.189164^{1.25} = 3.923066$

$CC_{11} = 0.487404,$
$CC_{12} = 0.412837,$
$CC_{13} = 0.543316,$
$CC_{21} = 0.554860,$
$CC_{22} = 0.459073\&$

$$CC_{23} = 0.477262$$

Thus, the coordinates of two clusters are found to be as follows:

$$C_1: (0.487404, \ 0.412837, \ 0.543316)$$
$$C_2: (0.554860, \ 0.459073, \ 0.477262)$$

We determine the modified values of membership grades like the following:

$$\mu_{11} = \frac{1}{1 + \left(\frac{d_{11}}{d_{12}}\right)^{\frac{2}{g-1}}}$$

Where

$$d_{11} = \sqrt{\begin{matrix}(0.487404 - 0.2)^2 \\ + (0.412837 - 0.4)^2 + (0.543316 - 0.6)^2\end{matrix}} = 0.293221526$$

$$d_{12} = \sqrt{(0.55486 - 0.2)^2 + (0.459073 - 0.4)^2 + (0.477262 - 0.6)^2} = 0.380105056$$

$$\mu_{11} = \frac{1}{1 + \left(\frac{d_{11}}{d_{12}}\right)^{g}} = 0.888564\,, corresponding\ to\ g = 1.25$$

The values of other elements of $[\mu]$ can be calculated in the same way and these are found to be as follows:

$$[\mu]=\begin{bmatrix} 0.888564 & 0.111436 \\ 0.909207 & 0.090793 \\ 0.376533 & 0.623467 \\ 0.142926 & 0.857074 \\ 0.217307 & 0.782693 \\ 0.922229 & 0.077771 \\ 0.060962 & 0.939038 \\ 0.562085 & 0.437915 \\ 0.788303 & 0.211697 \\ 0.232446 & 0.767554 \end{bmatrix}$$

It completed one iteration of the algorithm.

2.2. Application of Fuzzy C –Means Clustering for Prediction of Disease in Leaf

As we know, agriculture is the backbone of our economic system, so we have to take care of it. A better result leads us to a more financially stable country. So we need to get rid of the loss we are facing in the agricultural field. The farmers should be well provided with the farming techniques so that it will be easier for them in farming practice. Government should be more alert in giving proper training to the farmer and providing them with proper facilities so that they feel helpful. In this way, we can get over the agricultural loss. For disease-free agriculture, we should properly provide pesticides and fungicides. Sometimes it is difficult to recognize the disease in the pre-stage, so proper tools are provided to recognize it in the very first stage and overcome the situation using fruitful techniques.

The main purpose of this study is to identify the types of disease in pomegranate leaves using the SVM classification technique. Here we calculate the diseased part using C-means clustering. The image processing technique is very helpful for studying an image. It is very difficult to recognize the disease with the naked eye, and it varies from person to person, so we use this technique for better results. It is also a time-saving technique. Calculating the affected and non-affected parts, we finally calculate our overall disease area in percentage. Figure 5 describes the steps of image segmentation using the clustering algorithm [9], and the description of the figure is explained in algorithm 1. Figures 6-7 show the results of segmentation ROI (region of interest) of various image datasets [16].

Algorithm 1: Pomegranate leaf disease prediction

Input: Query Pomegranate leaf image

Output: Classification of disease, calculate the diseased area in percentage, evaluate accuracy

Start

Step 1: input the query pomegranate leaf image.

Step 2: image processing

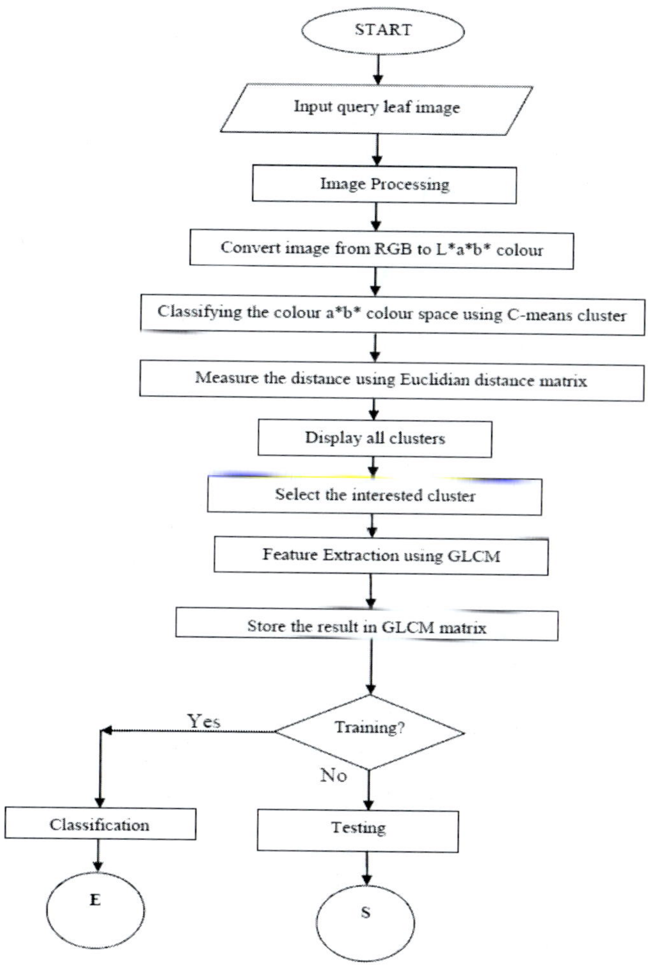

Figure 5. (Continued)

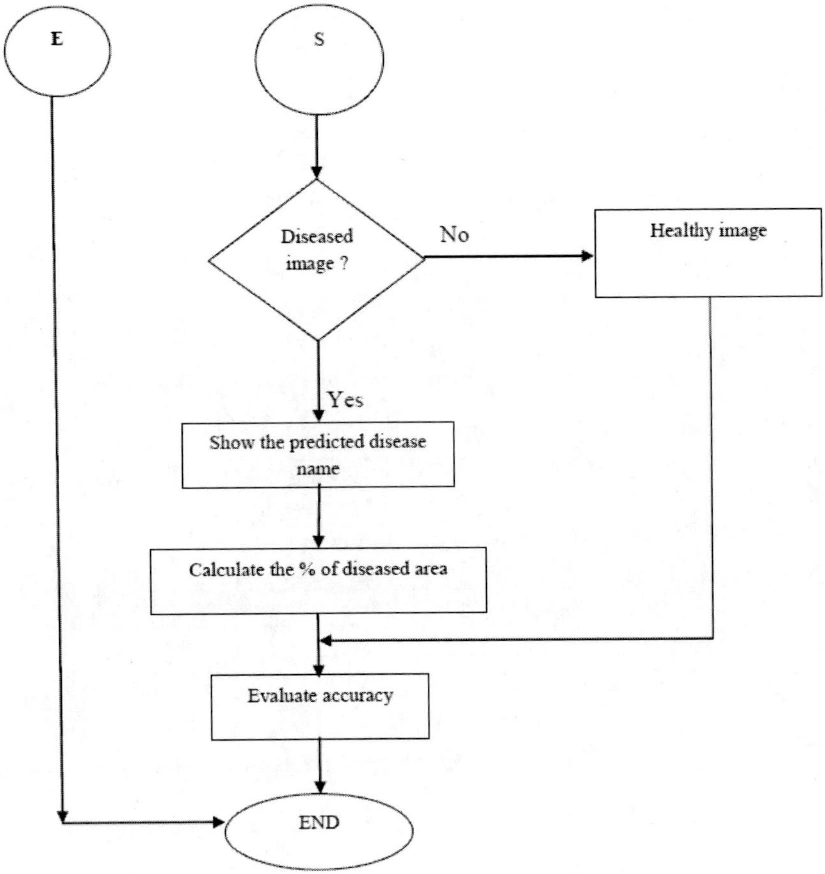

Figure 5. Image Segmentation Technique using clustering algorithm.

Step 2.1: Image resizing
Step 2.2: Image Enhancement
Step 3: Segmentation of Image.
Step 3.1: Convert image from RGB to l*a*b* colour space.
Step 3.2: Classifying the colour a*b* colour space using K-mean cluster.
Step 3.3: Measure Euclidian Distance.
Step 3.4: Display all cluster.
Step 4: Select the disease part cluster.
Step 5: Feature extraction using GLCM.
Step 5.1: Calculate all features.
Step 5.2: Store the feature in GLCM matrix.

Step 6: If training?
- Step 6.1: Yes, classification of image and stop.
- Step 6.2: No, Testing the image.

Step 7: If, disease image?
- Step 7.1: Yes, Show the predicted diseases name and go to step 8.
- Step 7.2: No, Healthy image and go to step 9.

Step 8: Calculate the percentage of diseased Area.

Step 9: Evaluating accuracy.

End

End

2.2.1. Result of Image Segmentation Using C- Means Clustering

Classification is done based on the training data sets. The new data sets or testing datasets are being checked according to the membership value of the training data sets. This research topic's classification is based on which type of disease class. Figures 6-7 show the simulation results of how much the disease affects the area.

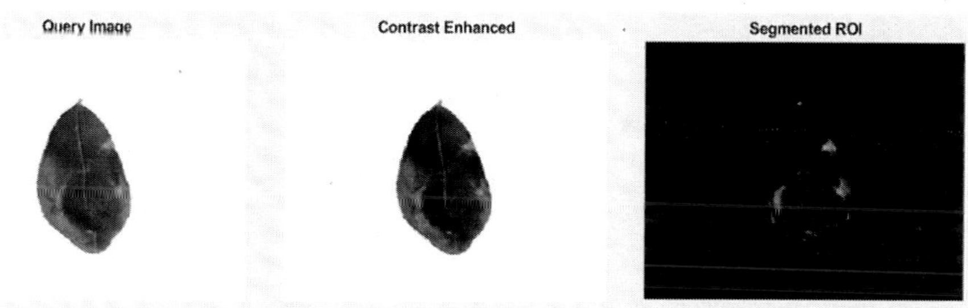

Figure 6. Clustering output 1.

Figure 7. Clustering output 2.

3. Role of Fuzzy Concept in the Field of Vehicular Ad Hoc Network

The Vehicular ad hoc network (VANET) is an emerging field in smart transportation. Figure 8 shows vehicle to vehicle (V2V) & vehicle to infrastructure communication (V2I). The clustering method is used to improve vehicular ad hoc network performance or quality of service (QoS). Performance is maximized by using clustering techniques. Moreover, can be a decrease in network energy and some collision. The multi-hop transmission increases network efficiency [17].

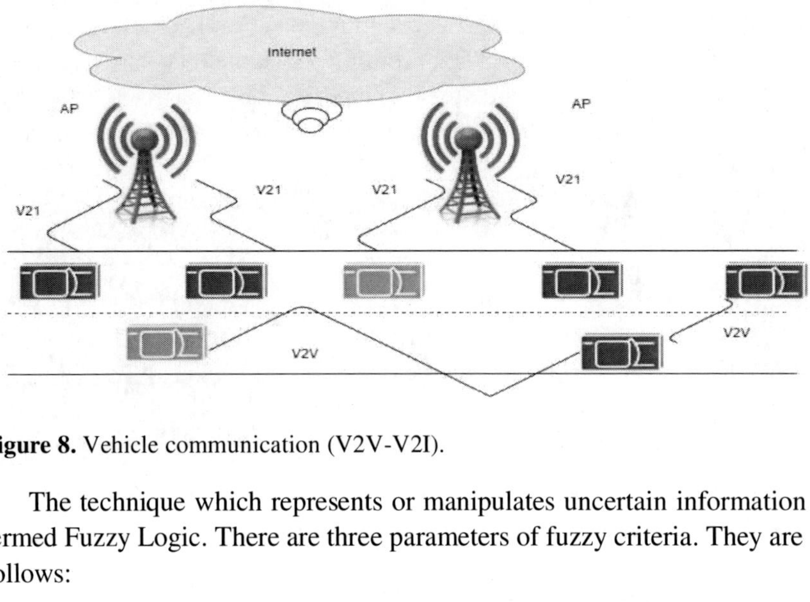

Figure 8. Vehicle communication (V2V-V2I).

The technique which represents or manipulates uncertain information is termed Fuzzy Logic. There are three parameters of fuzzy criteria. They are as follows:

(i) Node Distance
(ii) Remaining or Leftover Energy
(iii) Total number of neighboring nodes of each sensor node

Environment such as Vehicular ad hoc network (VANET) is susceptible to data transmission. Nodes fixed with sensors, actuators, software, and other connections that allow data to send and receive data through the internet and communication channel are known as IoT-based VANET. Transmission media such as - Bluetooth, wi-fi, and Zigbee can share data in two ways. The two ways are as follows:

(a) in ad hoc way or infrastructure less
(b) in infrastructures

Many sensor nodes occupy a large amount of data, leading to energy depletion (because nodes consist of batteries). The most common issue that is faced by all of us in today's generation is 'Traffic Jamming," which, as a result, leads to various serious issues. Increasing population leads to an increase in transportation. New computing technologies should be implemented in well-developed cities or smart cities. In fact, not only in this issue, new computing technologies should be implemented in other city services like - public safety, healthcare, education, and administration. The problem of road safety can be controlled by vehicular ad hoc networks (VANETs). It will also increase vehicle mobility. The different and multiple sets of nodes present among the neighbor is called sending set. Thus, an appropriate criterion is selected to select the rebroadcasting node to enhance the performance of inter-vehicle network characteristics. Moreover, the car following model is limited to highways or on roads. So, to avoid link disconnection, the car drivers should know the car or vehicle's position and direction. The beacon messages must be used in a periodic way between the neighbors. These updates are expensive in a different way.

The different ways are as follows:

(a) Delaying packet advancement
(b) One update causes an increase in the use of bandwidth
(c) In the media access layer, there will be an increase in the collision rule

Packet loss is a result of packet collision, which affects routing efficiency. Collision takes place, which leads to increased congestion in the network. Crowded conditions within a network result in an increase in pocket loss and pocket delay, which will decrease system efficiency. There is no use of beacon messages in an opportunistic routing protocol to solve this problem. They were found as a fault-tolerant protocol in vehicular networks. Wireless channel broadcast is used in vehicular networks. It upgrades the network performance in data forwarding. Moreover, for rebroadcasting nodes, packet advancement towards the destination, a number of duplicated packets, and density are used for prioritization criteria. This is the way where the packet is moved with the least consumed resources and gives proper performance considering the opportunistic routing scheme. In advance, it is not necessary to send beacon packets to keep the neighbor's status and network density illuminated. Hence, calculating this parameter is important, which leads to the reduction of congestion and packet loss. There are three input parameters that are used by the fuzzy protocol with the objective of proper decision-making for rebroadcasting. Three variables are chosen as the input to the fuzzy system for proper decision-making. These are as follows -:

Give the importance of the packet's advancement
Different and local density of the vehicles
Control of duplicate packets in the network

Routing techniques tried to pick the rebroadcasting nodes which lead towards the destination. Further, the local density of a parameter is dynamic in the network. Proper decisions while considering the density should make by routing. Duplication of packets results in congestion in the network with these three parameters; fuzzy decision making is used by routing and packet rebroadcasting for picking the right routes for retransmission; three concepts were made, which are merged in the fuzzy logic.

Following are the three concepts -:

1. Improvement in packets to reduce the delay between the links
2. Local control to manage fast communication changes, which results in minimizing transmission and maximizing throughput
3. Least number of received duplicate packets for choosing cluster head.

From the set of relays, the nearest best node is selected, which leads to the generation of routes with maximum throughput and lower delay of nodes from starting to end with the data delivery of productive rate. This part will tell us about fog computing, the Internet of Things (IoT), and edge computing that allow technologies for full deployment and management of VANET applications & services. The schedules points of many procedures support multiple rounds of data to the roadside unit (RSU) and continuously involve clustering in every round of single top data transmission in the roadside unit, but the clustering maximizes the number of control messages in every round. As a result, there will be the possibility of collision, and network energy will decrease. IoT is one of the most popular networking concepts and creates a smart environment. It expects the collection of information from the smart environment and makes daily life easy to make tough decisions. For this, an inter linkage between people, objects, and processes has to be done. There should be a measurable infrastructure to enable enough and full utility of the available resources. The two most important applications of IoT based on Vehicular ad hoc network (VANET) are smart cities and smart transportation, including intelligent traffic management. This data from the smart traffic light nodes and traffic information center infrastructure can be reached by any vehicle at any time or at any point. These application requires mobility support and are delay–sensitive. To satisfy all these demands, fog computing was introduced. Fog computing can be described as a highly computing-generated platform that provides storage, computing, and networking services between edge devices and old cloud computing data centers that offer low delaying and geographical distribution, mobility support, location awareness, and real-time interactions. All mentioned above makes fog computing an appropriate platform that fulfills all the VANETs term requirements. As the number of vehicles increases, the amount of data generated by these vehicles will also increase. Thus, cloud computing plays an important role in big data management and analytics. This section describes the frequently used protocols for the machine to machine (M2M) communication. The Message Queue Telemetry Transport (MQTT) is a client-server which accepts the message transport protocol. It is easy to implement, lightweight, open & simple. It runs over Internet Protocol (IP), Transmission control protocol (TCP), or other network protocols that give bi-directional and lossless connections. The feature of MQTT is the usage of an accepted method pattern that gives one too many message distributions. It has reduced transport overhead and protocol exchange to minimize network traffic and has a unique mechanism to inform interested parties when an unusual disconnection occurs. The Constraint Application Protocol (CoAP) is a special web transfer protocol using constrained nodes and networks. Generally, the nodes have an 8-bit microcontroller with few amounts of reading Only Memory (ROM) and Random Access Memory (RAM). Constraint networks generally have a high packet error rate; the typical throughput is ten kbps. This protocol is planned for M2M applications like building automation and smart city. The key concepts of CoAP are Universal Resource Locator (URLs) and Internet Media types. CoAP has a friendly interface with Hyper Text Markup Language (HTTP) for integration with the web. Many people still drive in the era of autonomous cars. The expensive driverless cars, trust issues, etc. one the reason why people will continue to drive their cars on their own. Many scientists and automotive engineers work on Driver Assistant Systems (DASs) and Driver Monitoring Systems (DMSs) as key features required to attain total points on a safety. Dass and DMSs are modern and intelligent systems. They are kept inside the vehicle and help drivers in a number of ways. These systems are based on comprehensive sensing network and artificial intelligence (AI) techniques and have made it possible to start the generation of connected cars. Algorithm 2 describes the energy-efficient scheme in the VANET. Table 2 describes the meaning of symbols used in the VANET.

Algorithm 2: Fuzzy system inspired energy efficient scheme for VANET

All vehicles (nodes) are homogenous in nature and have balanced initial energy.

Step 1: Randomly distribute the vehicles (nodes) in the network
Step 2: All the other devices (nodes) and road side units are static
Step 3: Euclidean method is used for calculating the distance
Step 4: Road side unit will receive data by cluster lead in a multi – hop way and in certain situation in single hope way
Step 5: The nodes which are placed inside R distance to a certain distance to a certain device are defined as its neighbors

Following is the energy consumption modes for sending L bit data package from transmitter to receives in d (distance) –

$$\{I^*E_{elec}+1^*\epsilon_{mp}{}^*d^4\ ;\ \text{if } d<d_0\} \quad (1)$$

where,

d = distance
E_{elec} = Amount of energy consumed for sending binary numbers that is 0 &1
ϵ_{fs} = Multiple Route Operation

Required Energy (E_{RX}):-

E_{Rx} is defined as energy required for the receiver to receive data.
E_{RX} can be get from the following relation –

$$d_0 = \sqrt{\frac{\epsilon fs}{\epsilon mp}} \quad (2)$$

$$E_{RX} = 1^*\ E_{elec} \quad (3)$$

Table 2. Symbols and their Descriptions

Symbols	Meaning
CH	Cluster head node which is responsible for collecting received environmental data and sending them to the road side unit.
SCH	The reserve cluster head is replaced by the cluster head if lack of energy occurs
CHL	The head of cluster head is responsible to send the gathered data to the road side unit
BS	Road Side Unit
R	Number of Rounds
Mod(r,3)	It shows the cluster method which is used by the division, remainder function
Compare	Used for the comparison of two values
Criteria	Criteria of suitability for each cluster head (distance energy)
Energy	Node residual energy in every moment
Terminologies CM	It is the cluster of member node which is responsible for receiving the environmental data
Competition Radius	It is used to get the cluster head leader
FND	First Dead Node
HND	Half Dead Node
LND	Last Dead Node
E_{elec}	Amount of energy consumed
ϵ_{fs}	Multiple Route Operation
E_{RX}	Energy Required
Elex	Energy Derived
ERC	One bit data received

Fuzzy Logic System:

Earlier we used to measure the distance by coordinate points or other different methods. In advance, longitude and latitude points measure the distance of neighbors. It is assumed that each vehicle is connected with Global Positioning System (GPS). GPS is used to calculate the latitude and longitude points. It figures out the neighbor vehicle's distance and according to these points the neighbor vehicle is calculated.

Let,

L_{A1} be the latitude point of vehicle V_1

L_{o1} be the longitude point of vehicle V_1

L_{A1} be the latitude point of vehicle V_2

L_{o1} be the longitude point of vehicle V_2

Formula for measuring distance between V_1 &$V_{2:}$- D = R*C

Where, R = radius of the circle

C = range of center of radius equal to the distance between two points

Speed:

Transmission of data can be distributed by speed of the vehicle in VANET. It can be caused by dynamic topology of changes in VANET. It is assumed that each vehicle is connected with the speedometer of the vehicle as it measures the speed of the vehicle in VANET.

Link Residual Time (LRT):

Link Residual Time (LRT) is defined as the duration of vehicle when they are in the range of communication. The LRT can be known by using physical layer and the distance parameters such as transmission power, receive power, Signal to Noise (SNR) ratio. Efficiency optimization of data is transmitted by using LRT, where SNR is the received signal and noise ratio.

SNR = The threshold value of signal to noise ratio
D= distance
^ (lambda) = path loss

$$LRT = -\sqrt{\frac{10\wedge(SNR(0)-SNR(TH)/5\wedge-d}{Relativespeed}} - (t_1 - t_2) \quad (4)$$

There are main three parts of the Fuzzy Logic Unit (FLS). These are as follows:-

1. Fuzzy Logic Engine (FLE)
2. Fuzzifier
3. Defuzzifier
4. There are three input parameters received by the Fuzzification System. They are as follows:-
   (1) Speed
   (2) Distance
   (3) Link Residual Time (LRT)

Member Function is answerable to convert into input data to fuzzy logic data. The distance value of member function is the speed value of member function is the LRT value in member function. Member function can be described inside the triangular function with the ranging value from 0 to 1.

- *Algorithm and Model*

  Fuzzy Control (FC) is the common and popular application. The industrial appliances since starting are the household appliances, process control and the automotive systems from various other fields.

- *Fuzzy Control in the Fuzzy Control Systems:*

  Fuzzy rules mechanism helps to decide certain operations for various modules of environment which is been represented by Fuzzy System (FS). The method, Fuzzy Control allows any type of model to control law which depends on the following stated law, "If...........then..............".
- *Linguistic Variables:*

  Linguistic Variables is the central idea of fuzzy logic. It plays the most important role in the application of fuzzy logic. We can represent numerous numerical values by a single linguistic variable. It is also called the form of a data compression. Linguistic variable data compression form is highly suggested.
- *Fuzzy Rules:*

  Let the input variable be I, j and k for the out of the process, Algorithm rule "if.........then........." is used here; If I is less than j then k is average and If I is greater and j is average then k is greater. These rules are known as Fuzzy Control Rules. The 'If' part is known as Antecedent and the 'Then' part is known as Consequent.
- *Control Knowledge Base (CKB):*

  The two main parameters is considered while designing Control Knowledge Base (CKB). They are as follows:-

  1. Linguistic Variable Set
  2. Mechanism of Control Knowledge Base

     Linguistic Variable Set tells the value for the entire process's main parameters. In this, both input and output parameters must be expressed in the form of Linguistic Variable. For the smoothness of process control, selection of granularity of input and output variables plays an important role. Mechanism of Control Knowledge Base is the key to develop for using the input and output linguistic parameter's description. To achieve this four different methods are there. They are as follows:

     a. Self-Organization
     b. Modeling of Operator's Control Operation
     c. Expert's Experience and Knowledge
     d. Modelling a Process
- *Defuzzification Method*

  This operation cause a non-fuzzy control action which shows a member function for a inferred Fuzzy control action. There are various different defuzzification approaches that have been found. Out of the four different approaches there are some selected approaches that is apply here-

1. Center of Area (COA)
2. Tsukamoto's Defuzzification Method (TDM)
3. Mean of Maximum Method (MOM)
4. De-Fuzzification once

Conclusion

This part presents the fundamental hypothesis of the fuzzy concept, which utilizes the idea of fuzzy sets. The compositional guideline is the thing that is applied for settling on derivations in dynamics related to the fuzzy concept. The idea of the Cartesian item can be straightforwardly reached out to multiple fuzzy sets. Scientist Zadeh presented the extension guideline to give a strategy for stretching out standard numerical ideas to their corresponding fuzzy item. The extension guideline is utilized to map one fuzzy set onto another set by taking the help of a crisp function.

The new pattern in the advancement of neuro-fuzzy frameworks has significantly accentuated the significance of collaboration between the essentials of fuzzy sets and neural organizations. The subsequent structures of the neuro-fuzzy frameworks exploited a variety of learning instruments, essentially beginning inside the hypothesis of neurocomputing and the utilization of fuzzy models (dominatingly rule-based frameworks) being grounded in the domain of fuzzy sets. In a perfect world, one can expect that neuro-fuzzy frameworks ought to completely take advantage of the linkages between these two advances while emphatically saving their clear characters (versatility or learning capacities to be shared by the straightforwardness and full interpretability of the subsequent neuro fuzzy builds). Curiously, this collaboration actually turns into an objective yet to be fulfilled. This examination is an endeavor to address the central interpretability challenge of neuro-fuzzy frameworks. Our hidden guess is that the straightforwardness of any neuro-fuzzy framework interfaces straightforwardly with the rationale texture of the framework, so the rationale essentials of the basic design happen to early stage pertinence. Having this as a primary concern, the advancement of neuro-fuzzy models depends on an assortment of rationale-driven preparing units named here fuzzy (rationale) neurons. These are reasonably straightforward rationale arranged components that accompany obvious semantics and versatility. Attributable to their variety, such neurons structure fundamental structure squares of the organizations. The examination returns to the current classes of rationale neurons, furnishes with their scientific classification, comprehends their useful provisions, and reveals insight into their conduct when being treated as computational segments of any fuzzy neuro design. The two principle classes of aggregative and reference neurons are profoundly established in the key tasks experienced in the innovation of fuzzy sets (counting rationale activities, semantic modifiers, and rationale reference tasks).

References

[1] Zadeh, L. A. (1996). Fuzzy sets. In *Fuzzy sets, fuzzy logic, and fuzzy systems: selected papers by Lotfi A Zadeh* (pp. 394-432).

[2] Bai, Y., Wang D. (2006) Fundamentals of Fuzzy Logic Control — Fuzzy Sets, Fuzzy Rules and Defuzzifications. In: Bai Y., Zhuang H., Wang D. (eds) Advanced Fuzzy Logic Technologies in Industrial Applications. *Advances in Industrial Control.* Springer, London. https://doi.org/10.1007/978-1-84628-469-4_2.

[3] Cox, E. (1992). Fuzzy fundamentals. *IEEE spectrum,* 29(10), 58-61.

[4] Aliev, R. A. (2013). *Fundamentals of the fuzzy logic-based generalized theory of decisions.* Springer.

[5] Pedrycz, W. (2004). Heterogeneous fuzzy logic networks: fundamentals and development studies. *IEEE Transactions on Neural Networks,* 15(6), 1466-1481.

[6] Kose, U. (2012). Fundamentals of fuzzy logic with an easy-to-use, interactive fuzzy control application. *International Journal of Modern Engineering Research,* 2, 1198-1203.

[7] De Silva, C. W. (2018). *Intelligent control: fuzzy logic applications.* CRC press.

[8] Gradojevic, N., &Gençay, R. (2013). Fuzzy logic, trading uncertainty and technical trading. *Journal of Banking & Finance,* 37(2), 578-586.

[9] Behura, A. (2021). The cluster analysis and feature selection: Perspective of machine learning and image processing. *Data Analytics in Bioinformatics: A Machine Learning Perspective,* 249-280.

[10] Behura, A., &Behura, A. (2020, July). Road Accident Prediction and Feature Analysis by Using Deep Learning. In *2020 11th International Conference on Computing, Communication and Networking Technologies (ICCCNT)* (pp. 1-7). IEEE.

[11] Behura, A. (2021). Congruence of Deep Learning in Medical Image Processing: Future Prospects and Challenges. *Technical Advancements of Machine Learning in Healthcare,* 197-221.

[12] Behura, A., Sahu, S., &Kabat, M. R. (2021). Advancement of Machine Learning and Cloud Computing in the Field of Smart Health Care. *Machine Learning Approach for Cloud Data Analytics in IoT,* 273.

[13] Behura, A., & Priyadarshini, S. B. B. (2021). Assessment of Load in Cloud Computing Environment Using C-means Clustering Algorithm. In *Intelligent and Cloud Computing* (pp. 207-215). Springer, Singapore.

[14] Behura, A. (2021). A Deep Learning Application for Prediction of COVID-19. In *Impact of AI and Data Science in Response to Coronavirus Pandemic* (pp. 127-148). Springer, Singapore.

[15] Behura, A., Sahu, S., &Kabat, M. R. (2021). Smart Technology Application for COVID-19 Detection, Control, Prediction and Analysis. *In Impact of AI and Data Science in Response to Coronavirus Pandemic* (pp. 209-234). Springer, Singa-pore.

[16] Behura, A., Srinivas, M., &Kabat, M. R. (2022). Giraffe kicking optimization algorithm provides efficient routing mechanism in the field of vehicular ad hoc networks. *Journal of Ambient Intelligence and Humanized Computing,* 1-20.

[17] Ozera, K., Bylykbashi, K., Liu, Y., &Barolli, L. (2018). A fuzzy-based approach for cluster management in VANETs: performance evaluation for two fuzzy-based systems. *Internet of Things,* 3, 120-133.

Chapter 16

Applications of Cognitive IoT for the Wellbeing of Human Life: An Empirical Study

Gurpreet Singh Chhabra[1], Umashankar Ghugar[1], G. N. V. Rajareddy[1] and Buddhadeb Pradhan[2*]

[1]GITAM School of Technology, GITAM Deemed to be University, Visakhapatnam, India

[2]Department of Computer Science and Engineering, University of Engineering and Management Kolkata, Kolkata, India

Abstract

Technology advancement is capable of managing daily human activities as well. The Internet of Things (IoT) is the most exciting term in the media world, and it has enticed many businesses to adopt it. On the other hand, cognitive computing refers to problem-solving systems that mimic human cognitive processes. The Cognitive Internet of Things (CIoT) technology combines cognitive systems and the Internet of Things (IoT). The primary advantage of this technology is to solve the myriad management challenges which CIoT does. (CIoT) has gained a lot of attention in recent years. IoT management algorithms are incredibly complicated, necessitating human thinking and automated cognitive processes for quick and precise decision-making. This book chapter has elaborated on how cognitive IoT has benefited for human beings.

Keywords: IoT, cognitive IoT, machine learning, artificial intelligence, smart city

[*]Corresponding Author's Email: buddhadebpradhan@gmail.com.

In: Computational Intelligence for Sustainable Development
Editor: Brojo Kishore Mishra
ISBN: 979-8-88697-198-9

Introduction

Cognitive IoT refers to the use of cognitive computing technologies to data generated by connected devices and the actions made by such devices. Researchers are already undoubtedly familiar with the Internet of Things, sensors, and actuators. Smart sensors provide a slew of possibilities for bettering the consumer experience. These sensors can learn the user's preferences and alter their settings accordingly. Smart home thermostats, for example, may adjust to multiple users' optimum temperature settings in the same property. There are three components to cognition as understanding, reasoning, and learning.

In a computer, system awareness entails generating meaning from large amounts of organized and unstructured data, creating a model of concepts, entities, and relationships. Reasoning is the process of applying a model to obtain answers or solve issues without explicitly programming the answers or solutions. And learning entails being able to automatically enter new information from data, which is a critical component of comprehending at scale. At scale, complex models of concepts and relationships can be time-consuming and expensive. Furthermore, because many relationships are unknown or unclear in advance, they can only be discovered in practice by having a machine look for patterns in massive data sets. Figure 1 depicts the various applications of cognitive IoT.

The notion of ubiquitous mobile learning -learning anywhere and anytime - indicates a pragmatic understanding of learning experiences, especially in an informal format that happens beyond classroom settings and resides in daily activities (Merriam and Bierema, 2014). The majority of adults engage in hundreds of hours of informal learning (Caffarella and Merriam, 2012); therefore, a unique aspect of mobile learning is that it is a "personal, contextual, and situated" learning process, especially emphasizing individuals' informal and unstructured experiences (Traxler, 2007). By recognizing the disruptive nature of mobile learning regardless of location and time, it is imperative to understand what design strategies and approaches are effective in mobile learning.

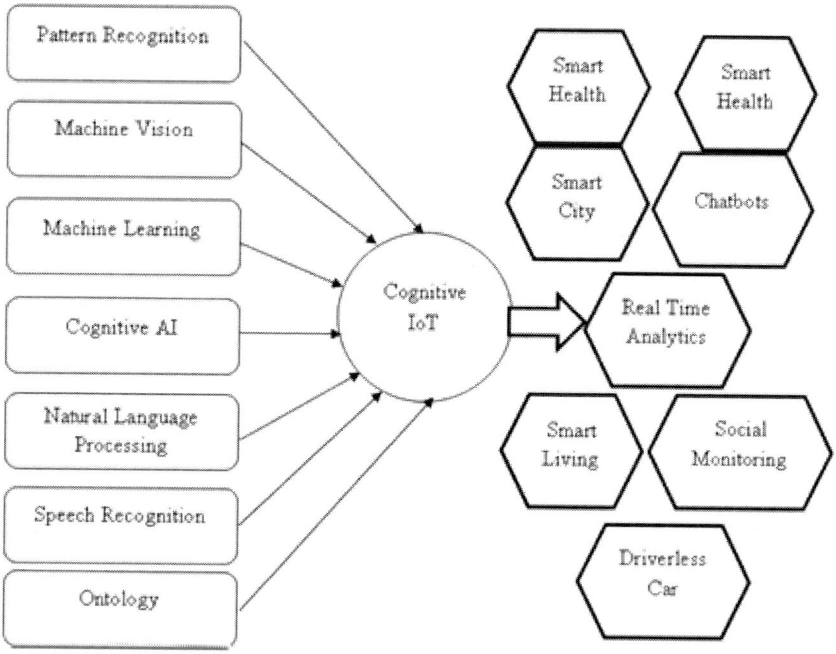

Figure 1. Cognitive IoT and Applications.

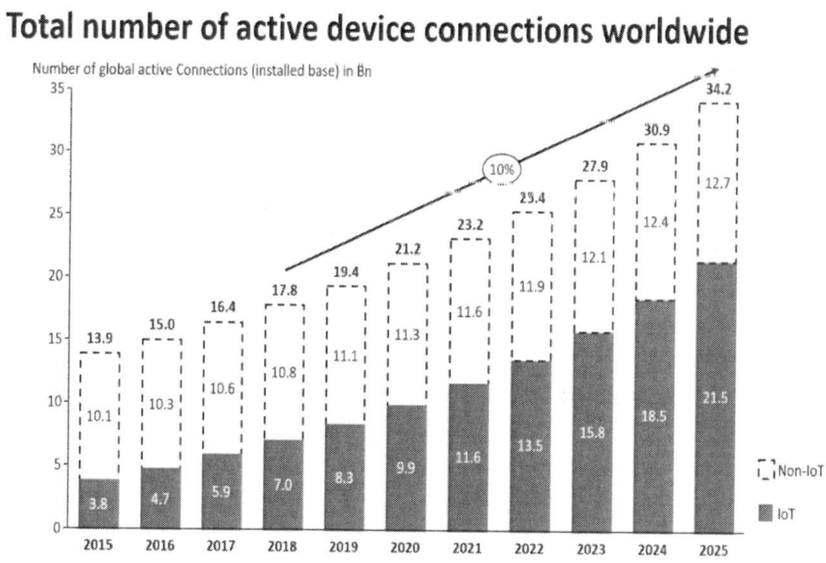

Figure 2. IoT Worldwide (Source: https://iot-analytics.com/state-of-the-iot-update-q1-q2-2018-number-of-iot-devices-now-7b/).

Let's look at an example to grasp the Cognitive IoT better. On our treadmill, let's employ Cognitive IoT. We start running on the treadmill. A sophisticated camera scans us and positively recognizes us. IoT devices start detecting our heart rate after measuring our weight. Our medical profiles, the report from our most recent cardiology appointment, and, yes, our designated contacts are all pulled up by AI. We begin to feel exhausted. The AI-controlled mind of the treadmill decided to stop the treadmill based on all available information and the output from the heart rate monitor and communicated with our doctor and caretakers with all relevant and confidential-authorized data. While the treadmill can call for help if necessary, it had already stopped the user from having a heart attack before the workout began. Figure 2 depicts the current and potential IoT applications around the world.

Internet of Things

In 1998, Kevin Ashton invented the phrase "Internet of Things" to define the Internet and pervasive computing future. This technological revolution will define the future of reachability and connectivity. 'Things' refer to any object on the globe, whether a participating device or a non-collaborating dumb object, in the Internet of Things. The Internet may include anything in the world. The items become connected nodes over the Internet via data communication pathways, usually using side-to-side Radio Frequency Identification (RFID) tags. IoT also includes intelligent devices. Smart Objects include physical and digital components that perform functions for humans and the environment. Apart from depicting IoT frameworks and objects, subsequent reviews emphasized that most IoT objects should know, referred to as "Smart Objects" (SO), and are expected to be recognized, detect, and communicate with others and make decisions independently of others. Figure 3 displays two potential IoT communication behaviors enabled by cell phones. Through direct engagement, the smartphone may ask the status of the IoT device in its area and then function as a bridge between low-level peer-to-peer protocols like Bluetooth or Wi-Fi and Internet protocols like HTTP and TCP. For example, the Fitbit fitness tracker uses a 4G network to transport a user's amount of steps from their phone. When searching the web on their mobile phone, consumers near an IoT-enabled device can examine linked data offered by interested persons via the Web service (Roy Want and colleagues, IEEE 2015). The RFID tool significantly improves embedded communication prototypes, enabling a microchip policy for WLAN data connection. Those function as electronic barcodes and help in automated detection. Because passive RFID tags do not have batteries, the ID is conveyed to the RFID reader using the power of the reader's debriefing signal.

M2M infrastructures are built on the Internet of Things (IoT), although it is not restricted to it. M2M stands for machine-to-machine communication that does not require human interaction. Even non-connected elements can join the Internet of Things by recognizing a data communication device, such as a barcode or an RFID tag, with a device (which might even be a smartphone) connected to the Internet.

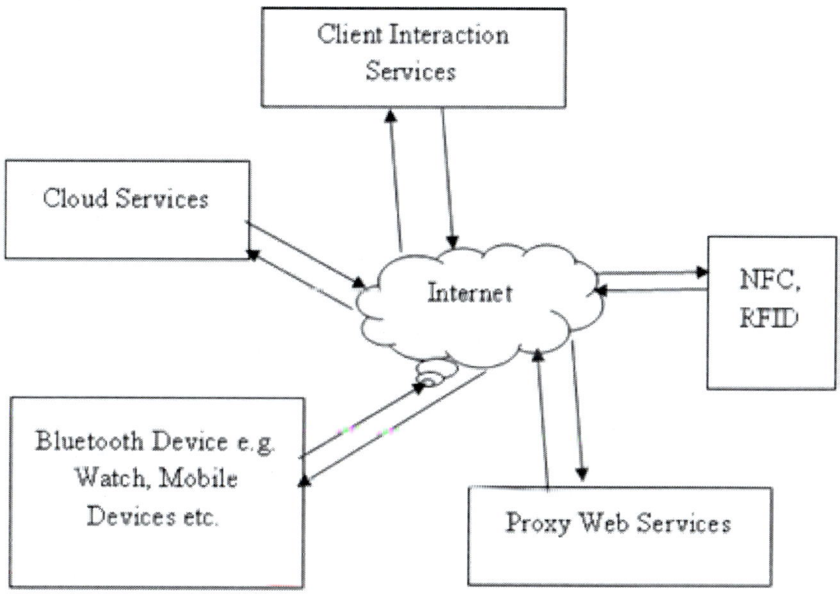

Figure 3. Method How Smart Objects Interact Through IoT.

AI and IoT

Artificial Intelligence (AI) and the IoT are closely connected. The IoT is all about connecting devices and utilizing the data they create. The practice of making robots look intelligent is known as artificial intelligence (AI). It seems like they are overlapping. Because IoT devices create large amounts of data, AI must deal with these massive volumes if the data is ever to be understood. It is only applicable when data leads to action. Data must be available with context and creativity. IoT and AI are coupled in this scenario, resulting in 'connected intelligence' rather than just linked devices. Traditional data analysis and action planning approaches are not developed to handle the large amounts of real-time data generated by IoT devices. AI-based analysis and reaction are required to extract the proper value from data.

Two very different real-time and post-event processing can benefit from AI:

- Post-event processing — finding trends in data sets and running predictive analytics on them, such as the relationship between traffic congestion, air pollution, and chronic respiratory illnesses in a city center.
- Real-time processing — reacts swiftly to circumstances and generates information about them, such as scanning license plates for parking fees with a remote video camera.

When we say artificial intelligence, we mean machine learning. Machine learning gives you the capacity to recognize patterns in data. It takes these trends into account when analyzing data or triggering actions. The goal of embedding AI into IoT systems is to provide another layer of intelligence throughout the whole IoT stack – from infrastructure to applications. AI is mainly used to create sensors that self-calibrate or self-heal when the IoT network or a single sensor fails, identifying them by proximity or even inventing new "virtual sensors." The definition is from EdyLiongosari, Global Managing Director and Chief Scientist of Accenture Labs

On the other hand, the IoT is a network of networked devices transmitting data. IoT gadgets have made their way into our daily lives and aim to provide high comfort. They rely on internet connectivity and create enormous volumes of data related to user behavior, personal information, and so on, which might not be ignored. However, many businesses have no idea how to store and manage such massive data. It revolves around IoT's development and capabilities. The combination of the rapidly expanding technologies is expected to radically disrupt the competitive landscape by forcing all businesses to adapt their product portfolios to emerging realities. As a result, new concepts will rise across all industries. The competitive climate will bring about four significant changes:

- Greater Revenues: In terms of gross profit and returns, the combined impact of AI and IoT will be highly beneficial to many businesses.

Manufacturers of IoT devices, IoT data providers, and companies and their services based on intelligent sensors.

- Augmented Safety Standards: The real-time monitoring systems can assist in maintaining a strict check, preventing any faults or disasters, and improving overall safety, security, and productivity. It can also help to reduce the danger of loss of life and property damage.
- Reduced Costs: Intelligent sensors, sensor-equipped domestic appliances, smart power meters, and other devices that monitor equipment will lower operational expenses for households and businesses.
- Customer Experience: Smart sensors provide a slew of possibilities for bettering the consumer experience. These sensors can learn the user's preferences and alter their settings accordingly. Smart home thermostats, for example, may adjust to multiple users' optimum temperature settings in the same property.

Cognitive IoTApplications

Smart sensors open up a slew of possibilities for bettering the consumer experience. These sensors can learn about user preferences and adjust their settings accordingly. Smart home thermostats, for example, may adapt to multiple users' preferred temperature settings.

- IoT in Manufacturing: Numerous manufacturing industries, viz. automobiles, aircraft, mining, food, beverages, etc., are integrated with intelligent sensors to perform different kinds of analysis to increase efficiency. An autonomous system having upgraded sensors must be deployed, and these sensors will help to protect the areas of the industry as well as maintain fault tolerances that lead to an entire proof safety system for the industry with a low-cost maintenance.
- Smart Houses: In IoT based smart home concept, all the associated devices are connected through a standard network combined with AI to perform smart decisions efficiently. Appliances can be accessed remotely by the blessings of smart technology to make life more straightforward and hassle-free. For instance, coffee preparation and TV program can be running only by clicking a switch or through a voice signal like the popular Alexa kind of intelligent device. Fire and

other emergency departments are also called through the automatic detection of fire and different emergencies to save lives.

- Smart buildings: Buildings can be equipped with smart sensors that can perform predictive analysis, enhancing their safety by preventing hazards and risks such as fire, short-circuit, flooding, etc. They additionally monitor the individual's movement and reduce costs by adjusting temperatures and improving energy capacity.
- Airlines: Smart sensors are integrated with aircraft to monitor the faults and errors within the aircraft. Maintenance teams will be alerted using those sensory signals, which also can predict the weather conditions and detect other hazardous obstacles quickly so that there is a chance of risk factor minimization. It also guarantees the safety of the passengers in an emergency.
- Smart cities: A smart city is a city that employs information and communication technology (ICT) and other allied technologies to improve the efficiency of daily urban operations and the quality (quality of service) of services provided to city residents. The demand for innovative ways of urban management that have minimum impact on the environment, public life, and administration has increased as the world's population, and urbanization continue to expand. The use of information and communication technology in municipal management has aided the integration of smart urbanization. The concept of smart cities has received a lot of attention due to the spread of smart gadgets and their recent advances. Smart cities use existing networks to connect everyday real-time artifacts. As a result of the Internet of Things, traditional networks that connect millions of linked objects have transformed (IoT). Continuous technical developments in pervasive computing, wireless networks, device communication, and the creation of cross-vertical applications are required to reinforce the IoT idea of Smart City further.
- Healthcare: Virtual medical devices are connected to the internet, gathering considerable additional data, providing insight into symptoms and patterns, providing remote care, and frequently allowing patients greater control over their health and therapy.
    - Insulin pens with smart continuous glucose monitoring (CGM): Diabetes, which affects one out of every ten individuals and needs continual monitoring and treatment administration, has become fertile ground for the growth of smart gadgets.

- Closed-loop (automatic) insulin delivery: The open-source effort OpenAPS, which stands for Open Artificial Pancreas System, is one of the most appealing areas of IoT medicine. OpenAPS is a closed-loop insulin administration device that, unlike a CGM, detects the quantity of glucose in a patient's circulation while also supplying insulin — thereby "closing the loop."
- Wearable technology doesn't necessarily have to be built with a medical purpose in mind to provide healthcare advantages. In 2017, Takeda Pharmaceuticals USA and Cognition Kit Limited, a mental health assessment platform, collaborated to test the usage of Apple Watch software to monitor and evaluate patients with Major Depressive Disorder (MDD).

Cognitive IoT and Covid19 Pandemic

The global world is fighting against a novel coronavirus, and healthcare officials are scrambling to provide the finest tools to protect citizens from infection and to treat those who have already been affected. Technology innovation is gradually migrating to healthcare systems in this case. Countries pool all available resources and employ modern technology to reduce COVID-19 danger and profile people. During times of crisis, the Internet of Things (IoT) and other technologies such as Cloud and AI have shown to be quite valuable. According to MIT researchers, overlaying a geographic information system (GIS) on IoT mobile data can enable epidemiologists to find patient zero and identify all who have come into contact with infected patients. This technology aids in monitoring high-risk individuals and can thus serve as a source of information for healthcare professionals to take appropriate action. Wearable technology has advanced quickly in recent years, firmly establishing itself as a rapidly evolving product category. Healthcare stakeholders are uncovering new techniques and drugs to help stop the flood of Covid-19 cases, which are increasing daily. Technology and healthcare technologies are now coming into their own and assisting us in the fight against the epidemic. Wearable technology, which is effective at monitoring numerous physical and physiological characteristics such as heart rate, body temperature, blood pressure, movement, sleep, and more, is critical in this circumstance.

Conclusion

As our lifestyle changes gradually, everyone wants in their life, every person wants a machine that works like a human being. Secondly, the world population has also increased, resources are less effective, and calculative use of resources is a foremost concern. Nowadays, drinking water is a significant problem in Asia due to waste of water and industrial pollution; all are based on human-dependent systems. Using IoT devices for water management, we can analyze water use and waste and monitor water pollution. Other than this, there are some open challenges in cognitive IoT as:

- Energy-efficient sensing
- Security of data
- Secure reprogrammable networks and privacy
- Protocol for sensor networks

References

Yin Y, Zeng Y, Chen X, Fan Y (2016). The internet of things in healthcare: an overview. *Journal of Industrial Information Integration* 1:3–13.

Chen M, Zhang Y, Qiu M, Guizani N, Hao Y (2018). SPHA: smart personal health advisor based on deep analytics. *IEEE Communications* 56(3):164–169.

Hu, L, Zhao, S, Xiong, X, Zheng, K, Chatzimisios, P, Hossain, M S, & Xiang, W. (2016) Internet of Things Cloud: Architecture and Implementation. *IEEE Commun.Mag*. 54(12-Supp):32–39.

Chen, M, Hao, Y, Hu, L, Hossain, M S, &Ghoneim, A. (2018).Edge-CoCaCo: Toward Joint Optimization of Computation, Caching, and Communication on Edge Cloud. *IEEE Wireless Commun* 25(3):21–27.

Muhammad G, Rahman SKMM, Alelaiwi A, Alamri A (2017). Smart health solution integrating IoT and cloud: a case study of voice pathology monitoring. *IEEE Commun.Mag*. 55(1):69–73.

Chen M, Herrera F, Hwang K (2018). Cognitive computing: architecture, technologies and intelligent applications. *IEEE Access* 6: 19774–19783.

Hossain, MS (2016). Patient State Recognition System for Healthcare Using Speech and Facial Expressions. *J.MedicalSystems* 40(12):272:1–272:8.

Hossain MS, Muhammad G (2016). Audio-visual emotion recognition using multi-directional regression and Ridgelet transform. *Journal on Multimodal User Interfaces* 10(4):325–333.

Cognitive IoT for Healthcare, https://www.ibmbigdatahub.com/blog/what-cognitive-iot.

Chapter 17

A Blockchain Network Simulation Tool - Hyperledger Fabric: Detailed Overview on Its Features, Process, and Implementation Steps

Priyanka Gaba[1,*], Urvashi Sugandh[2,†], Manju Khari[3,‡] and Arvind Panwar[4,§]

[1,2]Galgotias University, Greater Noida, India

[3]Jawaharlal Nehru University, New Delhi, India

[4]Dr. Akhilesh Das Gupta Institute of Technology and Management, New Delhi, India

Abstract

Blockchain is a decentralized, immutable, distributed ledger and consensus-based technology. From the time of invention when it was just used for just bitcoin till now, it has its widespread usage in almost all the fields like financial, healthcare, agriculture, IoT, cloud, supply chain, etc. These applications of blockchain could be of public or private or hybrid type blockchain and even maybe permissioned or permissionless. Based on the application's requirement for the type of blockchain it can be implemented on any of the platforms available from a range of platforms in today's market. The available platforms for blockchain are Hyperledger, Ethereum, Stellar, Quorum, Corda, etc. which are of significant interest to researchers working to improvise them as per the market's requirements. Hyperledger is an umbrella project which includes multiple projects for solving multiple purposes while implementation like Fabric, SawTooth, Besu, Indy, etc. Hyperledger also

* Corresponding Author's Email: priyanka.gaba2202@gmail.com.

† Corresponding Author's Email: urvashi5sugandh@hotmail.com.

‡ Corresponding Author's Email: manjukhari@yahoo.co.in.

§ Corresponding Author's Email: arvind.nice3@gmail.com.

In: Computational Intelligence for Sustainable Development
Editor: Brojo Kishore Mishra
ISBN: 979-8-88697-198-9

encompasses multiple tools like caliper, cello, burrow, explorer, grid, etc. Hyperledger Fabric is one of the most popular projects with a modular architecture to implement blockchain-based applications. Exploring the Hyperledger fabric platform in detail to have an understanding of using it for the application's implementation is the target of this chapter. Fabric components, their configuration files, the complete process, and the steps required to complete its implementation are explained in detail in this chapter to have a deep understanding of the complete implementation process of fabric for any application.

Keywords: blockchain, Hyperledger Fabric, Hyperledger Caliper

1. Introduction

Blockchain is a disruptive technology noticed in 2009 with its implementation in cryptocurrency named Bitcoin. Satoshi Nakamoto, an anonymous personality, gives his contribution to the implementation of blockchain technology. Bitcoin is the very first decentralized cryptocurrency that makes use of blockchain technology. In a centralized system, it is quite difficult to share data or any information between the parties because of trust factors. Also, a centralized system has some limitations which could be overcome by using blockchain technology. There is a possibility of single-point failure in a centralized system and storage of data is not possible at a third party due to privacy reasons. But to do so, we need a trusted entity but each entity has its different policies and may not be acceptable. With the implementation of blockchain, all the above-mentioned limitations can be overcome because of its distributed ledger system where a copy of each ledger is distributed or shared among all the entities or nodes (Androulaki et al. 2018). Blockchain is categorized into three major types: public, private, and consortium Blockchain. Further, the blockchain could be permissioned or permissionless. Blockchain technology is considered as a backbone for the current cryptocurrencies. Other than cryptocurrency, Blockchain has its application in many fields such as health care, banking systems, agriculture, etc. Being a new technology, blockchain encounters several challenges based on performance, scalability, and privacy. Some of the other issues that have been introduced by the researchers are latency, bandwidth, throughput, usability, and wastage of resources (Nasir et al. 2018). All these challenges or issues take the attention of researchers toward the platform of Blockchain. Blockchain because of its popularity in the market to develop a range of applications forced various researchers to create new platforms for its implementation. Various available platforms are available in the market to fulfill the need of the type of blockchain required, and the application's requirements. The platforms available are Hyperledger, Ethereum, Stellar, Quorum, Corda, etc. ("Hyperledger Caliper," n.d.).

Hyperledger Project is a collective effort to create an open-source distributed ledger-based and smart contract-based framework (Cachin 2017). This project was initiated by the Linux Foundation and includes several projects and frameworks that allow businesses to create their application on the blockchain.

Initially, Hyperledger Fabric was known as Hyperledger Fabric v0.6. It was introduced by IBM as an open-source and permissioned platform of blockchain. Later on, the market has an abundant number of blockchain platforms. Still, there is no perfect way to evaluate these platforms based on different facts such as scalability, latency, performance, and security.

In this book chapter, the main focus is on the basics of Hyperledger fabric. To get a better understanding, the basic concept of blockchain including its features is explained before the Hyperledger fabric. The next section includes the introduction of Hyperledger and Hyperledger fabric which further includes its components, features, and configuration files. As this book chapter is about Hyperledger fabric so it explains its implementation steps. In the last section, the paper is concluded.

2. Blockchain

Blockchain is comprised of two different words: Block and Chain. In the Blockchain, a block is a collection of multiple transactions and it may be of any type for example monetary transactions, or query data. These informative blocks connect with another block to form a chain known as Blockchain. The first block in the blockchain is known as the Genesis block. Block contains information in three parts: Block header, Block data, and metadata as shown in Figure 1. Block Header contains the current block number, block hash, and previous hash. Block data, comprises a list of transactions. In each transaction, there are two fields: Header, and data. The header encompasses the data related to the channel. The data fields encompass Transaction Proposal, Endorsements, and Proposal Response. Block Metadata, the section contains the details related to the Block creator, and its signature; the last configuration block number and its signature, and the last offset persisted and it's signature (Pilkington 2016).

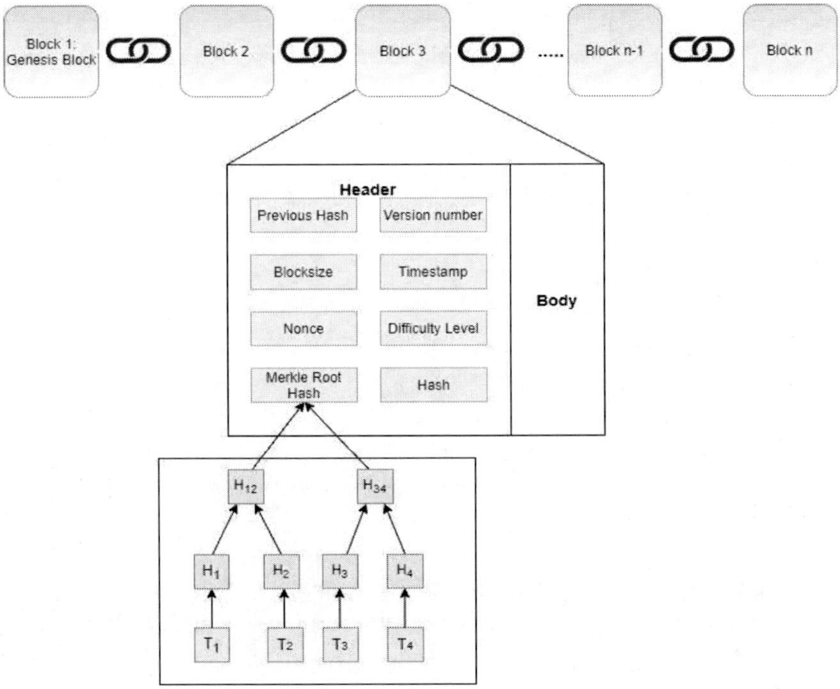

Figure 1. Elements of blockchain.

- *Previous hash:* A blockchain is made up of numerous linked nodes known as blocks, the previous hash holds the hashed value of the preceding node's address so the tracking of nodes becomes easy.
- *Blocksize:* It defines the size of the block.
- *Nonce:* Nonce is a number that is to be used only a single time. It is a random integer number and miners are always in search correct number. The range of this random number lies between 0 to 2^32.
- *Merkle root hash:* A Merkle root checks the data for corruption, hacking, or manipulation using mathematical procedures. As an example, if a block includes ten transactions, we'll need to merge those transactions into one Hash Value to identify it. To do so, it employs the concept of binary tree to construct the block's hash, which is referred to as the Merkle Root.
- *Version number:* This number defines the version which is used by the block. There are three types of versions v1.0, v2.0, v3.0, and v4.0.
- *Timestamp:* This field is used to have the knowledge of the timing on which the block is used and also it is used to authenticate the block.

- *Difficulty level:* It defines the complexity and processing power necessary to mine the network. If the difficulty level is high, it means that it will require costly computing equipment for mining. For example, target algorithms like RIPEMD, MD5, SHA-2, and SHA-3 are employed to raise the complexity.

2.1. Features of Blockchain

The trust required within a blockchain network in any of the applications is facilitated by the following features of the blockchain network even in fact without the presence of trusted mediators.

- *Decentralized*: A single authority is not responsible for the maintenance of the complete network rather multiple nodes who all are involved are responsible for the same. They can do this by storing everything on the blockchain and anyone can access it using their private key (Zheng et al. 2017). This feature makes the system highly fault-tolerant from accident failures and gives users control over their assets they don't have to rely on a third party and so no risk is involved.
- *Distributed Ledger*: A public ledger is open for all providing all the necessary information regarding transactions and users of the network except for private and federated Blockchain. The distributed ledger has 3 main features: Recorded, transparent, and decentralized (Panwar and Bhatnagar 2020). Every node has to maintain the ledger and participate in the validation of the block. To add a transaction to the ledger, it must be validated by the majority of nodes in the network. Update in the ledger is maintained fast so no malicious user can try to temper the network.
- *Immutable*: Immutable means not able to change or modify. In blockchain immutable means once the block is added to the ledger no one can change anything because the hash of the node will change and so all subsequent nodes will change (Meng et al. 2018). This change is easily identifiable in the network as the hash of that node is different from the rest of the network so no edit, or delete is possible in the blockchain network.

- *Secure:* The security of the blockchain network is maintained by using an encryption technique. To double-fold security, cryptography is also used to hash every piece of information. Every block has its unique hash and also contains the hash of the previous block as well. To access any information every node uses its private key and to make a transaction, the public key is used.
- *Consensus*: A commonly acknowledged view or conclusion between a group of nodes to agree on the state of the blockchain. To run a system smoothly, a consensus algorithm is essential where a huge number of nodes are validating a block. It's also necessary for the decentralization feature of every blockchain. Different types of consensus algorithms like Proof-of-Work, Proof-of-Stake, etc. (Wang et al. 2019)

3. Hyperledger

Hyperledger was evolved in 2015 by the Linux foundation hence it is an open-source platform ("Hyperledger – Open Source Blockchain Technologies" n.d.). Hyperledger offers features like scalability, secrecy, complexity, vital security necessities, and compliance; along with other features like distributed ledgers, smart contracts, and responsive interfaces (Ranjan et al. 2019). Therefore, Hyperledger can be applied to several fields and spread blockchain technologies for cross-industry applications (Priyanka and Raw 2020). Hyperledger popular use cases are Healthcare, supply chain, finance, insurance, VANET, etc. (Gaba and Raw 2020) It provides higher efficiency than other blockchain platforms. Hyperledger is an umbrella project consisting of multiple projects like Fabric, SawTooth, Besu, Indy, caliper, etc. Hyperledger also encompasses multiple tools like a caliper, explorer, cello, burrow, etc. The essential components of Hyperledger are Assets, Ledger, Privacy, Chaincode, consensus, and Membership services. This section further discusses the Hyperledger frameworks and tools in detail.

3.1. Hyperledger Frameworks

Hyperledger observes an umbrella strategy that includes several frameworks to implement business logic based on blockchain. These frameworks reflect the distinct tactics and challenges that would come in designing the application using Blockchain. Various supported frameworks are discussed below in detail and also shown its characteristics in Figure 2.

- *Hyperledger Aries:* Hyperledger Aries is a shared, reusable, and interoperable toolkit that is intended for efforts and solutions that are concerned with the creation, transmission, and storage of verified digital credentials, among other things. It is a platform for peer-to-peer interactions that are based on blockchain technology. Secure secret storage and decentralized key management features are offered by this project, which makes use of the cryptographic support given by Hyperledger Ursa.
- *Hyperledger Besu:* Hyperledger Besu is an Ethereum client that is intended to be enterprise-friendly for both publicly and privately permissioned network use cases. It is built on the Hyperledger Fabric platform. It may also be used on test networks, such as Rinkeby, Ropsten, and Görli, to demonstrate its capabilities. Hyperledger Besu provides several different consensus methods, including PoW and PoA, among others (IBFT 2.0, IBFT, Etherhash, and Clique). This software's sophisticated permissions methods are particularly developed for usage inside a consortium context.
- *Hyperledger Fabric:* To construct applications or systems with a modular design, Hyperledger Fabric is meant to serve as a basis for doing so. Hyperledger Fabric enables components such as agreement and membership services to be plug-and-play, making them more convenient to use. Its modular and adaptable architecture makes it suitable for a wide variety of industrial applications. It provides a one-of-a-kind method to a consensus that allows for high performance at a large scale while maintaining anonymity.
- *Hyperledger Indy:* Hyperledger Indy offers a variety of tools, modules, and reusable components for creating digital identities that are based on blockchains or other distributed ledgers that are compatible across administrative domains, apps, and any other siloed environment. It may be used in conjunction with other blockchains or as a stand-alone system, allowing for the decentralization of identification to be powered by Indy.

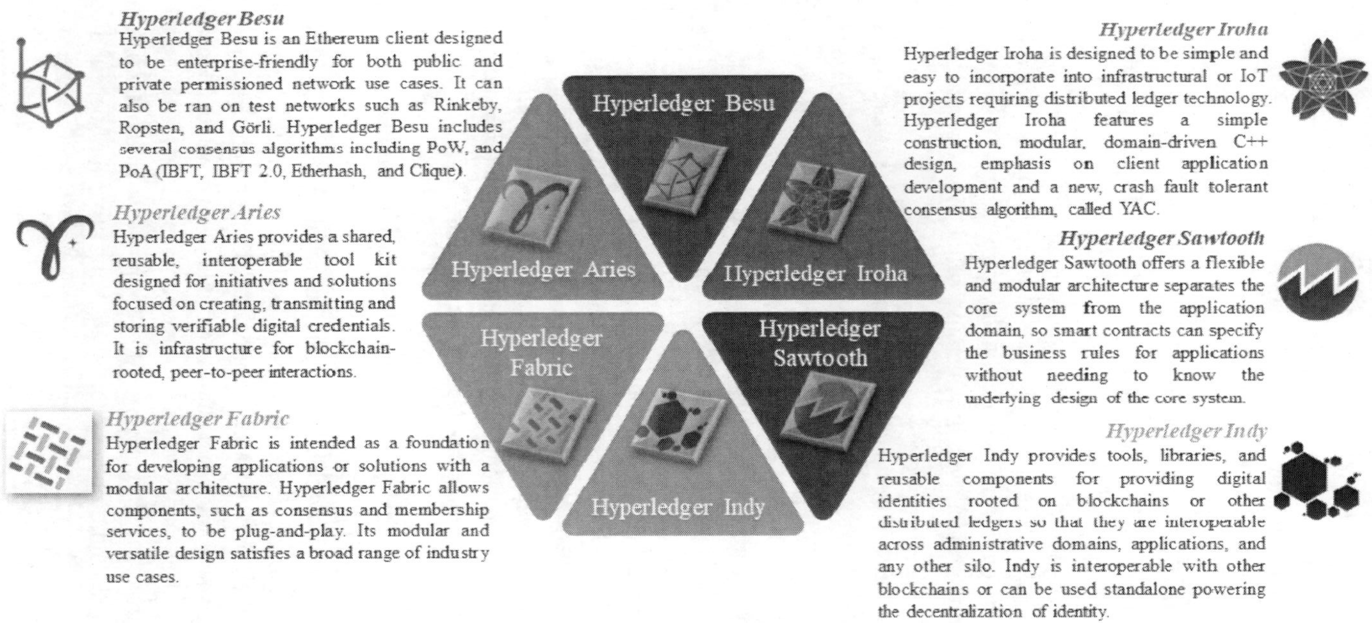

Figure 2. Hyperledger frameworks.

- *Hyperledger Iroha:* Hyperledger Iroha is intended to be straightforward to integrate into infrastructure and Internet of Things applications that need distributed ledger technology. It has a basic architecture, a flexible, domain-driven C++ design, a focus on client application development, and a novel, crash-fault-tolerant consensus algorithm called YAC. Hyperledger Iroha is a distributed ledger technology.
- *Hyperledger Sawtooth:* With Hyperledger Sawtooth, the core system and application domains are separated by a flexible and modular architecture, allowing smart contracts to express business logic for applications without having to be aware of the core system's underlying design. Proof of Elapsed Time (PoET) and Practical Byzantine Fault Tolerance (PBFT) are two of the consensus methods supported by Hyperledger Sawtooth.

3.2. Tools

Hyperledger comes with a variety of tools aiming to improvise blockchain-based implementation of the applications. These tools provide Hyperledger a capability to tackle complex challenges ranging from security, performance, complexity, interoperability, etc. Various tools under Hyperledger are explained below in detail and also shown in Figure 3.

- *Hyperledger Grid:* Businesses demand new technology and standards to improve supply networks, and Hyperledger is aware of the influence of blockchain on supply chains. Hyperledger Grid has been offered as a solution to the problem by Hyperledger. Blockchain-based supply chain systems with DLT features may be built and implemented using this platform.
  Libraries, Development Tools, and data models are all part of Grid's utility set. The primary goal of these tools is to ensure that the creation of smart contracts and user interfaces that have cross-industry validation is expedited.
  A single platform will make it easier for all stakeholders to interact and collaborate on sourcing and procurement, finance, inventory management, and distribution. One of the top Hyperledger development tools for supply chain management is Grid.

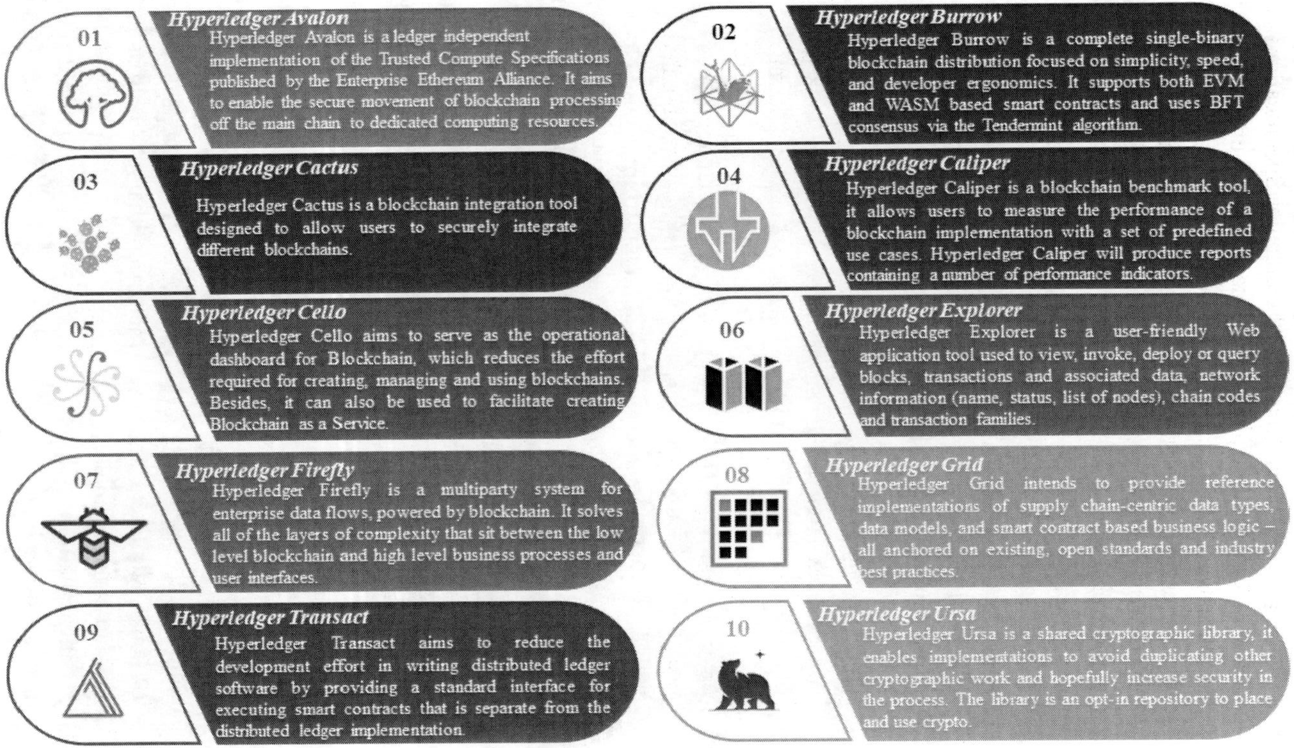

Figure 3. Hyperledger tools.

- *Hyperledger Avalon:* The Trusted Computer Specifications, released by the Enterprise Ethereum Alliance, are implemented using Hyperledger Avalon, a ledger-independent adaptation of the Trusted Computer Specifications. Its goal is to make it possible for blockchain operations to be moved securely off the main chain and onto specialized computer resources. Avalon is intended to assist developers in reaping the advantages of computational trust while also mitigating its shortcomings, according to the company.
- *Hyperledger Burrow:* In contrast to other blockchain distributions, Hyperledger Burrow is a single-binary blockchain distribution that is focused on developer ergonomics, simplicity, and speed. It supports smart contracts that are built on EVM and WASM, and it leverages BFT consensus using the Tender mint algorithm to do this. In addition to a complex event system, it can keep a relational database modeling of on-chain information. Integrated governance and permissions may be changed by on-chain proposal transactions, which are recorded on the blockchain. It is designed for public permissioned proof-of-stake use cases, but it may also be used for private/consortium networks, as well as for other applications.
- *Hyperledger Cactus:* Primarily, Hyperledger Cactus is a plugin-based platform that enables developers to safely combine numerous blockchains into a single application. The primary goal of developing this tool is to – increase interoperability across blockchain networks by addressing the fragmentation problem.

  One thing you should be aware of is that Hyperledger Cactus is a new project that is currently in the early stages of development. There are only connections available for the most common Hyperledger framework tools at this time (Fabric and Hyperledger Besu). However, the developers and contributors are working on other common protocols, including Corda, Geth, and Quorum, as well as their protocols.

  Although this tool is still in the introductory phase, Hyperledger has formally acknowledged it. Why? Hyperledger Cactus addresses two primary issues: blockchain fragmentation and the dangers associated with enterprise adoption.
- *Hyperledger Caliper:* The maximal efficiency of a blockchain system is not always achieved by all solutions. The effectiveness and performance of a system will always be dependent on a variety of external circumstances and application scenarios. Having said that,

there was no standard platform available to address this problem at the time until the Hyperledger Caliper is released ("Hyperledger Caliper," n.d.).

Specifically, Caliper covers the three most important reasons why the present blockchain development activities need the use of a performance benchmark tool. They're right here —

- The absence of the source codes for the current performance reports raises concerns about the validity of the assessments' validations.
- a deficiency in a pre-defined set of performance metrics
- There are currently no practical methods for benchmarking the use cases that are appropriate for a given blockchain.

It makes use of measurements and terminology developed by PSWG (the Hyperledger Performance and Scalability Working Group). All of these potential and perspectives have already piqued the attention of several corporate firms in Hyperledger caliper. This project is being worked on by developers from Huawei, IBM, Hyperchain, Oracle, Bitwise, and Soramitsu, among other companies and organizations.

The Hyperledger Caliper project seeks to quantify the performance of a specific blockchain implementation based on a set of use cases that have been predefined. Currently, Caliper employs four performance metrics to compare blockchain systems to present use cases, which are available on the Caliper website.

- *Hyperledger Cello:* It is the goal of Hyperledger Cello to act as the operational dashboard for blockchain technology, which will minimize the amount of time and effort necessary for the creation, management, and use of blockchains. Furthermore, it may be utilized to make the creation of Blockchain as a Service more convenient. Cello offers an operational interface for effectively managing blockchains that are operating on top of a variety of infrastructures, such as BareMetal, virtual machines, and multiple container platforms.
- *Hyperledger Explorer:* Hyperledger Explorer is a user-friendly web application tool that can be used to view, invoke, and deploy blocks, transactions and associated data, network information (such as name,

status, and a list of nodes), transaction families, and chain codes, as well as any other relevant information that is stored in the ledger.

Hyperledger Explorer is a Blockchain Website UI that provides users with pictorial representations of blockchain activity metrics, which may be useful while developing blockchain applications. This program also makes it possible for anybody to manage blockchains and distributed ledgers without having to be concerned with the technical details. As a result, Hyperledger Explorer provides tremendous commercial value to its users.

The interface of Explorer makes use of a variety of technologies, including ReactJS, Material-UI from Google, web sockets, and other similar technologies. Currently, Explorer is compatible with two of the most popular Hyperledger framework tools: Hyperledger Fabric and Hyperledger Iroha, respectively (alpha release).

- *Hyperledger Firefly:* Hyperledger FireFly is a multi-party system for business data flows that is driven by blockchain technology, according to the company. It deals with the layers of complexity that exist between the low-level blockchain and the high-level business operations and user interfaces, and it does it in one go. Because it allows developers to concentrate on business logic rather than infrastructure, FireFly allows them to create blockchain applications for enterprises far quicker.
- *Hyperledger Transact:* Hyperledger Transact intends to minimize the amount of time spent developing distributed ledger software by offering a common interface for running smart contracts that is independent of the distributed ledger implementations. To develop additional smart contract languages, Hyperledger Transact uses an extensible technique known as "smart contract engines," which are virtual machines or interpreters that process smart contracts and may be extended at any time.
- *Hyperledger Ursa:* It is a cryptography framework that is shared throughout the network, allowing developers to avoid duplicating work that has already been done in the cryptographic field.

  A public and collaborative repository of cryptographic protocols is becoming more necessary as the Hyperledger ecosystem grows in size and scope, according to the developers involved. This module would save time and effort by eliminating the need to develop distinct cryptographic protocols for each project. Another advantage of using

a shared library is that it enhances cross-platform compatibility and modularity.
Shortly said, Ursa makes it easy for developers across the ecosystem to collaborate on development projects, making it one of the most preferred Hyperledger development tools. These are the primary motivations for the Hyperledger Ursa project. —

- Keeping Unnecessary Duplications to a Minimum: Because cryptography implementation is exceedingly difficult, developers must spend a significant amount of effort improving it. Developers, on the other hand, frequently re-code the very same cryptographic algorithms in different ways. If they collaborate on the implementation protocols, they might save a significant amount of time and prevent having to recode the same protocol again.
- Improvements in cybersecurity and fewer bugs: Because the Ursa library is based on a coordinated partnership, all developers will contribute to making the basic cryptographic protocol safer, easier to maintain, and less prone to bugs.
- Interchange: Because all of the projects in the network are utilizing the same cryptographic libraries, there will be more interoperability between projects inside the library. Additionally, since they both use the same security standards, the cryptographic verification system will be more efficient.

Projects will automatically prefer keeping this common component if there is an agreement on the security protocols and if there has been a consensus. This will result in increased modularity throughout the whole development effort as a consequence of the library.

4. Hyperledger Fabric

Hyperledger Fabric is one of the most popular projects used to develop applications with modular architecture. It is a distributed ledger technology, that uses smart contracts, and a private, permissioned network that means allows organizations to use trusted Membership Service Provider (MSP) to choose who can enroll and access the network (cryptocurrencies and 2016 n.d.).

4.1. Features

The Hyperledger Fabric Model consists of six key design features.

- *Assets:* An asset can be any item that has some monetary value that can even be exchanged over the network like food, medicine, cryptocurrency, etc. In fabric, these are signified as a key-value pair, represented in binary or JSON format.
- *Ledger:* Ledger comprises components like world state and Blockchain. The World state outlines the current state of the ledger at any given time. The blockchain contains the transaction log that archives all previous transactions that have caused the present value of the world state. The transaction history of each channel is shared and immutable as the current transactions can't be changed, only new transactions can be added. Ledger supports a SQL-like query.
- *Privacy:* The privacy feature imposes restrictions on ledgers to be shared across the network or limits to some specific participants.
- *Smart contracts:* Fabric smart contracts are written code that contains the logic of the application, also called Chaincode (Androulakı et al. 2018). It aims to optimize the network by ensuring aspects like performance, scalability, verification, and trust. Chaincode mostly interacts with the world state not with transaction log to query out current data.
- *Consensus:* It verifies the transactions and other events within the blockchain network. Some prevalent consensus mechanisms are Proof-of-Work (PoW) and Proof-of-Stake (PoS) but Hyperledger uses Practical Byzantine Fault Tolerance (PBFT).
- *Security & Membership services:* Membership services manage all the identities of participants of the permissioned blockchain network to control the data access to make it secure.

4.2. Components

Three categories of Hyperledger Fabric components exist and those are "Core components", "Tools/Utilities," and "Fabric Certification Authority" as shown in Figure 4. Core components include peer, orderer, and clients. Tools/ Utilities include configtxgen, configtxlator, cryptogen. Fabric Certification

Authority includes fabric-ca-server and fabric-ca-client. These components are a required set of binaries for setting up Hyperledger Fabric Infrastructure for production or experiment. All of these binaries require configuration information to be provided in YAML format which is a data-serialization human-readable language.

"Peers," when launched as a process become a node of the network. Peers are used as a tool to manage network and channel configuration by hosting ledgers and chaincode (Thakkar, Nathan, and Viswanathan 2018). Each of the peers in the network maintains its copy of the ledger. They exchange data among themselves using the gossip data dissemination protocol (Gorenflo et al. 2019). Peers are of different types depending on their roles in the network like anchor peers, committing peers, endorsing peers, and leader peers. "Anchor peers" are the only peers who are discoverable. Inter-organization communication is possible via their anchor peers. Each organization must have at least one anchor peer to communicate with the network. "Leader peers" are the only nodes connected to the orderer for receiving the new block and then distributing the transactions of the block to the organization's committing peers. To become a leader, a peer could be statically assigned or dynamically assigned. Any peer which has a smart contract installed could behave as an "endorsing peer." These peers simulate the transactions on receiving its proposal and based on simulation results prepare a transaction proposal and send that to the Ordering peers. Each peer in the channel can be a "committing peer." These peers receive blocks from the orderer and validate them and then that change is appended to the peer's copy of the ledger.

"Orderer peers" are also known as ordering services that provide a communication channel for fabric. They can be configured using any of these consensus methods: Solo, Kafka, or Raft (Yusuf and Surjandari 2020). Orderer establishes consensus for the complete Blockchain network. Consensus assistance in governing whether to commit a transaction or not. These transactions are being submitted by Client nodes to endorsers and broadcast transaction proposals to the orderers. Orderers ensure the order of transactions, create a block from it, and ensure a consistent ledger state across the network (Androulaki et al. 2018). Clients will use this application by sending updates or querying the database. Only the orderer peers keep records of valid and invalid transactions, whereas the other peers just keep only valid transactions.

Tools/utilities required for Hyperledger fabric implementation are configtxgen, configtxlator, and cryptogen. "Configtxgen" tool is used for managing a network or channel configuration ("Configtxgen," n.d.). "Configtxlator" is used to translate protobuf (a binary format used by GRPC) to JSON and vice-versa ("Cryptogen," n.d.). "Cryptogen" generates crypto material for testing.

Two fabric certification authorities are fabric-ca-server, and fabric-ca-client who are responsible for creating identities by producing a key-pair of public key and private key and also distributing certificates to different actors ("Fabric-CA Commands," n.d.). The mechanism that allows identities to be trusted and a complete network will recognize them without disclosing their private key is called Membership Service Provider (MSP). MSP also defines the roles and permission of every member within the organization. The digitally signed certificates by the CA, bind together the actor with the actor's public key. "Fabric-ca-server" launched as a standalone process that exposes services for managing identities and certificates. "Fabric-ca-client" works as a command-line interface tool that is used for executing services on fabric-ca-server i.e., for managing identities and certificates ("Fabric-CA Commands," n.d.).

Hyperledger Fabric consists of multiple organizations. Each Organization can have multiple peers and client nodes. Organizations or peers may join multiple channels. Each channel has its own distributed ledger and state DB. Channel is a group of organizations for the same purpose to maintain the privacy of transactions by just making the transactions available for the members of the channel. Chaincode is also deployed on channels, not on the network.

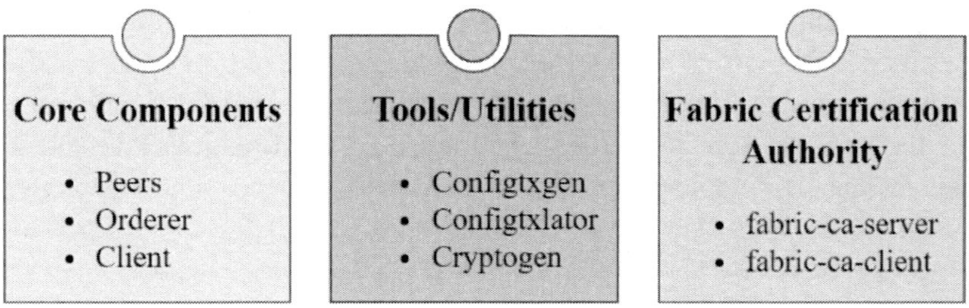

Figure 4. HLF Components.

4.3. Configuration Files

Various configuration files created for creating the binaries of the above components in HLF implementation are configtx.yaml, crypto-config.yaml, orderer.yaml, docker-compose-base.yaml, core.yaml.

4.3.1. Orderer.yaml

"Orderer.yaml" file is to define the orderer configurations. This file is to provide the access required by the orderer for the genesis block to initialize the orderer channel. This file contains 8 sections. 1st section is the "General" section to specify general properties of the orderer which includes, LocalMSPID, LocalMSPDir, ListenAddress, ListenPort, Cluster parameters, Keepalive, TLS (if enabled), BootstrapMethod (value of this could be "provisional" if genesis is generated dynamically or "file" if genesis block is provided as a file), BootstrapProfile or BootstrapFile depending on BootstrapMethod value, and BCCSP Setup. The BlockChain Cryptographic Service Provider (BCCSP) is the package that offers the implementation of cryptographic standards and algorithms. The 2nd section is "LedgerType" whose value could be RAM, JSON, or file. "FileLedger" is the section created if the value is "file", this manages blocks in LevelDB i.e., filesystem, and the location of the file needs to be specified in the orderer.yaml file. The value "JSON" means blocks are managed in a JSON file. The value "file" means blocks are managed in LevelDB. The 3rd Section is "operations" where transport layer security can be enabled for server-level authentication; for data encryption of messages exchanged between peers and GRPC server; for client authentication. This can be done by specifying properties like PrivateKey, Certificate, RootCAs, and ClientAuthRequired. The 4th section is "Metrics" for configuring the metrics collection for the orderer. The properties which are specified in this section are network type which is TCP or UDP; server address and write interval. The 5th section is "debug" to control the debugging options for the orderer. Two properties specified here are BroadcastTraceDir and DeliverTraceDir which specify the directory where a file is stored for Broadcast service or Deliver service request.

4.3.2. Configtx.yaml

"Configtx.yaml" file is to specify configuration elements for various artifacts. This file contains sections for organizations, orderers, applications, channels, capabilities, and profiles (Hua et al. 2020). The "Organization" section contains a list of member organizations. One organization is for the orderer and others are generally consisting of various peers. Each organization is defined by an anchor (&) symbol to avoid repetitions. It contains details like name, id, list of anchor peers, MSPDir which is a path to a folder containing the crypto materials for the organization, and policies that define rules for readers, writers, Admins, and for endorsement. The "Orderer" section specifies the configuration for the orderer and is defined by the anchor (&) symbol so that it can be referred from other files within configtx.yaml file ("Configtxgen," n.d.). This contains details like orderertype, Address, Batchtimeout, Batchsize (specifies maximum message count, absolute and preferred maximum bytes), and policies which define rules for readers, writers, Admins, and block validation. Orderertype could be SOLO or Kafka. The amount of time to wait before creating a batch is defined in Batchtimeout. Batchsize contains three arguments: MaxMessageCount, AbsoluteMaxBytes, PreferredMaxBytes. These values can be decided by the admin or developer depending upon requirements. "Application" section specifies Access Control Lists (ACL) for Lifecycle System chaincode (LSCC) functions, Quering System chaincode (QSCC) functions, Configuration System Chaincode (cscc) functions, and some miscellaneous peer functions. The "Channel" section defines the default set of policies which defines rules for readers, writers, and Admins for accessing a channel. The Channel name is also specified by the anchor (&) symbol. 'Capabilities' are binary version management that defines the application, orderer, and channel capabilities by using the anchor (&) symbol. 'Profiles' are needed for the generation of configuration components. Attributes in the profile depend on the config component type. It contains the section for generating the Genesis block and also the application channel. Three network configuration artifacts generated by the cryptogen tool are Genesis Block, Channel Transaction, and Anchor Peer Update Transaction.

4.3.3. Crypto-config.yaml

The "Crypto-config.yaml" file contains the list of organizations managing orderers and the list of member organizations managing peers (Hua et al. 2020). While creating an orderer organization, one needs to specify the name, domain, and hostname of the orderer. While creating peers of the organization, need to specify the name, the domain of the organization, and several peers that are to be created for this organization. Admin is created by default for every organization by the cryptogen tool. Multiple peers are created by specifying the count field under the template section and are assigned names by default like peer0, peer1, and so on. Multiple users can also be created by specifying the count field under the users' section. The names of different users are assigned like user0, user1, and so on.

4.3.4. Docker-compose-base.yaml

"Docker-compose-base.yaml" file is available under the folder "docker/config." This file contains these sections. 1st section is Version; for specifying the version of the composeyaml file. 2nd section is Network; which is used by services defined in a docker-compose file for intercommunication in isolated networks. The 3rd section is Volumes; which defines the volumes created on the host machine file system. On the host machine, when the container is launched it has a read/write file system. When a container is shut down or killed all of the data written within the container file system is lost. To prevent this lost data needs to use the host machine file system for reading/writing ledgers as well as configuration data. The volume section creates this linkage between containers and folders on the host machine file system. Examples of volumes are those created on host machines of different organization peers. 4th section is services; each service definition is identified by a unique container name, an image that is used for the container, a default or landing directory after the launch of the container, and a command executed once the container has started.

4.3.5. Core.yaml

The "Core.yaml" file is available in various organization folders under the docker/config folder. The 1st section contains "Peer" which is used as a utility for managing peer and network configuration. On launch, the peer reads the location of the core.yaml file; reads its configuration and overrides any configuration for which the environment variables have been set. Then it reads the MSP and connects to the orderer. Data is then written to the storage location specified in the config or environment variable. The properties which are set in this section are listenAddress, chaincodeListenAddress, chaincodeAddress, keepalive, gossip, tls, and BCCSP (Blockchain Crypto Service Provider). To enable transport layer security, one needs to specify file path in properties like a key, cert, rootcert, clientRootCAs, clientKey, and clientCert. BCCSP, exposes cryptographic functions like: encrypt, decrypt, key pair generation, message digest, private key security, etc. It's default parameter specifies its types which could be sw or PKCS11. For either of these properties need to be set: Hash, Security, FileKeyStore. Other properties that need to be set for BCCSP are mspconfigpath, localMspId, the client (CLI common client config options), deliveryclient, localMspType, profile, adminService, handlers, and discovery (to check which peers have joined a certain channel). The 2nd section is "logging," to specify logging levels for different modules like gossip, grpc, ledger, policies, MSP, and cauthdsl. The log level values could be Debug, Warning, Error, Info, Notice, and critical. The 3rd section is "Ledger", to specify ledger configuration that encompasses both the blockchain and the state Data. The transaction log of a blockchain is always stored in LevelDB data which is stored in the file system. State Data can be configured to use LevelDB or CouchDB (Zhou et al. 2019). CouchDBConfig sets credentials like username, and password and sets parameters like maxRetries, maxRetriesOnStartup, requestTimeout, internalQueryLimit, maxBatchUpdateSize, warmIndexesAfterNBlocks, and create GlobalChangesDB. 4th section is 'Metrics,' to specify the configuration from one of the metrics providers: statsd, Prometheus, or disabled. The parameters included in this are network, address, writeinterval, and prefix. The 5th section is "chaincode" which contains different sections for golang, node, and java languages. The properties set under the chaincode section are id, which is used by the Chaincode stub to register the executing Chaincode; builder, for setting the docker image used for building the chaincode. In all the three languages sections, the runtime property is set to specify the base image for that language. Other tuning parameters in the chaincode section are installTimeout, startuptimeout, executetimeout, keepalive, system (for enabling various types of chaincodes), and logging. The 6th section is "VM," for managing VM systems. Settings for docker VM's include. tls, if enabled need to specify the file path for ca, cert, key; attachStdout, to enable or disable the standard out/err from chaincode containers for debugging purposes; hostConfig, to specify parameters on creating docker container like NetworkMode, Dns, LogConfig, Memory.

4.4. HLF Process

Two possible interactions of application with peers are ledger- query and ledger-update. The request is generated by the application to peers. The fabric SDK allows programmers to use any language like java, python, go, or nodejs to create an API for the users to interact with the Blockchain network. For ledger query requests, peers need not consult other peers of the network as the required information is already there in the peer's copy of the ledger. The ledger-update request works initially in the same way as the ledger-query with a few extra steps. The peer cannot itself update a ledger instead other peers need to agree to this change.

For ledger-query requests, the Application sends requests to Peer. Peer invokes chaincode to query. Chaincode generates the response from the ledger which through peer is sent to the application back as a response of query.

The process of ledger-update is divided into three parts *Endorsement, Ordering, and Validation* as shown in Figure 5. A client generates a request for the transaction. The application generates a transactional proposal and forwards it to the endorsing peer nodes for endorsement. After that, the peers would validate if the proposed transaction is valid or not, follows the defined rules, and whether the client has appropriate permission for this or not (Sousa, Bessani, and Vukolic 2018). Next, the endorsing peer invokes the chaincode for the transaction proposals to query the current state of the ledger.

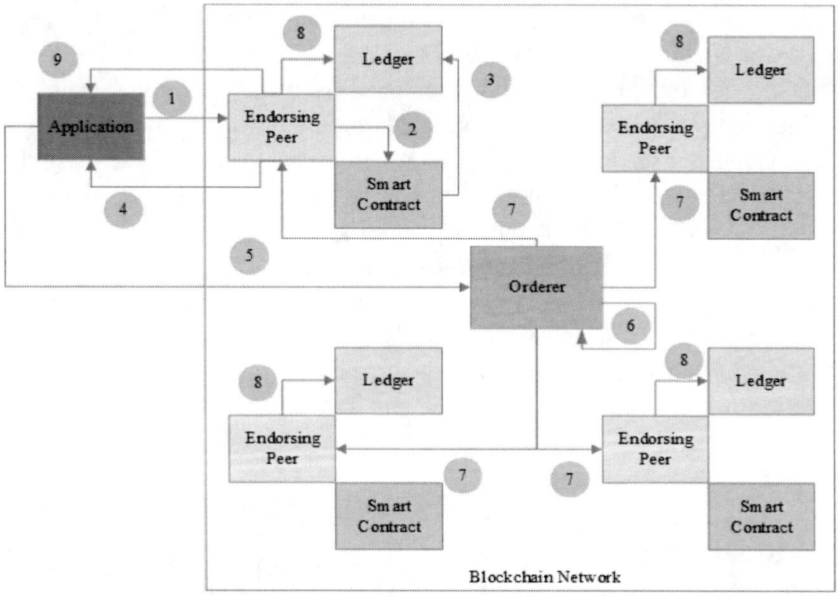

1. Transaction Proposal Request
2. Invokes Chaincode for transaction proposal.
3. Generates Proposal Response.
4. Sends Proposal response to client.
5. Sends transaction responses and endorsers signature for ordering.
6. Apply consensus to create block from Transactions.
7. Delivers new block to all Endorsing peers.
8. Block validation and updating the ledger.
9. Sends Transaction status to client.

Figure 5. HLF transaction flow process.

Further chaincode generates the proposal response for the same, signs it, and sends it back to the application. The application needs to collect responses from various peers, validate the signatures, compare proposal responses, and pack them into a request. The generated transaction request with responses and signatures are being sent to the ordering nodes. Orderer verifies all cryptographic material, policies, and read/write sets are correct (Androulaki et al. 2018). The policy could be that all peers must agree to it or few. Depending upon policy orderer decides whether to validate the transaction or discard it. The validated transactions are bundled together to create a block when either the timeout occurs or the block size limit is reached or the block contains the maximum number of required transactions (Androulaki et al. 2018). The newly created block is sent to all peers for validation and then updating their ledger. By this, all the peers have the same updated copy of the ledger. Finally, the transaction status, that is whether it is validated or not, is sent to the client application who has submitted the transaction.

5. Hyperledger Fabric Implementation Steps

The steps in Hyperledger fabric implementation involve first initializing and launching the network. The second is Set context for organization1. The third is to initialize the Chaincode. The fourth is to Instantiate the Chaincode. The fifth is to Update or query the Chaincode from Organization1. The sixth is to Change the context to Organization2 and query the chaincode to verify the data on chaincode. The seventh is to Shut down the system. These all are explained below.

5.1. To Initialize and Launch the Network

5.1.1. Generate Configuration

The first step is to generate crypto-materials for different organization identities. In our example, we are considering five areas as different organizations so five folders are created containing their certificates and key stores. This step is not meant to be used in production as real identities are issued by the government or some trusted organization. In the case of the test setup, the cryptogen tool is used to carry out this step in HLF ("Cryptogen," n.d.). This tool uses Public-key Infrastructure (PKI) for premising and issuing certificates to all identities of the network. Identities include participants, orderers, peers, and members. Certificates are managed by Certification

Authorities (CA) which issue and revoke identities. This process is decentralized means each organization CA is responsible for issuing certificates to their own organization identities. The cryptomaterial is generated in a crypto-config.yaml file in a folder named crypto-config. The cryptogen command used for creating configuration is:

Cryptogen generate --config =<<config file path>>
--output =<<output folder name>>

Various organization details like a unique name, their domain, and several peers and other users need to be mentioned in the file for completing the above command action ("Cryptogen," n.d.). All the peers are named peer1, peer2, and so on. Similarly, different users are named as user1, user2, etc.

5.1.2. Specifying Artifacts Configuration and Creating Genesis Block

The next step is to specify configuration elements for various artifacts in configtx.yaml file. This file contains sections for organizations, orderer, application, channel, capabilities, and profiles. All these are explained in the above section. Three network configuration artifacts generated by the cryptogen tool are Genesis Block, Channel Transaction, and Anchor Peer Update Transaction. The Genesis block and channel transaction artifacts are generated at this step. Anchor Peer Update Transaction step is used in the next step.

a. Creating a Genesis Block

To create a Genesis Block certain configuration elements are required from configtx.yaml file like configuration for the orderer, MSP of the orderer, and other organizations. The command used for creating the Genesis block in the current folder is:

Configtxgen	-outputBlock	<<genesis block file name>>
	-profile	<<profile name>>
	-channelId	<<Channel name>>

If channel id is not specified then this command would generate the genesis block for the default channel. The generated file is a binary format file that is used for launching the orderer. As this file is in binary format so can't see the content but can inspect it by dumping the content of the block in JSON format using the command.

Configtxgen -inspectBlock <<genesis block file name>>

b. Creating Channel Transaction:
To create channel transactions certain configuration elements are required from configtx.yaml file like channel id, configuration for the application, configuration for the organizations, and their associated MSP. The command used for creating Channel Transaction in the current folder is:

Configtxgen -outputCreateChannelTx << file name>>
-profile <<profile name>>
-channelId <<Channel name>>

This generated transaction file is used by peers for submitting transactions for the creation of application channels. The file is in binary format so can't see the content but can inspect it by dumping the content in JSON format using the command.

Configtxgen -inspectChannelCreateTx<<Tx file name>>

5.1.3. Launching Docker Containers

The next step is by docker-compose to launch the services which are specified in the yaml file named as docker-compose.yaml. The up command used for launching services is as follows:

docker-compose -f <<file name>> up -d

-f flag is to override the default file with a different name file and the -d flag is to launch the container in detached mode or background mode. Other docker commands are:

To shut down the services:	docker-compose down
To stop the services:	docker-compose stop [container]
To start the services:	docker-compose start [container]
To pause the services:	docker-compose pause
To restart the services:	docker-compose restart

5.1.4. Setting Up Anchor Peers for Different Organizations

To set up an anchor peer for 1st organization steps that need to be followed are given below. To set up anchor peer for the next organization step a, c, and d are followed as the channel is just created once.

- Set the organization context for native peer commands
  To run the further commands from peers of a particular organization, context needs to be set. This step will set all parameters like "LOCALMSPID", which is set as organization MSP id; "FABRIC_LOGGING_SPEC" whose value is set to "INFO"; "FABRIC_CFG_PATH" set to point to "config/organization_name"; "CORE_PEER_ADDRESS" set to "peer1.oraganization_name.com"; "ORDERER_ADDRESS" is also set to point to orderer address. Apart from these other properties related to chaincode are also set in this step, and those are chaincode constructor, name, path, version, channel id, language, package name, package folder, and sequence.

- Submit the create channel transactions
  Channel as discussed above is joined by specific organizations for maintaining the privacy of the transaction. Various commands that can be carried out using peer channel commands are to create, join, list, fetch, getinfo, signConfigTx, and update ("Peer Channel," n.d.). The syntax for create command is:

```
Peer channel create     -f <<path to file>>
                        -c <<channel ID>>
                        -t <<time out, optional>>
                        -o <<orderer address>>
                        --outputBlock<<channel id block>>
```

This command is executed by the admin of the organization and requires the following values. A channelTx file is specified by using the –f flag, channel Id is specified by the –c flag. By default, the timeout is 5 sec still can specify it using the –t flag. Orderer address is specified by –o flag, although this command is executed by peer actual command is executed on Orderer so it is required. Channel ID block file returned by orderer is the 0th block or called as genesis block for the channel. Any intended peer who wishes to join must have access to the genesis block. This command needs to be executed only once and the one who is executing this command will get the genesis block as a result of this command. Other organizations can get a copy of this block by using the fetch command.

- Join an organization to the channel
  This command is also executed by the admin of the organization so that the peer binary initializes the ledger or state DB ("Peer Channel," n.d.). The syntax of this command is:

  ```
  Peer channel join        -b <<path of genesis block file>>
                           -o <<orderer address>>
  ```

  This command requires the Genesis block file path which is specified by the –b flag.

- Update anchor peer in channel
  - Creating anchor peer update transaction:

  The anchor peer update transaction file created here is used by peers for submitting transactions. The command used to create this transaction file is:

```
Configtxgen -outputAnchorPeersUpdate     << file name>>
                     -profile            <<profile name>>
                     -channelId          <<Channel name>>
                     -asOrg              <<Organization name>>
```

The transaction includes only those changes for which the specified organization has the privilege. This generated transaction file is in binary format so can't see the content but can inspect it by dumping the content of the block in JSON format using the command.

```
Configtxgen -inspectChannelCreateTx      <<Tx file name>>
```

- Update the channel

This enables the admin to submit configTx to the orderer. The working behind this is that the peer binary receives the updated configTx file as part of the peer channel update command. It submits the transaction to the Orderer.

Orderer then validates the transaction and sends out the same to all the peers in the network ("Peer Channel," n.d.). The syntax for this command is:

Peer channel update –f <<transaction file>>
-c <<channel ID>>
-o <<Orderer address>>

5.1.5. Set the Context for Organization1

This step is the same as already done above in step 4 (a). If the complete network is run on a single machine so to switch between various organizations and execute the commands from their end, one needs to switch context using this command.

5.1.6. Initialize the Chaincode

Application chaincode is used here, which defines the application behavior. It is launched by peers and instantiated in docker containers. Application chaincode must be consistent across all endorsing peers and it is bound to a channel. Application chaincode lifecycle executes multiple operations. Initially, it starts with the Developer who writes the code for chaincode and "installs" it on to the peer. The developer then "instantiates" the chaincode and once it is successful, the application chaincode is in the instantiated state. Further for upgrading the chaincode the developer installs the newer version of it to the peer and then invokes the "upgrade" command.

- Checks if the package already exists then no need to create the package again and just come out of this block otherwise creates the chaincode package. The Chaincode package is used for achieving consistency across the network. The syntax of command for creating a package is:

  Peer chaincode package –flags <<package_filename>>

Different flags that could be used in this command are "–n" for name, "-v" for version, "-l" for language, "-p" for path, and "-i" for instantiation policy.

- Install the Chaincode

This step is executed by the admin on the peer. It needs copying of chaincode to the file system on peer which is typically known as "Installed on peer". The peer validates the credentials to make sure that the caller is authorized to install chaincode on the peer.

These steps happen when an admin executes the install command on the peer. 1st step is Peer launches a build container, which has all the tools for building source code for the chaincode. Next step peer copies the chaincode source to build containers. Then initiate the build process where source code is compiled and packaged. Once the build process executes successfully, it creates a package within the container which is read by the peer. Peer then writes this package on the local file system. Hence after compilation of this, one can see the files in the local file system to see the package in chaincode. The image used for the build container can be configured by parameter under the core.yaml. The syntax for the install command is ("Peer Chaincode," n.d.):

Peer chaincode install
-n <<Unique name of chaincode assigned by admin>>
-v <<Version of chaincode>>
-l <<language in which chaincode is written, default is golang>
-p <<path to chaincode>>

The above install command is used if the package is not used but if the package is used then the install command syntax would be:

Peer chaincode install <<package_filename>>

- Approving the chaincode

The chaincode created above needs to be approved before use. The syntax of the command with various flags is given below.

peer chaincode approve for myorg. -channelID <<Channel ID>>
-name <<CC_Name>>
-version <<CC_version>>
--package-id <<Package_ID>>
--sequence <<CC_Sequence>>
-o <<Orderer_address>>
--waitForEvent

5.1.7. Instantiate the Chaincode

This step is done after installing the chaincode as till the install command, the chaincode is not associated with any channel we launch the container on a specific channel. Admin executes the instantiate command. The peer validates the credentials to make sure that the caller is authorized and then passes the control to instantiation logic.

These steps happen when an admin executes an instantiate command on a peer. The 1st step is to Peer read the package from the local file system and based on the type of chaincode language used it selects the base image for the chaincode container. Then builds an image for the chaincode container, at this point the package for the chaincode is part of the chaincode image. Peer then launches the chaincode container for the chaincode image it has created. At this point, the interaction between the chaincode container and the peer occurs over the gRPC interface. The base image used for the chaincode image is configured by way of the core.yaml parameters. The syntax for instantiating command is ("Peer Chaincode," n.d.):

Peer chaincode instantiate
-n <<Unique name of chaincode assigned by admin>>
-v <<Version of chaincode>>
-c <<Constructor parameter for chaincode in JSON format>>
-C <<Channel ID>>
-p <<Endorsement policy for the chaincode>>
-E <<Name of the Endorsement system Chaincode, default escc>>
-V <<Name of the verification system chaincode, default is vscc>>

5.1.8. Update or Query the Chaincode from Organization1

The required operations before updating or querying the Chaincode are to initialize the orderer binary and launch it, initialize the peer binary and launch it, and install, and instantiate the chaincode from any organization.

The Chaincode query command is executed on a local peer, who reads the state of the chaincode from the local ledger and sends it back to the peer binary. As query commands just read the state from the local machine so there is no state change and no transaction log maintained ("Peer Chaincode," n.d.). The syntax of the query command is:

Peer chaincode query -n <<Name of the chaincode>>
-C <<Channel ID>>
-c <<Arguments in JSON format>>

For updating the chaincode invoke-command is used. For this command, the peer binary creates the transaction proposal and sends it to the endorsing peer. Endorsing peer simulates the execution of the chaincode, endorses the transaction, and sends it back to the peer binary. Peer binary sends out the endorsing Transaction to the orderer. Orderer creates the Block and delivers it to various peers in the network. This command may lead to state change and so transactions are also recorded in the ledger ("Peer Chaincode," n.d.).

Peer chaincode invoke -n <<Name of the chaincode>>
-C <<Channel ID>>
-c <<Arguments in JSON format>>

5.1.9. Change the Context to Organization2 and Query the Chaincode to Verify the Data on Chaincode

To change the context to organization 2 some properties need to be set which is mentioned in step no 4 (a). After that to check the current state of the ledger in this organization, a query command can be executed as mentioned in step E above. If any change is needed from this organization can be done using invoke-command as mentioned in step E above.

5.1.10. Shutdown the System

This step is just to down the docker when need not do any step further. The command for this is:

Docker-compose –f <<docker-compose-base.yaml file>> down

Conclusion

Blockchain, a collection of multiple blocks is a distributed, decentralized, consensus-based technology. Blockchain technology is the need of most of the applications in the market because of its capability to make the system more secure and reliable. A variety of blockchain frameworks are available in the market for its implementation. One can choose any framework or tool as per the suitability of their application logic. Hyperledger Fabric, one of the popular frameworks of Hyperledger is the target of this chapter. The distinctive capabilities of Fabric make it an adaptable system for permissioned blockchains having flexible trust expectations that allow the fabric to support a diversity of business use cases like government, finance, VANET, IoT, agriculture, healthcare, and many more. The features, components, configuration files, and implementation steps are discussed in detail in this chapter. The fabric consists of various components like core components, tools, and fabric certification authority to set up fabric infrastructure. To create binaries of these components, various configuration files are required and those are configtx.yaml, crypto-config.yaml, orderer.yaml, docker-compose-base.yaml, core.yaml. The complete process of implementation in fabric includes endorsement, ordering, and validation. The seven steps for the complete implementation include initializing and launching the network, setting the context for organization1, initialize the Chaincode, Instantiate the Chaincode, Update or querying the Chaincode from Organization1, Change the context to Organization2, querying the chaincode to verify the data on chaincode, Shut down the system. The complete process and implementation steps are explained in detail in this chapter which can be considered by a developer to understand the development of a blockchain application.

References

Androulaki, Elli, Artem Barger, Vita Bortnikov, Srinivasan Muralidharan, Christian Cachin, Konstantinos Christidis, Angelo De Caro, et al. 2018. "Hyperledger Fabric: A Distributed Operating System for Permissioned Blockchains." In *Proceedings of the 13th EuroSys Conference, EuroSys 2018*. https://doi.org/10.1145/3190508.3190538.

Cachin, Christian. 2017. "Architecture of the Hyperledger Blockchain Fabric." *Leibniz International Proceedings in Informatics, LIPIcs* 70: 24.1-24.16. https://doi.org/10.4230/LIPIcs.OPODIS.2016.24.

"Configtxgen." n.d. https://hyperledger-fabric.readthedocs.io/en/release-1.4/commands/configtxgen.html.

cryptocurrencies, C Cachin - Workshop on distributed, and undefined 2016. n.d. "Architecture of the Hyperledger Blockchain Fabric." *Pdfs.Semanticscholar.Org*. Accessed October 2, 2020. https://pdfs.semanticscholar.org/f852/c5f3fe649f8a17ded391df0796677a59927f.pdf.

"Cryptogen." n.d. https://hyperledger-fabric.readthedocs.io/en/release-1.4/commands/cryptogen.html.

"Fabric-CA Commands." n.d. https://hyperledger-fabric.readthedocs.io/en/release-1.4/commands/fabric-ca-commands.html.

Gaba, Priyanka, and Ram Shringar Raw. 2020. "Vehicular Cloud and Fog Computing Architecture, Applications, Services, and Challenges." In *IoT and Cloud Computing Advancements in Vehicular Ad-Hoc Networks*, edited by Ram Shringar Rao, Vishal Jain, Omprakash Kaiwartya, and Singh Nanhay, 268–96. IGI Global. https://doi.org/10.4018.

Gorenflo, Christian, Stephen Lee, Lukasz Golab, and Srinivasan Keshav. 2019. "FastFabric: Scaling Hyperledger Fabric to 20,000 Transactions per Second." *ICBC 2019 - IEEE International Conference on Blockchain and Cryptocurrency*, 455–63. https://doi.org/10.1109/BLOC.2019.8751452.

Hua, Song, Shenbin Zhang, Bingfeng Pi, Jun Sun, Kazuhiro Yamashita, and Yoshihide Nomura. 2020. "Reasonableness Discussion and Analysis for Hyperledger Fabric Configuration." *ArXiv*.

"Hyperledger – Open Source Blockchain Technologies." n.d. Accessed October 2, 2020. https://www.hyperledger.org/.

"Hyperledger Caliper." n.d. https://hyperledger.github.io/caliper/.

Meng, Weizhi, Jianfeng Wang, Xianmin Wang, Joseph Liu, Zuoxia Yu, Jin Li, Yongjun Zhao, and Sherman S. M. Chow. 2018. "Position Paper on Blockchain Technology: Smart Contract and Applications." In *Lecture Notes in Computer Science (Including Subseries Lecture Notes in Artificial Intelligence and Lecture Notes in Bioinformatics)*. Vol. 11058 LNCS. https://doi.org/10.1007/978-3-030-02744-5_35.

Nasir, Qassim, Ilham A. Qasse, Manar Abu Talib, and Ali Bou Nassif. 2018. "Performance Analysis of Hyperledger Fabric Platforms." *Security and Communication Networks* 2018. https://doi.org/10.1155/2018/3976093.

Panwar, Arvind, and Vishal Bhatnagar. 2020. "Distributed Ledger Technology (DLT): The Beginning of a Technological Revolution for Blockchain." In *2nd International Conference on Data, Engineering and Applications, IDEA 2020*. https://doi.org/10.1109/IDEA49133.2020.9170699.

"Peer Chaincode." n.d. https://hyperledger-fabric.readthedocs.io/en/release-1.4/commands/peerchaincode.html.

"Peer Channel." n.d. https://hyperledger-fabric.readthedocs.io/en/release-1.4/commands/peerchannel.html.

Pilkington, Marc. 2016. "Blockchain Technology: Principles and Applications." *Research Handbooks on Digital Transformations*, 225–53. https://doi.org/10.4337/9781784717766.00019.

Priyanka, and Ram Shringar Raw. 2020. "The Amalgamation of Blockchain with Smart and Connected Vehicles: Requirements, Attacks, and Possible Solution." *Proceedings - IEEE 2020 2nd International Conference on Advances in Computing, Communication Control and Networking, ICACCCN 2020*, 896–902. https://doi.org/10.1109/ICACCCN51052.2020.9362906.

Ranjan, Shishir, Amit Negi, Himansh Jain, Bhuvnesh Pal, and Himanshu Agrawal. 2019. "Network System Design Using Hyperledger Fabric: Permissioned Blockchain Framework." *2019 12th International Conference on Contemporary Computing, IC3 2019*, 1–6. https://doi.org/10.1109/IC3.2019.8844940.

Sousa, Joao, Alysson Bessani, and Marko Vukolic. 2018. "A Byzantine Fault-Tolerant Ordering Service for the Hyperledger Fabric Blockchain Platform." *Proceedings - 48th Annual IEEE/IFIP International Conference on Dependable Systems and Networks, DSN 2018*, no. Section 4: 51–58. https://doi.org/10.1109/DSN.2018.00018.

Thakkar, Parth, Senthil Nathan, and Balaji Viswanathan. 2018. "Performance Benchmarking and Optimizing Hyperledger Fabric Blockchain Platform." *Proceedings - 26th IEEE International Symposium on Modeling, Analysis and Simulation of Computer and Telecommunication Systems, MASCOTS 2018*, 264–76. https://doi.org/10.1109/MASCOTS.2018.00034.

Wang, Wenbo, Dinh Thai Hoang, Peizhao Hu, Zehui Xiong, Dusit Niyato, Ping Wang, Yonggang Wen, and Dong In Kim. 2019. "A Survey on Consensus Mechanisms and Mining Strategy Management in Blockchain Networks." *IEEE Access* 7: 22328–70. https://doi.org/10.1109/ACCESS.2019.2896108.

Yusuf, H, and I Surjandari. 2020. "Comparison of Performance Between Kafka and Raft as Ordering Service Nodes Implementation in Hyperledger Fabric." *International Journal of Advanced Science and Technology* 29 (7): 3549–54.

Zheng, Zibin, Shaoan Xie, Hongning Dai, Xiangping Chen, and Huaimin Wang. 2017. "An Overview of Blockchain Technology: Architecture, Consensus, and Future Trends." *Proceedings - 2017 IEEE 6th International Congress on Big Data, BigData Congress 2017*, no. June: 557–64. https://doi.org/10.1109/BigDataCongress.2017.85.

Zhou, Ence, Haoli Sun, Bingfeng Pi, Jun Sun, Kazuhiro Yamashita, and Yoshihide Nomura. 2019. "Ledgerdata Refiner: A Powerful Ledger Data Query Platform for Hyperledger Fabric." *2019 6th International Conference on Internet of Things: Systems, Management and Security, IoTSMS 2019*, 433–40. https://doi.org/10.1109/IoTSMS48152.2019.8939212.

About the Editor

Dr. Brojo Kishore Mishra
Professor
Department of Computer Science & Engineering
Associate Dean (International Affairs)
GIET University, Gunupur, Odisha, India
Email: brojomishra@gmail.com

Index

D

E

F

G

H

I

R

S

T

U

V

W

Y